Extreme QCD 2012

(xQCD)

Journal of Physics: Conference Series Volume 432

Washington DC, USA
21 – 23 August 2012

ISBN: 978-1-62748-309-4
ISSN: 1742-6588

Printed from e-media with permission by:

Curran Associates, Inc.
57 Morehouse Lane
Red Hook, NY 12571

Some format issues inherent in the e-media version may also appear in this print version.

Printed by Curran Associates, Inc. (2013)

Institute of Physics
Dirac House, Temple Back
Bristol BS1 6BE UK

Phone: 44 1 17 929 7481
Fax: 44 1 17 920 0979

techtracking@iop.org

Additional copies of this publication are available from:

Curran Associates, Inc.
57 Morehouse Lane
Red Hook, NY 12571 USA
Phone: 845-758-0400
Fax: 845-758-2634
Email: curran@proceedings.com
Web: www.proceedings.com

TABLE OF CONTENTS

DAY 1

DAY 2

DAY 3

FOREWORD

The Extreme QCD 2012 conference, held at the George Washington University in August 2012, celebrated the 10th event in the series. It has been held annually since 2003 at different locations: San Carlos (2011), Bad Honnef (2010), Seoul (2009), Raleigh (2008), Rome (2007), Brookhaven (2006), Swansea (2005), Argonne (2004), and Nara (2003).

As usual, it was a very productive and inspiring meeting that brought together experts in the field of finite-temperature QCD, both theoretical and experimental. On the experimental side, we heard about recent results from major experiments, such as PHENIX and STAR at Brookhaven National Laboratory, ALICE and CMS at CERN, and also about the constraints on the QCD phase diagram coming from astronomical observations of one of the largest laboratories one can imagine, neutron stars. The theoretical contributions covered a wide range of topics, including QCD thermodynamics at zero and finite chemical potential, new ideas to overcome the sign problem in the latter case, fluctuations of conserved charges and how they allow one to connect calculations in lattice QCD with experimentally measured quantities, finite-temperature behavior of theories with many flavors of fermions, properties and the fate of heavy quarkonium states in the quark-gluon plasma, and many others.

The participants took the time to write up and revise their contributions and submit them for publication in these proceedings. Thanks to their efforts, we have now a good record of the ideas presented and discussed during the workshop. We hope that this will serve both as a reminder and as a reference for the participants and for other researchers interested in the physics of nuclear matter at high temperatures and density. To preserve the atmosphere of the event the contributions are ordered in the same way as the talks at the conference.

We are honored to have helped organize the 10th meeting in this series, a milestone that reflects the lasting interest in this research area and the steady progress in resolving the outstanding issues. Many challenges remain and we are confident that this series will continue well into the future. Next year, the meeting will be organized at the Albert Einstein Center for Fundamental Physics at the University of Bern. We expect to have another interesting and lively meeting and look forward to meeting you there.

February 23, 2013
Andrei Alexandru
Alexei Bazavov
Keh-Fei Liu

Journal of Physics: Conference Series **432** (2013) 012001 doi:10.1088/1742-6596/432/1/012001

Constraints on physics of neutron stars from X-ray observations

Craig O. Heinke
Dept. of Physics, U. of Alberta, CCIS 4-183, Edmonton, Alberta T6G 2E1, Canada
E-mail: heinke@ualberta.ca

Abstract. I summarize some constraints on the physics of neutron stars arising from X-ray observations of the surfaces of neutron stars, focusing on using models of low-magnetic-field neutron star atmospheres to interpret their X-ray spectra. I discuss observations of spectral lines, pulsation profiles, X-ray bursts, radius measurements of transiently accreting neutron stars in quiescence, crust and core cooling measurements of transiently accreting neutron stars, and cooling of young neutron stars. These observations have constrained the neutron star mass and radius (and thus the internal composition, and dense matter equation of state), the superfluidity and neutrino emissivity properties of the core, and the composition and superfluid state of the crust.

1. Introduction

The behavior of matter at high densities (several times nuclear density) and relatively low temperatures ($kT << mc^2$) cannot be probed in terrestrial experiments. The nonlinear aspects of quantum chromodynamics ensure that the equation of state cannot be extrapolated unambiguously from well-understood lower-density matter. The interiors of neutron stars (NSs) are composed of high-density, relatively cold matter, and the physics of these interiors affect the observable properties (e.g. mass, radius, surface temperature) of the NSs. Therefore, astrophysical observations of NSs permit constraints on the physics of dense matter, as have been summarized in detailed reviews such as [1, 2]; see also Lattimer (in these proceedings) for a complementary discussion.

NSs can be observed in a variety of manifestations [3], many of which offer possibilities for constraints on their internal physics. In this short review, I focus on a few recent advances involving thermal X-ray radiation from the surface of the NS [4, 5]. Other constraints arise from, for instance, timing of radio pulses [6, 7] which can provide measurements of the spin rate [8] of NSs; neutrino emission from the formation of NSs in supernovae [9]; gravitational waves as two NSs spiral together and coalesce [10]; high-energy radiation produced by the decay of extremely high magnetic fields in the enigmatic NSs known as magnetars [11]; and studies of binary companion stars and the accretion disks [12]. A key result is NS mass measurements in binaries with a radio pulsar, often using tests from general relativity (GR), which suggest that the majority of NSs are around 1.4 solar masses ($M_\odot$) [13], with some ranging as high as 1.97 $M_\odot$[14].

Here I discuss NSs where we observe thermal radiation from the surface (specifically, those with low magnetic fields, e.g. $B < 10^{11}$ G). As they cool, isolated NSs radiate heat from their surfaces which may have nonuniform temperature distributions, leading to flux variations with the NS rotation phase. NSs accreting material from a companion star experience occasional thermonuclear fusion events on their surfaces, "X-ray bursts", which often provide sufficient radiation to briefly expand the NS's atmosphere. Accreting NSs often experience periods of quiescence during which material piles up in an accretion disk rather than reaching the NS, thus allowing one to study NS surface emission.

1.1. Thermal surface emission

NSs are born extremely hot, and their extreme gravity and magnetic fields can accelerate particles that heat their surfaces as NSs become older. Thus, many NSs have temperatures of order a million degrees over part or all of their surface, and thus glow in X-rays. This X-ray radiation is observed by X-ray telescopes on orbiting satellites (as X-rays cannot penetrate Earth's atmosphere); the findings reported below have principally come from NASA's *RXTE* and *Chandra*, and ESA's *XMM-Newton* satellites.

The surface radiation from NSs must pass through the thin (few cm) NS atmospheres (if present), which modify the radiation through free-free absorption, spectral lines, and photoionization jumps. Strong magnetic fields heavily distort the structure of electronic orbitals, and thus alter the ionization energies and opacities [15]. Thus, it is easier to get constraints on NS physics when studying those NSs which are expected to have relatively low surface magnetic fields ($B<10^6$ T), as the number of free parameters affecting the spectrum is reduced. Accretion of material from a companion appears to reduce NS magnetic fields by diamagnetic screening, from typical birth fields of 10^8 T to $\sim10^4$ T [16].

Due to the strong surface gravity, NS atmospheres stratify by element within < 100 s [17]. Thus, at times of little or no accretion, the part of NS atmospheres that produce the spectra is usually assumed to be pure hydrogen (during X-ray bursts, the composition should be complex, and one young NS appears to have a carbon atmosphere [18]). Low-B-field hydrogen atmosphere models give observable X-ray spectra similar to blackbodies, but shifted to higher energies due to the inverse dependence of the electron free-free opacity on photon energy [19, 20]. The intense gravity of NSs shifts the observed spectrum to longer wavelengths, decreasing the observed temperature. Although they complicate the spectral interpretation, gravitational effects permit constraints on the compactness (mass divided by radius) of the NS.

2. Constraints

2.1. Spectral lines

An unambiguous identification of a clear spectral line due to an atomic transition from a NS surface would permit, by measuring how the energy of the line is decreased by escaping from the gravitational potential of the NS, a straightforward measurement of the compactness of the NS. A tentative identification of narrow absorption lines in a high-resolution X-ray spectrum of X-ray bursts from one accreting NS (EXO 0748-676) suggested a gravitational redshift of z=0.35 [21], consistent with theoretical expectations [1], and sparking many attempts to find similar lines in other accreting NSs. Unfortunately, deeper observations of this NS with multiple X-ray telescopes have failed to confirm these lines [22]. Furthermore, the NS rotation period, later measured at 552 turns per second [23], would broaden any lines produced on the NS surface [24, 25], implying that the observed absorption lines do not arise from the NS surface. Searches for absorption lines in high-resolution spectra of other accreting NSs have also met with no success.

Other spectral absorption lines have been observed in isolated NSs, both in nearby NSs [26, 27] and in very young NSs [28, 29]. These spectral lines appear to all be consistent with variants of proton or electron cyclotron absorption (or emission) lines [30, 31], thus their study can tell us the surface magnetic field strength but not the surface redshift of the NS. Recently, a time-variable absorption line has been claimed in a normal pulsar, which shows a magnetic field strength incompatible with the line's production by cyclotron absorption [32]. It is difficult to understand how a line produced by an atomic transition would have a variable energy. However, the apparent lines could instead be due to an intrinsically complicated multiple-blackbody spectrum (cf. [33]), which when fit by a single blackbody gives the appearance of absorption lines [32].

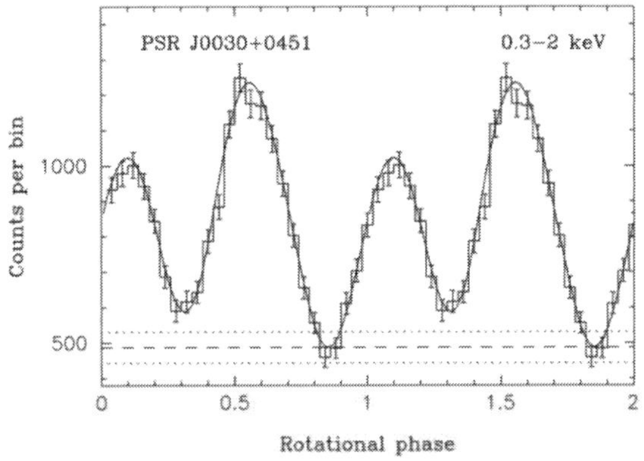

Figure 1. X-ray pulsation profile of the radio pulsar PSR J0030+0451 folded on the NS pulse period (repeated twice for clarity), as observed by the XMM-Newton telescope and modeled by [34]. The two peaks indicate the maximum visibilities of the two polar caps.

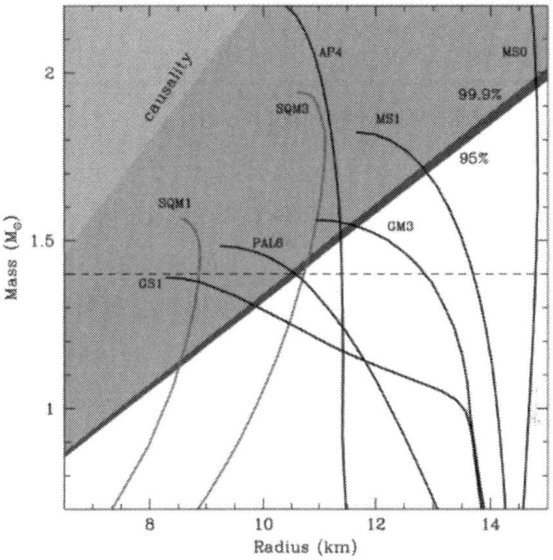

Figure 2. Constraints on the mass and radius of the NS in PSR J0030+0451 from its pulsation profile [34]. The cyan region at upper left is not allowed due to causality (the requirement that pressure waves must propagate at speeds below c). The region ruled out by fits to this NS is the green shaded region (at 99.9% confidence); the dark blue region is also included at 95% confidence. Several loci of mass and radius for different NS models are also plotted [35]. The dashed line is the canonical NS mass, 1.4 $M_\odot$.

2.2. Pulsation profiles

Polar caps of NSs may be heated either by accretion, funneled onto the caps by the NS's magnetic field, or by radio pulsar activity, which produces energetic (relativistic) positrons and electrons that collide with and heat the polar caps [36]. Either situation produces temperature inhomogeneities over the NS surface, which combine with the angular dependence of intensity from the atmosphere and ray-tracing in a general relativity metric to produce a complicated pulse profile [37]. Generally, the Schwarzschild metric combined with appropriately oblate (due to their rotation) neutron stars, Doppler effects, and light-travel time delays provides a good approximation to the full GR lightcurve calculation [38].

Recent work has focused on accreting millisecond pulsars and radio millisecond pulsars, which are both believed to have relatively low magnetic fields. The accreting millisecond pulsars, due to continuing accretion of a range of elements, may have spectra reasonably described by blackbodies, while radio millisecond pulsars, without continuing accretion, should typically have hydrogen atmospheres. The simplest constraint from modeling the pulsation profiles is that a relatively sharp, peaked profile requires a larger star, since radiation from a more compact star suffers more gravitational bending and thus produces a smoother pulse profile. A strong constraint on the radius of the NS in the radio pulsar PSR J0030+0451 was calculated by [34], giving $R > 10.4$ km (99.9% confidence) for an assumed mass of 1.4 $M_{\odot}$, due to its highly peaked pulsation profile (Fig. 1).

More complicated modeling of the emission from accreting millisecond pulsars (which have components of their spectra that do not come from the surface, e.g. Comptonized radiation from above the surface, and reflected light) produces constraints in mass and radius space. For the well-studied object SAX J1808.4-3658, [39] find a relatively low mass below 1.5 $M_{\odot}$ (using a reasonable constraint on the inclination). Although this modeling involves more parameters, the mass constraint is consistent with optical studies of the companion star [40].

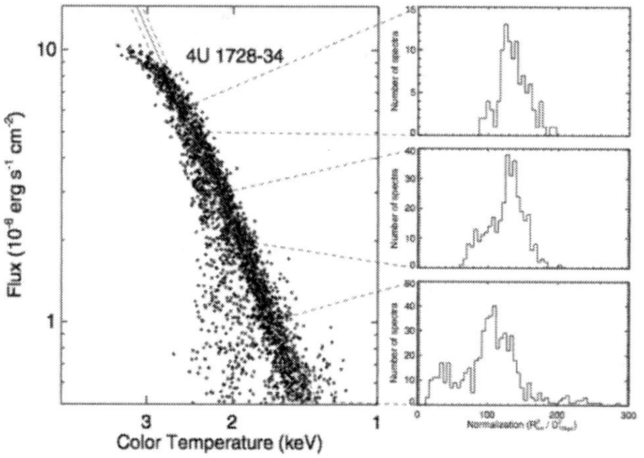

Figure 3. Flux vs. temperature for X-ray spectra taken at various times during multiple X-ray bursts by the same NS [41]. Diagonal lines indicate the best-fit blackbody normalization and its uncertainty. The majority of bursts are consistent with one normalization.

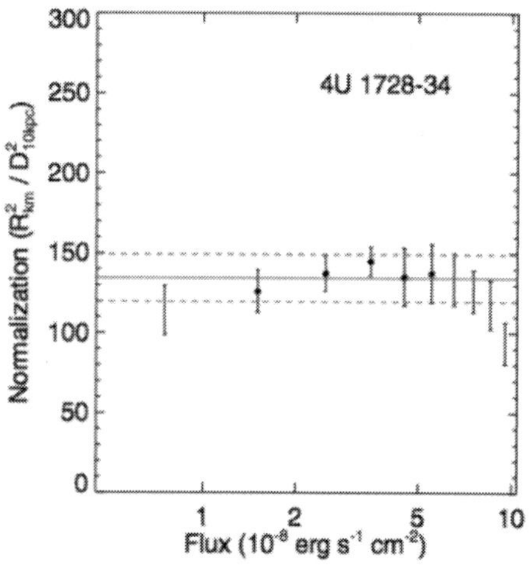

Figure 4. Inferred blackbody normalization vs. X-ray flux for the X-ray burst spectra of Fig. 3 [41].

2.3. X-ray bursts

Accreting NSs often experience unstable nuclear burning on their surfaces, leading to rapid X-ray brightenings known as "bursts" [42]. Several authors have calculated models for NS atmospheres during bursts, showing how the measured temperature (defined by fits to a blackbody spectrum) relates to the effective surface temperature; this relation is parametrized as a "color correction factor" [43, 44]. The burst may expand the NS atmosphere; the radiative flux necessary to overcome gravity and push the atmosphere away, the "Eddington limit", depends on the NS mass, but the observed flux will be decreased by the NS's gravity, and thus its compactness. As the radiation from the burst comes from the NS surface (rather than the physically complicated accretion flow that affects the radiation at other times), measurement of the temperature and flux of these bursts (and knowledge of their distances) determines the emitting area, and thus (after including GR corrections) the NS radius [45]. Similarly, measuring the radiation flux during bursts that expand the atmosphere tells us about the surface gravity of the NS [46].

Using time-resolved spectra of several bursts showing atmospheric expansion, [47, 48, 49] calculated constraints on the mass and radius of three NSs, finding relatively small NSs (with radii of 8-11 km). These calculations have been criticized by [50] on their choices of color correction factors, distances, and surface chemical compositions, and for not being internally self-consistent; the key problem is that the best-fit values of the observable properties produce complex (non-real-valued) values for the NS mass and radius. [50] present alternative calculations tending toward slightly larger (~12 km) NS radii, favoring the interpretation that the observed radius of the atmosphere remains larger than the true NS radius throughout the burst. [51] use their color correction factors to derive a large radius (>14 km) for a burst from one NS, though they have also been criticized [41] for the relatively poor fits of their spectra to blackbodies.

Several uncertainties surround the interpretation of X-ray bursts. A critical question is whether the NS radii measured during the course of a burst, and between bursts, are identical and equal to the NS radius [52, 41]. Fig. 3 (from [41]) shows temperature vs. flux at different times across multiple bursts from one NS, showing consistency of most burst spectra with one inferred radius (Fig. 4), which suggests that this uncertainty is manageable. Other questions include whether the burst fluxes consistently reach the same Eddington limit [53, 54], whether they show significant anisotropy [55], and whether they vary in their chemical composition [56].

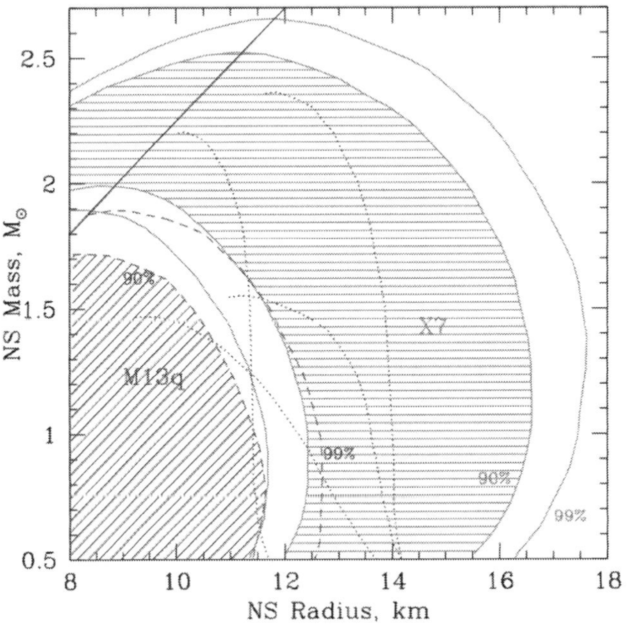

Figure 5. Constraints on the mass and radius of two transiently accreting NSs at known distances, as inferred from fitting their X-ray spectra with hydrogen atmosphere models. Hatched areas indicate the 90% confidence regions for NSs in 47 Tuc (red) [57] and M13 (blue) [58], with outside contours (solid for 47 Tuc, dashed for M13) marking the 99% confidence regions of each. Some sample NS equations of state are indicated by dotted black lines [35]. The 90% confidence contour regions for these two NSs do not overlap.

2.4. Quiescent accreting neutron stars: radius

Many accreting NSs show transient behavior, with periods of little or no accretion, in which the surface of the heated NS may be studied. Assuming a low-magnetic-field hydrogen atmosphere without anisotropies, and a distance known by other means (e.g. by studying the other stars in its stellar cluster), tight constraints may be placed on the NS mass and radius by measuring the temperature and flux, and thus inferring the radius [59]. Due to the gravitational redshift, curved constraints in mass and radius are produced (e.g. Fig. 5).

Deep X-ray observations of globular clusters containing transient accreting NSs have placed tight constraints on the mass and radius of several NSs [57, 58, 60, 61]. Not all these constraints are consistent with each other (see Fig. 5), which may be due to instrumental effects [57], and/or variations in atmospheric composition between NSs [61]. Combining results from these NSs with X-ray burst data (see above) gives robust constraints on the NS radius (between 10.4 and 12.9 km for 1.4 $M_\odot$ NSs), and thus on the dense matter equation of state [62].

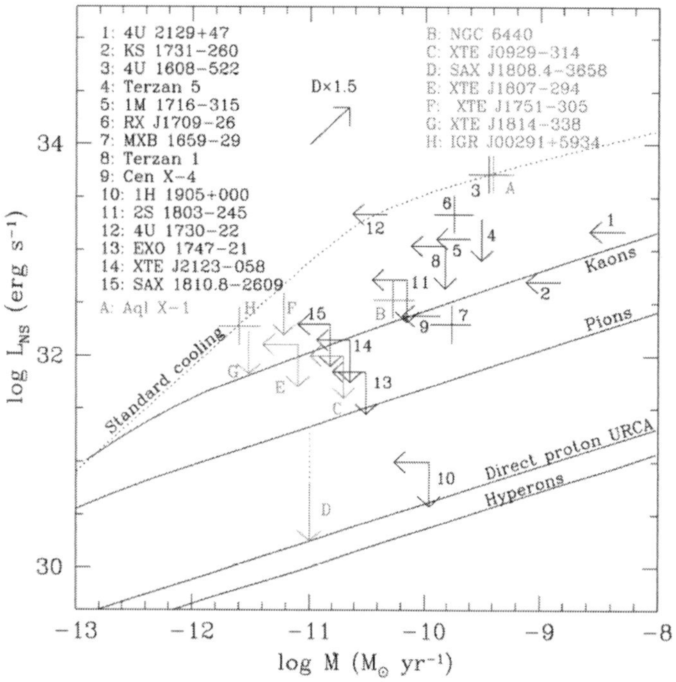

Figure 6. Cooling curves for various neutrino emission scenarios [63], compared with measurements or upper limits of the NS luminosity outside accretion, vs. the time-averaged mass transfer rate [64]. The dotted line relates heat input (via mass transfer) to the NS luminosity between accretion episodes for "standard" (slow) cooling, while the other diagonal lines suggest the maximal cooling achievable by specific enhanced cooling processes. The effect of a factor of 1.5 error in distance is indicated. SAX J1808-36's cold temperature (D) rules out most neutrino emission scenarios.

2.5. Quiescent accreting neutron stars: core temperature

Observations of transient NSs between periods of accretion measure the temperature that the NS has been heated to during accretion episodes. The kinetic energy of the infalling material is radiated away immediately, but the added material compresses the deep crust, leading to nuclear reactions that deposit heat deep in the crust, some of which flows into the core [65, 66]. Averaged over long periods including many accretion episodes, the heat flowing into and out of the core will reach equilibrium, and the NS surface (between outbursts) will reflect the core temperature. Neutrino emission from the NS core dominates the NS's energy loss, either through "standard" cooling (neutron-neutron neutrino bremsstrahlung processes) or enhanced neutrino cooling mechanisms such as direct URCA (involving protons or hyperons) or URCA-like processes involving (e.g.) pion or kaon-like Bose-Einstein condensations in the core, which will decrease the NS surface temperature [63, 67]. Differences in the surface temperature of NSs with the same average mass transfer rate (and thus heating history) can be explained by different NS masses, as this will change the central NS density and perhaps composition, and thus alter which cooling mechanisms may be accessed.

Strong evidence for enhanced neutrino cooling of the cores of transient NSs was provided by relatively cold NSs whose accretion histories suggested higher NS temperatures [68, 69, 70]. Perhaps the clearest evidence for enhanced neutrino cooling is the cold NS SAX J1808.4-3658, where the mass transfer rate is very well-known (Fig. 6); this object may require direct URCA processes involving protons or hyperons [71, 64]. An important question is whether the observed accretion history accurately reflects the long-term averaged mass transfer rates (see [72] and [73] for differing views on this NS).

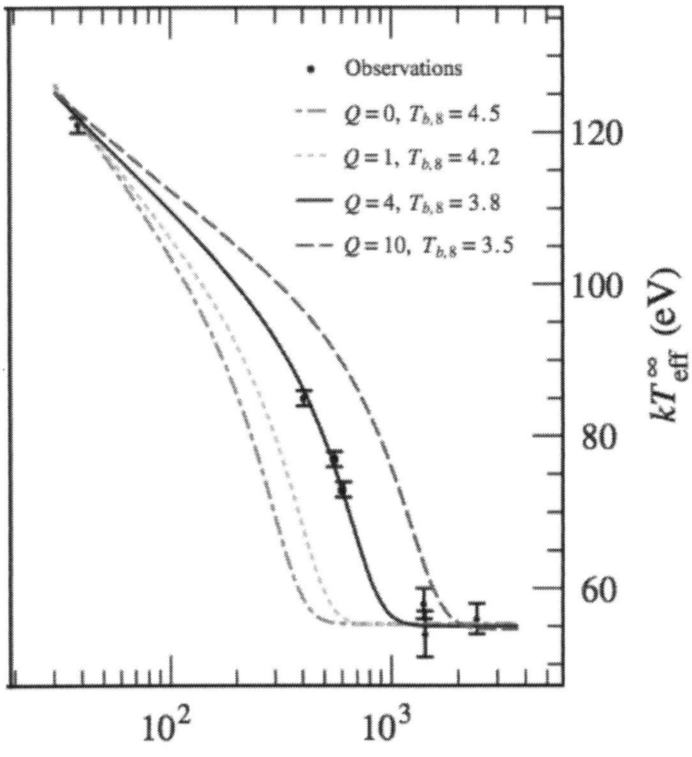

Figure 7. Surface temperature vs. time for the transiently accreting NS MXB 1659-29 after accretion stopped, vs. models with different choices of crustal impurities (the initial temperature after accretion is adjusted to match the first point) [74]. X-axis is days since accretion stopped, y-axis is NS surface temperature. The level of impurities (elements that disrupt the crystalline structure) in the crust is tightly constrained.

2.6. Quiescent accreting neutron stars: crust temperature

Observations of transient NSs immediately after long periods of accretion allow measurements of the cooling behavior of the crust, which radiates the heat produced by nuclear fusion within the crust within a few years, coming back to equilibrium with the core [75]. Crustal cooling has now been tracked for five transients over several years [76, 77, 78, 79, 80].

The crustal cooling behavior is dominated by the crust's thermal conductivity, which is set by the level of defects in the crystalline crust, and its heat capacity, which is strongly reduced if the neutrons are superfluid. Thus, measuring crustal cooling constrains the composition of the crust, and thus how nuclear reactions occur in the crust [81]. Comparisons of the crust cooling curves with models (Fig. 7) show that the crust is thin, highly conductive (so crystalline with low impurities), and that the neutrons are indeed superfluid [82, 74]. A key remaining uncertainty is whether the measured NS surface temperature is affected by continuing, low-level accretion [78], which might appear to increase the rate of the temperature decay.

2.7. Young cooling neutron stars

The thermal behavior of NSs can also be studied by measuring the temperatures of young NSs of known ages, from 300 to roughly a million years old [63]. Here there is less uncertainty about the heat input and timescales (compared to transiently accreting NSs), but there is also a smaller range of temperatures. Allowing for different compositions of the outer layers of the crust, [84] showed that the known young NSs were consistent with a "minimal" (standard) cooling model, without the enhanced cooling mechanisms suggested to explain some of the transient NSs. However, selection effects may affect this, as identifying young NSs often requires relatively high temperatures. There are several supernova remnants which may contain young NSs, but sensitive, unsuccessful X-ray searches for them have constrained any such NSs to relatively low temperatures. Considering the young ages of these supernova remnants, such cold NSs so early

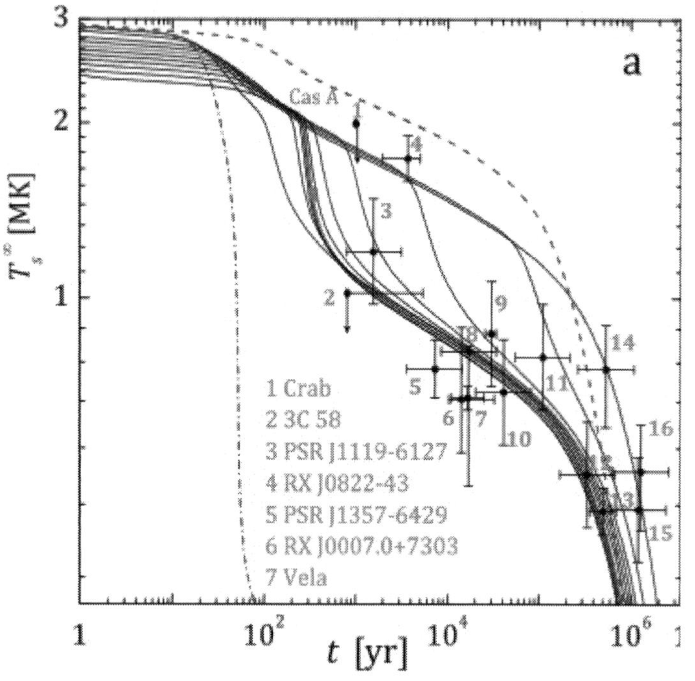

Figure 8. Cooling curves (solid lines) for NSs with a range of masses (varying other parameters can reach the blue dashed lines), compared with observations (crosses) of the temperatures of young NSs of different ages. The drop in model curves at around 100 years is due to the neutron superfluid transition. The slope of the model curve passing through the Cas A NS data point (red) also matches the observed temperature drop of the Cas A NS [83].

may require enhanced neutrino cooling [85].

The youngest known NS, in the 330-year-old supernova remnant Cassiopeia A (Cas A), shows evidence of rapid cooling, by ~3% over 10 years [86]. It is too hot now to have experienced enhanced cooling, as those mechanisms would operate throughout its history and make it very cold already. But its current temperature drop is too fast for normal cooling, requiring a recent transition to more rapid cooling. This can be explained if the neutrons in the NS core are currently transitioning to a superfluid state; as the neutrons bind into the quantum pairs required to enter the superfluid state, they release the "gap" energy of the pairing in the form of neutrinos [87]. Models of the temperature decline from such a transition match the data, for a critical temperature of the neutron superfluid of $5 - 9 \times 10^8$ K [88, 83]; see Fig. 8. The key question here is how well the calibration of the X-ray telescope (always a complex question) is known. This is the first observationally driven estimate of the superfluid critical temperature for any superfluid other than the helium-3 and -4 isotopes on Earth, and a beautiful illustration of the potential NSs can offer for constraining properties of the superdense matter.

Acknowledgments

I thank my collaborators in the research I have done in these areas, particularly the theorists Wynn Ho, George Rybicki, Peter Shternin, Ron Taam, and Dima Yakovlev. I am supported by an Ingenuity New Faculty Award and by an NSERC Discovery Grant.

References

[1] Lattimer JM, Prakash M. Phys. Rep.. 2007 Apr;442:109–165.
[2] Ozel F. Reports on Progress in Physics, in press. 2012 Oct;arXiv:1210.0916.
[3] Kaspi VM. Proceedings of the National Academy of Science. 2010 Apr;107:7147–7152.
[4] Pavlov GG, Zavlin VE. In: Bandiera R, Maiolino R, Mannucci F, editors. Texas in Tuscany. XXI Symposium on Relativistic Astrophysics; 2003. p. 319–328.
[5] Kaspi VM, Roberts MSE, Harding AK. Isolated neutron stars. In: Compact stellar X-ray sources. Lewin WHG, van der Klis M, editors. Cambridge Univ. Press; 2006. p. 279–339.

[6] Lyne AG, Graham-Smith F. Pulsar astronomy. Cambridge University Press (Cambridge Astrophysics Series, No. 16), 1990, 285 p.; 1990.
[7] Stairs IH. Science. 2004 Apr;304:547–552.
[8] Hessels JWT, Ransom SM, Stairs IH, Freire PCC, Kaspi VM, Camilo F. *Science*, 2006 Mar;311,1901–1904.
[9] Demorest PB, Pennucci T, Ransom SM, Roberts MSE, Hessels JWT. Nature. 2010 Oct;467:1081–1083.
[10] Kiziltan B, Kottas A, Thorsett SE. 2010 Nov., arXiv:1011.4291
[11] Kotake K, Sato K, Takahashi K. Reports on Progress in Physics. 2006 Apr;69:971–1143.
[12] Read JS, Markakis C, Shibata M, Uryū K, Creighton JDE, Friedman JL. Phys. Rev. D. 2009 Jun;79(12):124033.
[13] Mereghetti S. Astron. & Astroph. Rev.. 2008 Jul;15:225–287.
[14] Bhattacharyya S. Advances in Space Research. 2010 Apr;45:949–978.
[15] Lai D, Salpeter EE. ApJ. 1997 Dec;491:270.
[16] Cumming A, Zweibel E, Bildsten L. ApJ. 2001 Aug;557:958.
[17] Romani RW. ApJ. 1987 Feb;313:718–726.
[18] Ho, WCG, Heinke CO. Nature. 2009 Nov;462:71-73.
[19] Rajagopal M, Romani RW. ApJ. 1996 Apr;461:327.
[20] Zavlin VE, Pavlov GG, Shibanov YA. A&A. 1996 Nov;315:141–152.
[21] Cottam J, Paerels F, Mendez M. Nature. 2002 Nov;420:51–54.
[22] Cottam J, Paerels F, Méndez M, Boirin L, Lewin WHG, Kuulkers E, et al. ApJ. 2008 Jan;672:504–509.
[23] Galloway DK, Lin J, Chakrabarty D, Hartman JM. ApJ Lett. 2010 Mar;711:L148–L151.
[24] Chang P, Bildsten L, Wasserman I. ApJ. 2005 Aug;629:998–1007.
[25] Lin J, Özel F, Chakrabarty D, Psaltis D. ApJ. 2010 Nov;723:1053–1056.
[26] Haberl F. Astroph. & Space Sci.. 2007 Apr;308:181–190.
[27] van Kerkwijk MH, Kaplan DL. Astroph. & Space Sci.. 2007 Apr;308:191–201.
[28] Sanwal D, Pavlov GG, Zavlin VE, Teter MA. ApJ Lett. 2002 Jul;574:L61–L64.
[29] de Luca A, Salvetti D, Sartori A, Esposito P, Tiengo A, Zane S, et al. MNRAS. 2012 Mar;421:L72–L76.
[30] Suleimanov VF, Pavlov GG, Werner K. ApJ. 2010 May;714:630–635.
[31] Suleimanov V, Hambaryan V, Potekhin AY, van Adelsberg M, Neuhäuser R, Werner K. A&A. 2010 Nov;522:A111.
[32] Kargaltsev O, Durant M, Misanovic Z, Pavlov GG. Science. 2012 Aug;337:946–.
[33] Gotthelf EV, Perna R, Halpern JP. ApJ. 2010 Dec;724:1316–1324.
[34] Harding AK, Muslimov AG. ApJ. 2002 Apr;568:862–877.
[35] Pechenick KR, Ftaclas C, Cohen JM. ApJ. 1983 Nov;274:846–857.
[36] Morsink SM, Leahy DA, Cadeau C, Braga J. ApJ. 2007 Jul;663:1244–1251.
[37] Bogdanov S, Grindlay JE. ApJ. 2009 Oct;703:1557–1564.
[38] Lattimer JM, Prakash M. ApJ. 2001 Mar;550:426–442.
[39] Morsink SM, Leahy DA. ApJ. 2011 Jan;726:56.
[40] Elebert P, Reynolds MT, Callanan PJ, Hurley DJ, Ramsay G, et al. MNRAS. 2009 May;395:884–894.
[41] Lewin WHG, van Paradijs J, Taam RE. X-ray bursts. In: X-ray Binaries. W H G Lewin, J van Paradijs, & E P J van den Heuvel, editor. 1995. p. 175–232.
[42] van Paradijs J. ApJ. 1979 Dec;234:609–611.
[43] Damen E, Magnier E, Lewin WHG, Tan J, Penninx W, van Paradijs J. A&A. 1990 Oct;237:103–109.
[44] Güver T, Psaltis D, Özel F. ApJ. 2012 Mar;747:76.
[45] Majczyna A, Madej J, Joss PC, Różańska A. A&A. 2005 Feb;430:643–654.
[46] Suleimanov V, Poutanen J, Werner K. A&A. 2011 Mar;527:A139.
[47] Özel F, Güver T, Psaltis D. ApJ. 2009 Mar;693:1775–1779.
[48] Güver T, Özel F, Cabrera-Lavers A, Wroblewski P. ApJ. 2010 Apr;712:964–973.
[49] Güver T, Wroblewski P, Camarota L, Özel F. ApJ. 2010 Aug;719:1807–1812.
[50] Steiner AW, Lattimer JM, Brown EF. ApJ. 2010 Oct;722:33–54.
[51] Suleimanov V, Poutanen J, Revnivtsev M, Werner K. ApJ. 2011 Dec;742:122.
[52] Galloway DK, Lampe N. ApJ. 2012 Mar;747:75.
[53] Boutloukos S, Miller MC, Lamb FK. ApJ Lett. 2010 Sep;720:L15–L19.
[54] Güver T. Özel F, Psaltis D. ApJ. 2012 Mar;747:77.
[55] Zamfir M, Cumming A, Galloway DK. ApJ. 2012 Apr;749:69.
[56] Bhattacharyya S, Miller MC, Galloway DK. MNRAS. 2010 Jan;401:2–6.
[57] Rutledge RE, Bildsten L, Brown EF, Pavlov GG, Zavlin VE. ApJ. 2002 Oct;578:405–412.
[58] Heinke CO, Rybicki GB, Narayan R, Grindlay JE. ApJ. 2006 Jun;644:1090–1103.
[59] Webb NA, Barret D. ApJ. 2007 Dec;671:727–733.
[60] Guillot S, Rutledge RE, Brown EF. ApJ. 2011 May;732:88.

[61] Servillat M, Heinke CO, Ho WCG, Grindlay JE, Hong J, et al. MNRAS. 2012 Jun;423:1556–1561.
[62] Steiner AW, Lattimer JM, Brown EF. ApJL in press, 2012 May;arXiv:1205.6871.
[63] Haensel P, Zdunik JL. A&A. 1990 Jan;227:431–436.
[64] Brown EF, Bildsten L, Rutledge RE. ApJ Lett. 1998 Sep;504:L95.
[65] Yakovlev DG, Pethick CJ. ARAA. 2004 Sep;42:169–210.
[66] Levenfish KP, Haensel P. Astroph. & Space Sci.. 2007 Apr;308: 457–465.
[67] Heinke CO, Jonker PG, Wijnands R, Deloye CJ, Taam RE. ApJ. 2009 Feb;691:1035–1041.
[68] Colpi M, Geppert U, Page D, Possenti A. ApJ Lett. 2001 Feb;548:L175–L178.
[69] Wijnands R, Miller JM, Markwardt C, Lewin WHG, van der Klis M. ApJ Lett. 2001 Oct;560:L159–L162.
[70] Jonker PG, Bassa CG, Nelemans G, Juett AM, Brown EF, Chakrabarty D. MNRAS. 2006 Jun;368:1803–1810.
[71] Heinke CO, Jonker PG, Wijnands R, Taam RE. ApJ. 2007 May;660:1424–1427.
[72] di Salvo T, Burderi L, Riggio A, Papitto A, Menna MT. MNRAS. 2008 Oct;389:1851–1857.
[73] Patruno A, Bult P, Gopakumar A, Hartman JM, Wijnands R, et al. ApJ Lett. 2012 Feb;746:L27.
[74] Rutledge RE, Bildsten L, Brown EF, Pavlov GG, Zavlin VE, Ushomirsky G. ApJ. 2002 Nov;580:413–422.
[75] Cackett EM, Wijnands R, Miller JM, Brown EF, Degenaar N. ApJ Lett. 2008 Nov;687:L87–L90.
[76] Cackett EM, Brown EF, Cumming A, Degenaar N, Miller JM, Wijnands R. ApJ Lett. 2010 Oct;722:L137–L141.
[77] Fridriksson JK, Homan J, Wijnands R, Méndez M, Altamirano D, et al. ApJ. 2010 May;714:270–286.
[78] Díaz Trigo M, Boirin L, Costantini E, Méndez M, Parmar A. A&A. 2011 Apr;528:A150.
[79] Degenaar N, Wolff MT, Ray PS, Wood KS, Homan J, Lewin WHG, et al. MNRAS. 2011 Apr;412:1409–1418.
[80] Brown EF, Cumming A. ApJ. 2009 Jun;698:1020–1032.
[81] Horowitz CJ, Caballero OL, Berry DK. Phys. Rev. E. 2009 Feb;79(2):026103.
[82] Shternin PS, Yakovlev DG, Haensel P, Potekhin AY. MNRAS. 2007 Nov;382:L43–L47.
[83] Page D, Geppert U, Weber F. Nuclear Physics A. 2006 Oct;777:497–530.
[84] Kaplan DL, Frail DA, Gaensler BM, Gotthelf EV, Kulkarni SR, et al. ApJ Supp. 2004 Jul;153:269–315.
[85] Shternin PS, Yakovlev DG, Heinke CO, Ho WCG, Patnaude DJ. MNRAS. 2011 Mar;412:L108–L112.
[86] Heinke CO, Ho WCH ApJ Lett. 2010 Aug;719:L167–L171.
[87] Flowers E, Ruderman M, Sutherland P. ApJ. 1976 Apr;205:541–544.
[88] Page D, Prakash M, Lattimer JM, Steiner AW. Physical Review Letters. 2011 Feb;106(8):081101.

Journal of Physics: Conference Series **432** (2013) 012002 doi:10.1088/1742-6596/432/1/012002

Detailed measurements of bottomonium suppression in PbPb collisions at 2.76 TeV with CMS

Guillermo Breto Rangel (on behalf of the CMS collaboration)
University of California at Davis, One Shields Ave, Davis, CA 95616
E-mail: guillermo.breto.rangel@cern.ch

Abstract. The three Υ states can be separated using the CMS experimental apparatus via their dimuon decays in both pp and heavy-ion collisions. A suppression of the $\Upsilon(1S)$, $\Upsilon(2S)$ and $\Upsilon(3S)$ mesons is observed in PbPb collisions at $\sqrt{s_{NN}}$ = 2.76 TeV, compared to the yield in pp collisions scaled by the number of inelastic nucleon-nucleon collisions. Furthermore, a suppression of the excited Υ states has been measured with respect to the Υ ground state, expressed as a double ratio $[\Upsilon(nS)/\Upsilon(1S)]_{PbPb}$ / $[\Upsilon(nS)/\Upsilon(1S)]_{pp}$ with n = 2, 3, and 2 + 3. The centrality dependence of the double ratio, as well as the nuclear modification factors (R_{AA}) of the $\Upsilon(1S)$ and $\Upsilon(2S)$ states are presented, based on the analysis of the full data sample collected during the 2011 PbPb run and the 2010 pp run, which corresponds to an integrated luminosity of 150 μb^{-1} and 231 nb^{-1} respectively. The three Υ states suppression pattern is ordered as $R_{AA}^{\Upsilon(1S)} > R_{AA}^{\Upsilon(2S)} > R_{AA}^{\Upsilon(3S)}$.

1. Introduction

Quarkonia, bound states of a heavy quark and its antiquark, are very different kind of hadrons. For the ground states J/ψ and Υ the binding energies are around 0.6 and 1.2 GeV, respectively, and thus much larger than the typical hadronic scale $\Lambda \sim 0.2$ GeV; as a consequence, they are also much smaller, with radii of about 0.1 and 0.2 fm. It is therefore expected that they can survive in a quark-gluon plasma through some range of temperatures above T_c. The higher excited quarkonium states are less tightly bound and hence larger. Since deconfinement is related to colour screening [1], the critical quantity for dissociation of a bound state is the relation of binding energy of the state to the Debye screening radius of the medium. We therefore expect that the different quarkonium states have different "dissociation temperatures" in a quark-gluon plasma. Hence the spectral analysis of in-medium quarkonium dissociation should provide a QGP thermometer [2].

In particular, the Υ in heavy-ion collisions was suggested to be a theoretically and experimentally cleaner probe of the deconfined medium due to the larger mass of the bottom quark. For the bottomonium the effects of initial state nuclear suppression are expected to be reduced. Moreover, heavier quarks allow for cleaner theoretical treatments based on potential models because these models rely on taking the non-relativistic limit; and the heavier the quark the more reliable the treatment [4]. Furthermore, since bottom $Q\bar{Q}$ pairs are relatively rare within the plasma, the probability for regeneration of bottomonium states through recombination is much smaller than for charm quarks [3].

Finally, the three $\Upsilon(nS)$ states, characterized by similar decay kinematics and production mechanisms but distinct binding energies, further enable the measurement of relative state suppression, where common experimental and theoretical factors, and their respective uncertainties, cancel.

2. Muon reconstruction

Quarkonia are identified through their dimuon decay. The silicon pixel and strip tracker measures charged-particle trajectories for the range $|\eta| < 2.5$. The tracker consists of 66M pixel and 10M strip detector channels, providing a vertex resolution of $\sim 15\ \mu$m in the transverse plane. Muons are detected for the $|\eta| < 2.4$ range, with detection planes based on three technologies: drift tubes, cathode strip chambers, and resistive plate chambers. A detailed description of the CMS detector can be found elsewhere [5]. Its central feature is a superconducting solenoid of 6 m internal diameter, providing a magnetic field of 3.8 T. Due to the strong magnetic field and the fine granularity of the silicon tracker, the transverse momentum measurement based on information from the silicon tracker alone has a resolution between 1 and 2% for a typical muon in this analysis. CMS is therefore very well suited to measure dimuons. In this analysis the pp data was reconstructed using the heavy ion algorithm and the same nominal cuts (muons with $|\eta| < 2.4$ and $p_T > 4.0$ GeV/c). Incidentally, these muon cuts coincide with the ones used in our previous analysis [6]. .

3. Quantifying the observation of the relative suppression through the double ratio

In order to extract the $\Upsilon(nS)$ yields we performed an extended unbinned maximum likelihood fit to the invariant mass spectra shown in Fig. 1, similar to our previous analysis described in Refs. [6, 7]. The measured mass lineshape of each $\Upsilon(nS)$ state is parameterized by a "Crystal Ball" (CB) function. The mass differences between the states are fixed to their world average values and the mass resolution is forced to scale with the resonance mass.

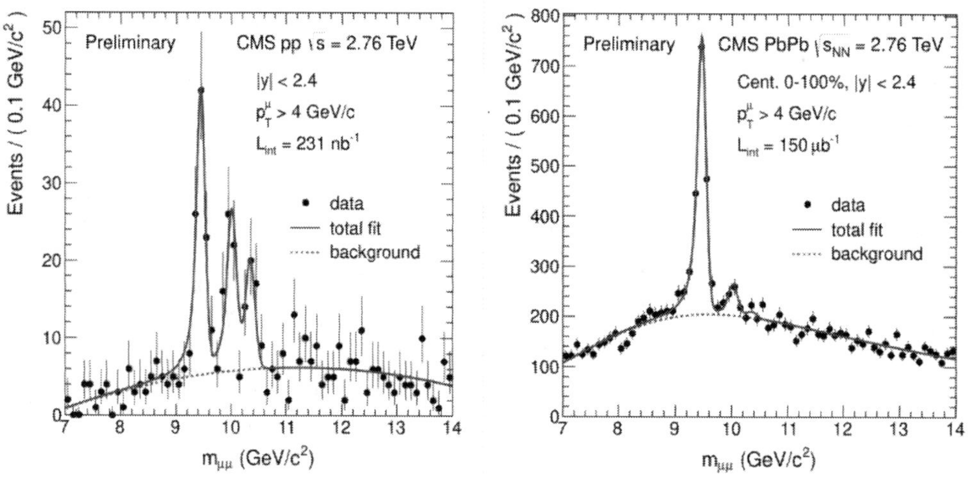

Figure 1. Simultaneous fit to the pp *(left)* and the much larger statistics 2011 *(right)* PbPb datasets.

The background model for the PbPb dataset consists of an exponential function multiplied by an error function describing the low-mass turn-on due to the kinematics cuts of the analysis.

This nominal model accurately describes the mass sidebands in the opposite-sign muon signal sample, as well as the alternative estimates of the shape of the combinatorial background obtained from like-sign muon pairs or via a "track-rotation" method. In the latter method [11] the azimuthal angular coordinate of one of the muon tracks is rotated by 180 degrees. The comprehensive studies performed on the background described the opposite sign background accurately and the minor differences were used to estimate systematic uncertainties.

The systematic uncertainties from the fitting procedure are evaluated by varying the fit function. An additional systematic uncertainty (1%), estimated from MC simulation, is included to account for

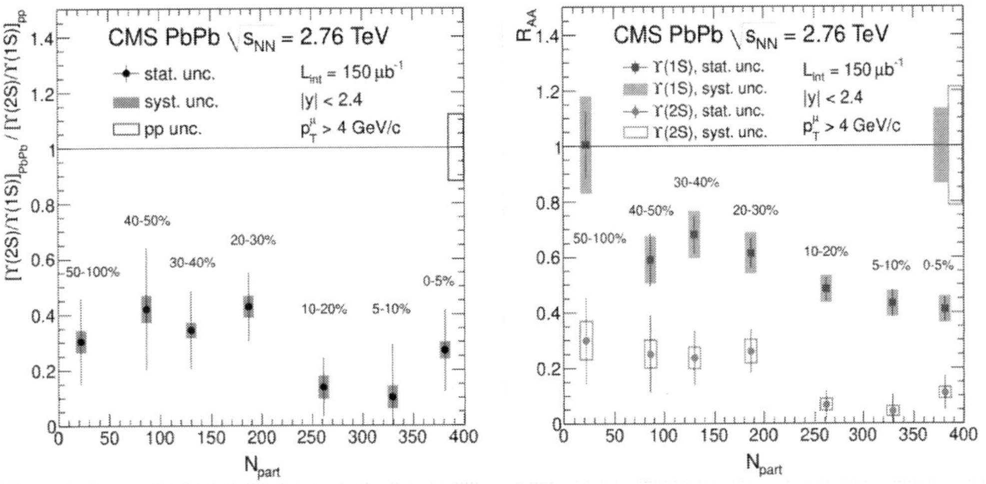

Figure 2. Centrality dependence of the double ratio *(left)* and of the nuclear modification factors *(right)* for the Υ(1S) and Υ(2S) states. The relative uncertainties from N_{part}-independent quantities (pp yields and, for the R_{AA}, also integrated luminosity) are represented by the boxes at unity, and are not included in the data points as these uncertainties do not affect the point-to-point trend. The event centrality bins used are indicated by percentage intervals.

possible imperfect cancellations of acceptance and efficiency.

The measurement of the ratio of the Υ(nS)/Υ(1S) ratios in PbPb and pp collisions, referred as the double ratio ($[\Upsilon(nS)/\Upsilon(1S)]_{PbPb}$ / $[\Upsilon(nS)/\Upsilon(1S)]_{pp}$), benefits from an almost complete cancellation of possible acceptance or efficiency differences among the reconstructed resonances. The simultaneous fit to the PbPb and pp mass spectra gives the double ratio Υ(2S)/Υ(1S) $= 0.21 \pm 0.07(stat) \pm 0.02(syst.)$ and for the Υ(3S) we set an upper limit at 95% confidence level (CL), based on the Feldman-Cousins statistical method [12], yielding a double ratio Υ(3S)/Υ(1S) < 0.17

Assuming the absence of suppression of the excited states relative to the Υ(1S) state hypothesis, the probability of observing a double ratio as discrepant or more to the value measured is very small corresponding to a significance that exceeds 5 σ in a one-sided gaussian tail test.

4. Describing the absolute suppression through the nuclear modification factor

Using both data sets, the production measured in PbPb collisions is compared to expectations from an independent superposition of nucleon-nucleon collisions typically expressed in terms of the nuclear modification factor:

$$R_{AA} = \frac{\mathscr{L}_{pp}}{T_{AA}N_{MB}} \frac{N_{PbPb}(Q\bar{Q})}{N_{pp}(Q\bar{Q})} \cdot \frac{\varepsilon_{pp}}{\varepsilon_{PbPb}}. \quad (1)$$

Here T_{AA} is the nuclear overlap function (ratio of the number of binary nucleon-nucleon collisions N_{coll} calculated from a Glauber model of the nuclear collision geometry [8, 9]) and the inelastic nucleon-nucleon cross section $\sigma_{inel}^{NN} = (64 \pm 5)$ mb at $\sqrt{s}$ = 2.76 $TeV, \mathscr{L}_{pp}$ is the pp luminosity, N_{MB} is the measured number of equivalent minimum bias events in PbPb, $\frac{N_{PbPb}(Q\bar{Q})}{N_{pp}(Q\bar{Q})}$ is the raw yield ratio, and $\frac{\varepsilon_{pp}}{\varepsilon_{PbPb}}$ the multiplicity dependent fraction of the efficiency ($\frac{\varepsilon_{pp}}{\varepsilon_{PbPb}} \sim 1.1$ for the most central bin).

The centrality integrated R_{AA} values are $0.56 \pm 0.08(stat.) \pm 0.07(syst.)$, $0.12 \pm 0.04(stat.) \pm 0.02(syst.)$, and lower than 0.10 (at 95% confidence level), for the Υ(1S), Υ(2S), and Υ(3S) states, respectively. A significant suppression of the Υ(nS) states in heavy-ion collisions compared to pp collisions is observed and this suppression is higher for the less strongly bound states. In other words, the $\Upsilon(1S)$ is the least suppressed and the $\Upsilon(3S)$ is the most suppressed of the three states.

The $\Upsilon(1S)$ and $\Upsilon(2S)$ suppressions are observed to increase with collision centrality. The suppression of $\Upsilon(2S)$ is stronger than that of $\Upsilon(1S)$ in all centrality ranges, including the most peripheral bin. The observed $\Upsilon(nS)$ yields contain contributions from decays of heavier bottomonium states and, thus, the measured suppression is affected by the dissociation of these states. This feed-down contribution to the $\Upsilon(1S)$ state was measured to be of the order of 50% [13], albeit in different kinematic ranges and center of mass energy than used here.

In addition to QGP formation, differences between quarkonium production yields in PbPb and pp collisions can also arise from cold-nuclear-matter effects. However, such effects should have a small impact on the double ratios reported here. Initial-state nuclear effects are expected to affect similarly each of the three Υ states, thereby canceling out in the double ratio. Final-state "nuclear absorption" becomes weaker with increasing energy [14] and is expected to be negligible for the $\Upsilon(1S)$ at the LHC [15].

5. Summary

In summary, the high statistics accumulated at the LHC allows the detailed studies of bottomonia. CMS has performed the measurements of the Υ(nS) mesons via their decay into dimuon pair in PbPb and pp collisions at 2.76 TeV. Based on a PbPb dataset of 150 μb^{-1}, CMS managed to separate the three Υ resonances and measured the double ratio $\Upsilon(2S)/\Upsilon(1S)$ in PbPb relative to pp collisions For the $\Upsilon(3S)$ double ratio a 95% CL upper limit was set. No clear centrality dependence of the double ratios can be inferred from the data, while such a dependence is observed when inspecting the individual $\Upsilon(1S)$ and $\Upsilon(2S)$ nuclear modification factors, R_{AA}. The nuclear modification factors for the three Υ states follow the sequential suppression hypothesis $R_{AA}^{\Upsilon(1S)} > R_{AA}^{\Upsilon(2S)} > R_{AA}^{\Upsilon(3S)}$. Aditionally, an upper limit on the $\Upsilon(3S)$ nuclear modification factor was set at 95% CL resulting in $R_{AA}(3S) < 0.1$. Finally, the quarkonia results seem qualitatively consistent with the presence of a deconfining medium. However, a quantitative understanding needs better statistics, in order to correct for possible cold nuclear matter effects (absorption/shadowing) at LHC with data from the upcoming pPb run, and quantitatively take into account the contributions of feed-down from the various higher lying quarkonia resonance states.

References

[1] T. Matsui and H. Satz, *Phys. Lett.* **B178** (1986) 416,doi:10.1016/0370-2693(86)91404-8.
[2] S. Digal *et al.*, *Phys. Rev. D* **64** (2001) 094015.
[3] X. Zhao and R. Rapp, *Nucl. Phys. A* **859** (2011) 114
[4] M. Strickland and D. Bazow, Nucl. Phys. A **879**, 25 (2012) [arXiv:1112.2761 [nucl-th]].
[5] CMS Collaboration, *JINST* **0803** (2008) S08004.
[6] CMS Collaboration, *Phys. Rev. Lett.* **107** (2011) 052302.
[7] CMS Collaboration, *Phys. Rev. D* **83** (2011) 112004, arXiv:1012.5545.
[8] CMS Collaboration, Phys.Rev.**C84:**024906,(2011).
[9] B. Alver *et. al.*, *Phys. Rev.* **C77** (2008) 014906.
[10] PDG Collaboration, *J. Phys.* **G37** (2010) 07502, 10.1088/0954-3899/37/7A/075021.
[11] STAR Collaboration, (2012). arXiv:1204.4244. Submitted to *Phys. Rev. D.*
[12] G. J. Feldman and R. D. Cousins, Phys. Rev. D **57** (1998) 3873.
[13] CDF Collaboration, *Phys. Rev. Lett.* **84** (2000) 2094.
[14] C. Lourenco, R. Vogt, and H. K. Woehri, *JHEP* **02** (2009) 014
[15] Z.-W. Lin and C. M. Ko, *Phys. Lett. B* **503** (2001) 104

Journal of Physics: Conference Series **432** (2013) 012003 doi:10.1088/1742-6596/432/1/012003

Static quark correlators and quarkonium properties at non-zero temperature

A. Bazavov and P. Petreczky

Physics Department, Brookhaven National Laboratory, Upton, NY 11973

E-mail: `bazavov@bnl.gov`, `petreczk@bnl.gov`

Abstract. We discuss different static quark correlators, including Wilson loops in 2+1 flavor QCD at non-zero temperature and their relation to in-medium quarkonium properties. We present lattice results on static correlation functions obtained with highly improved staggered fermion action and their implications for potential models.

1. Introduction

At sufficiently high temperatures strongly interacting matter undergoes a transition to a new state, often called quark-gluon plasma that is characterized by chiral symmetry restoration and color screening (see e.g. Ref. [1] for a current review). Experimentally such state of matter can be studied in relativistic heavy collisions. There was considerable interest in the properties and the fate of heavy quarkonium states at finite temperature since the famous conjecture by Matsui and Satz [2]. It has been argued that color screening in medium will lead to quarkonium dissociation above deconfinement, which in turn can signal quark-gluon plasma formation in heavy ion collisions. The basic assumption behind the conjecture by Matsui and Satz was the fact that medium effects can be understood in terms of a temperature-dependent heavy-quark potential. Color screening makes the potential exponentially suppressed at distances larger than the Debye radius and therefore it cannot bind the heavy quark and anti-quark once the temperature is sufficiently high. Based on this idea potential models at finite temperature with different temperature dependent potentials have been used over the last two decades to study quarkonium properties at finite temperature (see e.g. Refs. [3, 4] for recent reviews). It is not clear if and to what extent medium effects on quarkonium binding can be encoded in a temperature dependent potential. Effective field theory approach, namely the so-called thermal pNRQCD, can provide an answer to this question [5]. The notion of the potential can be defined using EFT approach both at zero and non-zero temperature. Thermal pNRQCD that will be discussed in the next section is based on the weak-coupling techniques. To understand the non-perturbative aspects of color screening as well as to test the reliability of the weak-coupling approach lattice calculations of the correlation functions of static quarks are needed. The correlation functions of static quarks that propagate around the periodic time direction $\tau = 1/T$ are related to the free energy of a static quark anti-quark pair. We will see that pNRQCD is a useful tool in understanding the temperature dependence of the static correlators. We also consider Wilson loops evaluated at time extent $\tau < 1/T$. They are naturally related to the static energy at non-zero temperature.

Journal of Physics: Conference Series **432** (2013) 012003 doi:10.1088/1742-6596/432/1/012003

In principle, it is possible to study the problem of quarkonium dissolution without any use of potential models. In-medium properties of different quarkonium states and/or their dissolution are encoded in spectral functions. Spectral functions are related to Euclidean meson correlation functions which can be calculated on the lattice. Reconstruction of the spectral functions from the lattice meson correlators turns out to be very difficult, and the corresponding results remain inconclusive. We will discuss the calculation of the spectral functions using potential models in the light of lattice calculations of Wilson loops.

2. pNRQCD at finite temperature

There are different scales in the heavy quark bound state problem related to the heavy quark mass m, the inverse size $\sim mv \sim 1/r$ and the binding energy $mv^2 \sim \alpha_s/r$. Here v is the typical heavy quark velocity in the bound state and is considered to be a small parameter. Therefore it is possible to derive a sequence of effective field theories using this separation of scales (see Refs. [6, 7] for recent reviews). Integrating out modes at the highest energy scale $\sim m$ leads to an effective field theory called non-relativistic QCD or NRQCD, where the pair creation of heavy quarks is suppressed by powers of the inverse mass and the heavy quarks are described by non-relativistic Pauli spinors [8]. At the next step, when the large scale related to the inverse size is integrated out, the potential NRQCD or pNRQCD appears. In this effective theory the dynamical fields include the singlet $\mathrm{S}(\mathrm{r},\mathrm{R})$ and octet $\mathrm{O}(\mathrm{r},\mathrm{R})$ fields corresponding to the heavy quark anti-quark pair in singlet and octet states respectively, as well as light quarks and gluon fields at the lowest scale $\sim mv^2$. The Lagrangian of this effective field theory has the form

$$\begin{aligned}\mathcal{L} = \; & -\frac{1}{4}F^a_{\mu\nu}F^{a\,\mu\nu} + \sum_{i=1}^{n_f}\bar{q}_i\, i\not{D} q_i + \int d^3r\, \mathrm{Tr}\left\{\mathrm{S}^\dagger\left[i\partial_0 + \frac{\nabla_r^2}{m} - V_s(r)\right]\mathrm{S}\right. \\ & \left. +\mathrm{O}^\dagger\left[iD_0 + \frac{\nabla_r^2}{m} - V_o(r)\right]\mathrm{O}\right\} + V_A\,\mathrm{Tr}\left\{\mathrm{O}^\dagger\vec{r}\cdot g\vec{E}\,\mathrm{S} + \mathrm{S}^\dagger\vec{r}\cdot g\vec{E}\,\mathrm{O}\right\} \\ & +\frac{V_B}{2}\mathrm{Tr}\left\{\mathrm{O}^\dagger\vec{r}\cdot g\vec{E}\,\mathrm{O} + \mathrm{O}^\dagger\mathrm{O}\vec{r}\cdot g\vec{E}\right\} + \dots \, .\end{aligned} \tag{1}$$

Here the dots correspond to terms which are higher order in the multipole expansion [6]. The relative distance r between the heavy quark and anti-quark plays a role of a label, the light quark and gluon fields depend only on the center-of-mass coordinate R. The singlet $V_s(r)$ and octet $V_o(r)$ heavy quark potentials appear as matching coefficients in the Lagrangian of the effective field theory and therefore can be rigorously defined in QCD at any order of the perturbative expansion. At leading order

$$V_s(r) = -\frac{4}{3}\frac{\alpha_s}{r},\; V_o(r) = \frac{1}{6}\frac{\alpha_s}{r} \tag{2}$$

and $V_A = V_B = 1$. One can generalize this approach to finite temperature. However, the presence of additional scales makes the analysis more complicated [5]. The effective Lagrangian will have the same form as above, but the matching coefficients may be temperature-dependent. In the weak coupling regime there are three different thermal scales: T, gT and g^2T. The calculations of the matching coefficients depend on the relation of these thermal scales to the heavy quark bound-state scales [5]. To simplify the analysis the static approximation has been used, in which case the scale mv is replaced by the inverse distance $1/r$ between the static quark and anti-quark. The binding energy in the static limit becomes $V_o - V_s \simeq N\alpha_s/(2r)$. When the binding energy is larger than the temperature the derivation of pNRQCD proceeds in the same way as at zero temperature and there is no medium modifications of the heavy quark potential [5]. But bound state properties will be affected by the medium through interactions with ultra-soft gluons, in particular, the binding energy will be reduced and a finite thermal width will appear due to medium induced singlet-octet transitions arising from the dipole interactions in the pNRQCD Lagrangian [5] (c.f. Eq. (1)). When the binding energy is smaller than one of the thermal scales the singlet and octet potential will be temperature-dependent and will acquire an imaginary part [5]. The imaginary part of the potential arises because of the singlet-octet transitions induced by the dipole vertex as well as due to the Landau damping in the plasma, i.e. scattering of the gluons with space-like momentum off the thermal excitations in the plasma. In general, the thermal corrections to the potential go like $(rT)^n$ and $(m_D r)^n$ [5], where m_D denotes the Debye mass. Only for distances $r > 1/m_D$ there is an exponential screening. In this region the singlet potential has a simple form

$$V_s(r) = -\frac{4}{3}\frac{\alpha_s}{r}e^{-m_D r} + i\frac{4}{3}\alpha_s T\frac{2}{rm_D}\int_0^\infty dx\frac{\sin(m_D r\,x)}{(x^2+1)^2} - \frac{4}{3}\alpha_s(m_D + iT), \tag{3}$$

The real part of the singlet potential coincides with the leading order result of the so-called singlet free energy [9]. The imaginary part of the singlet potential in this limit has been first calculated in [10]. For small distances the imaginary part vanishes, while at large distances it is twice the damping rate of a heavy quark [11]. This fact was first noted in Ref. [12] for thermal QED.

The effective field theory at finite temperature has been derived in the weak-coupling regime assuming the separation of different thermal scales as well as Λ_{QCD}. In practice, the separation of these scales is not evident and one needs lattice techniques to test the approach. Lattice QCD is formulated in Euclidean time. Therefore the next section will be dedicated to the study of static quarks at finite temperature in Euclidean time formalism.

3. Static meson correlators in Euclidean time formalism

Consider static (infinitely heavy) quarks. The position of heavy quark anti-quark pair is fixed in space and propagation happens only along the time direction. With respect to the color a static quark anti-quark ($Q\bar{Q}$) pair can be in a singlet or in an octet state. Therefore we can define the following $Q\bar{Q}$ (meson) operators

$$J(\vec{x},\vec{y};\tau) = \bar{\psi}(\vec{x},\tau)U(\vec{x},\vec{y};\tau)\psi(\vec{y},\tau), \tag{4}$$
$$J^a(\vec{x},\vec{y};\tau) = \bar{\psi}(\vec{x},\tau)U(\vec{x},\vec{x}_0;\tau)T^aU(\vec{x}_0,\vec{y};\tau)\psi(\vec{y},\tau), \tag{5}$$

for singlet and octet state respectively. Here $U(\vec{x},\vec{y};\tau)$ are the spatial gauge transporters connecting $\vec{x}$ and $\vec{y}$, $\vec{x}_0$ is the coordinate of the center of mass of the meson and T^a are the $SU(3)$ group generators. We consider the correlation function of singlet and octet meson operators at maximal Euclidean time $\tau = 1/T$:

$$G_1(r,T,\tau=1/T) = \langle J(x,y,\tau=1/T)J^\dagger(x,y;0)\rangle, \tag{6}$$
$$G_8(r,T,\tau=1/T) = \tfrac{1}{8}\langle J^a(x,y,\tau=1/T)J^{a\dagger}(x,y;0)\rangle. \tag{7}$$

Integrating out the static quark fields ψ and replacing the quark propagators by temporal Wilson lines $L(\vec{x}) = \prod_{\tau=0}^{N_\tau-1}U_0(\vec{x},\tau)$ with $U_0(\vec{x},t)$ being the temporal links, we get the following expression for the above correlators:

$$G_1(r,T) = \frac{1}{3}\langle\mathrm{Tr}\left[L^\dagger(\vec{x})U(\vec{x},\vec{y};0)L(\vec{y})U^\dagger(\vec{x},\vec{y},1/T)\right]\rangle, \tag{8}$$
$$G_8(r,T) = \frac{1}{8}\langle\mathrm{Tr}L^\dagger(x)\mathrm{Tr}L(y)\rangle - \frac{1}{24}\langle\mathrm{Tr}\left[L^\dagger(x)U(x,y;0)L(y)U^\dagger(x,y,1/T)\right]\rangle, \tag{9}$$
$$r = |\vec{x}-\vec{y}|.$$

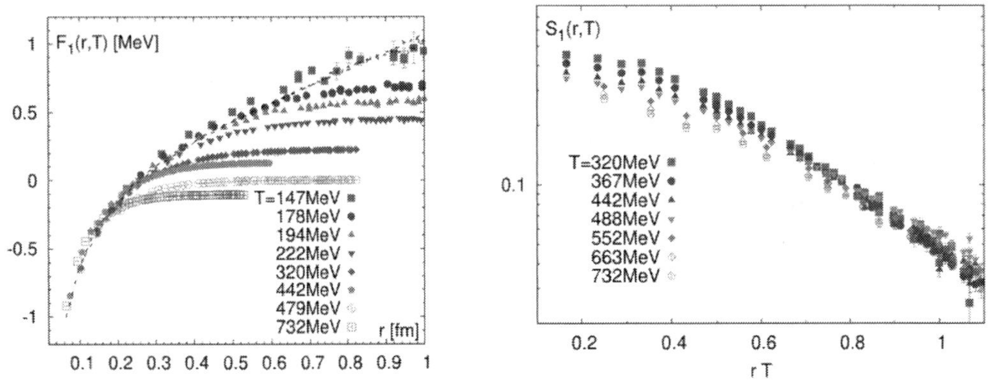

Figure 1. The singlet free energy (left) and the screening function (right) as function of the distance r at different temperatures calculated with the HISQ action.

The correlators depend on the choice of the spatial transporters $U(\vec{x}, \vec{y}; \tau)$. Typically, a straight line connecting points $\vec{x}$ and $\vec{y}$ is used as a path in the gauge transporters, i.e. one deals with time-like rectangular cyclic Wilson loops. In the special gauge, where $U(\vec{x}, \vec{y}; \tau) = 1$ the above correlators give the standard definition of the singlet and octet free energies of a static $Q\bar{Q}$ pair

$$\exp(-F_1(r,T)/T) = \frac{1}{3}\langle \mathrm{Tr}[L^\dagger(x)L(y)]\rangle, \tag{10}$$

$$\exp(-F_8(r,T)/T) = \frac{1}{8}\langle \mathrm{Tr}L^\dagger(x)\mathrm{Tr}L(y)\rangle - \frac{1}{24}\langle \mathrm{Tr}\left[L^\dagger(x)L(y)\right]\rangle. \tag{11}$$

One can also fix the Coulomb gauge and define the interpolating meson operators without the spatial transporters, and use the above expression to define the singlet and octet correlators.

The correlator $G(r,T) = \frac{1}{9}\langle \mathrm{Tr}L^\dagger(x)\mathrm{Tr}L(y)\rangle$ gives the free energy $F(r,T) = -T\ln G(r,T)$ of static quark anti-quark pair separated by distance r [13]. It can be expressed in terms of energy levels $E_n(r)$ of static quark anti-quark pair at $T = 0$ [14]

$$G(r,T) = \sum_{n=1}^{\infty} e^{-E_n(r)/T}. \tag{12}$$

It is tempting to rewrite Eq. (9) or Eq. (11) as

$$\exp(-F(r,T)/T) = \frac{1}{9}\exp(-F_1(r,T)/T) + \frac{8}{9}\exp(-F_8(r,T)/T) \tag{13}$$

and interpret this expression as the decomposition of the free energy of static $Q\bar{Q}$ pair into singlet and octet contributions [13, 15, 16, 17]. This decomposition is intuitively very appealing and should be valid in perturbation theory. However, it is problematic as $G_1(r,T)$ is path- or gauge- dependent. The problem is also evident if one writes the spectral decomposition of $G_1(r,T)$ [14]:

$$G_1(r,T) = \sum_{n=1}^{\infty} c_n(r) e^{-E_n(r,T)/T}. \tag{14}$$

The coefficients $c_n(r)$ are different from unity and are path- or gauge- dependent. The EFT approach can help to resolve this puzzle. One can use pNRQCD also in Euclidean time formulation [18] and study the Polyakov loop correlator in this framework. The Polyakov loop correlator can be written in terms of correlation function of singlet and octet fields [18]

$$G(r,T) = Z_s(r)\langle S(r,\tau = 1/T)S^\dagger(r,0)\rangle + Z_o(r)\langle O^a(r,\tau = 1/T)O^{a\dagger}(r,0)\rangle. \tag{15}$$

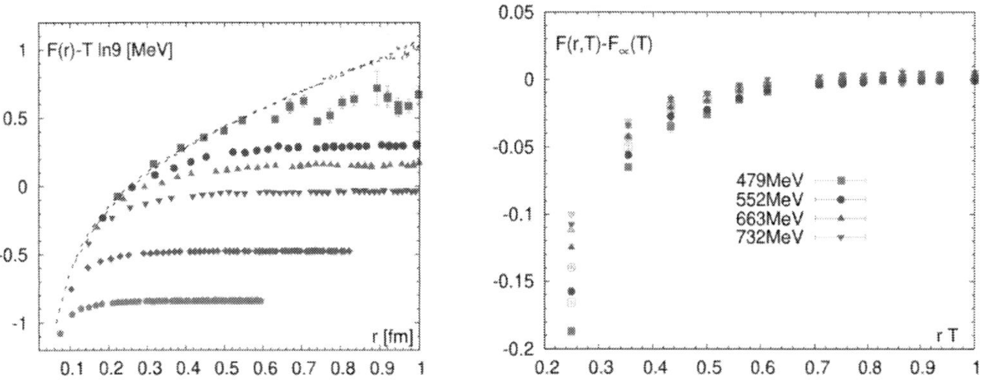

Figure 2. The free energy of static $Q\bar{Q}$ pair (left) and the difference $F(r,T) - F_\infty(T)$ (right) calculated with HISQ action as function of the distance r at different temperatures. In the right panel the filled symbols correspond to the lattice data, while the open symbols correspond to the values reconstructed from the singlet free energy. The legend in the left panel is the same as in Fig. 1 (left).

For $rT \ll 1$ one can use the zero temperature version of pNRQCD where the singlet and octet potentials are known up to 2-loop order. One can then show that $Z_s = Z_o = 1/9$ and thus in this limit the conjectured decomposition of the Polyakov loop correlator in terms of singlet and octet contribution is justified [18]

$$G(r,T) = \frac{1}{9}\exp(-V_s(r)/T) + \frac{8}{9}\exp(-V_o(r)/T). \tag{16}$$

The singlet and octet contributions are gauge-independent in this framework. When the binding energy $E_{bin} \sim \alpha_s/r$ is the largest scale in the problem the free energy of $Q\bar{Q}$ pair is dominated by the singlet contribution and is equal to the zero temperature potential [18] up to the term $T\ln 9$ coming from the normalization constant. When the temperature is much larger than the binding energy, i.e. $\alpha_s/(rT) \ll 1$ the exponentials in Eq. (16) can be expanded and we get

$$F(r,T) = \frac{\alpha_s^2}{(r^2T)}. \tag{17}$$

We see that despite no T-dependence of the potential in this limit, the free energy is strongly temperature dependent and it is very different from the potential. The complete next-to-leading order result can be found in Ref. [18].

Similarly, for the singlet correlator one can write

$$G_1(r,T) = \tilde{Z}_s(r)\langle S(r,\tau = 1/T)S^\dagger(r,0)\rangle \tag{18}$$

At leading order $\tilde{Z}_s(r) = 1$ and $F_1(r,T) \simeq V_s(r)$. At very high temperatures for $r \sim 1/m_D$ with $m_D = gT\sqrt{3/2}$ being the leading order Debye mass, the singlet and octet correlators can be calculated in the hard thermal loop (HTL) approximation [9]

$$F_1(r,T) = -\frac{4}{3}\frac{\alpha_s}{r}\exp(-m_D r) - \frac{4\alpha_s m_D}{3}, \quad F_8(r,T) = \frac{1}{6}\frac{\alpha_s}{r}\exp(-m_D r) - \frac{4\alpha_s m_D}{3}. \tag{19}$$

The singlet and octet free energies are gauge-independent at this order. At large distances the singlet and octet free energies approach a constant value $-\frac{4\alpha_s m_D}{3}$. This constant corresponds to the leading order result for the free energy of two isolated static quarks F_∞, which has been also calculated to next-to-leading order [19, 18]. The next-to-leading corrections are small and do not change the qualitative behavior of $F_\infty(T)$ which decreases with increasing temperatures. At leading order we have $(F_1(r,T) - F_\infty(T))/(F_8(r,T) - F_\infty(T)) = -8$.

The free energy of static $Q\bar{Q}$ pair was calculated at leading order long time ago [13, 15, 16]

$$F(r,T) = -\frac{1}{9}\frac{\alpha_s^2}{r^2}\exp(-2m_D r) - \frac{4\alpha_s m_D}{3}. \tag{20}$$

The above expression can also be obtained by inserting Eqs. (19) into Eq. (13) and expanding the exponentials to order α_s^2 thus confirming the validity of the decomposition and the partial cancellation of the singlet and octet contributions at leading order. The free energy was calculated at next-to-leading order for $r \simeq 1/m_D$ [16] but the decomposition into singlet and octet contributions was not shown. Because of the partial cancellation of the singlet and octet contributions for $r \sim 1/m_D$ we expect that $F(r,T) - F_\infty(T) \ll F_1(r,T) - F_\infty(T)$ at sufficiently high temperatures. In summary, the partial cancellation of the singlet and octet contribution to the free energy happens both at short and long distances and leads to its strong temperature dependence.

4. Lattice results on the free energy and the singlet free energy

We calculated Polyakov loop correlators as well as singlet correlators on the lattice in 2+1 flavor QCD using Highly Improved Staggered Quark (HISQ) action [20] on $24^3 \times 6$ and $16^3 \times 4$ lattices. The strange quark mass m_s was fixed to its physical value, while for the light quark masses we used $m_l = m_s/20$ that corresponds to the pion mass of about 160 MeV. The detailed choice of the lattice parameters is discussed in Ref. [20]. To calculate the singlet free energy we used the Coulomb gauge. The free energy and the singlet free energy have an additive divergent part that has to be removed by adding a normalization constant determined from the zero temperature potential. We used the normalization constants from Ref. [20]. The numerical results for the singlet free energy are shown in Fig. 1. At short distances the singlet free energy agrees with the zero temperature potential calculated in Ref. [20], while at large distances it approaches a constant value $F_\infty(T)$ equal to the excess free energy of two isolated static quarks. As the temperature increases the deviation from the zero-temperature potential shows up at shorter and shorter distances as the consequence of color screening. To explore the screening behavior in Fig. 1 we also show the combination $S(r,T) = r \cdot (F(r,T) - F_\infty(T))$ which we call the screening function. The screening function should decay exponentially. We indeed observe the exponential decay of this quantity at distances larger than $1/T$. Thus at high temperatures the behavior of the singlet free energy expected from the weak-coupling calculations seems to be confirmed by lattice QCD, at least qualitatively. Let us also mention that at high temperatures the behavior of the singlet free energy is similar to that observed in pure gauge theory [21, 22].

In Fig. 2 we show our results for the free energy of static $Q\bar{Q}$ pair as function of the distance at different temperatures. At short distances and low temperatures the free energy is expected to be dominated by the singlet contribution and we expect it to be equal to the zero temperature potential up to the term $T \ln 9$ coming from the normalization, see the discussion in the previous section. Therefore in the figure the numerical results have been shifted by $-T \ln 9$. Indeed, for the smallest temperature and the shortest distances $F(r,T) - T \ln 9$ is equal to the zero temperature potential shown as the dashed black line. At higher temperature $F(r,T)$ is very different from the zero-temperature potential. At large distance the free energy approaches a constant value $F_\infty(T)$ that decreases with increasing temperatures as expected (see discussions above). The temperature dependence of $F(r,T)$ is much larger than that of the singlet free energy. This is presumably due to the partial cancellation of the singlet and octet contribution discussed above. To verify this assertion we calculated $F(r,T) - F_\infty(T)$ using the numerical data for

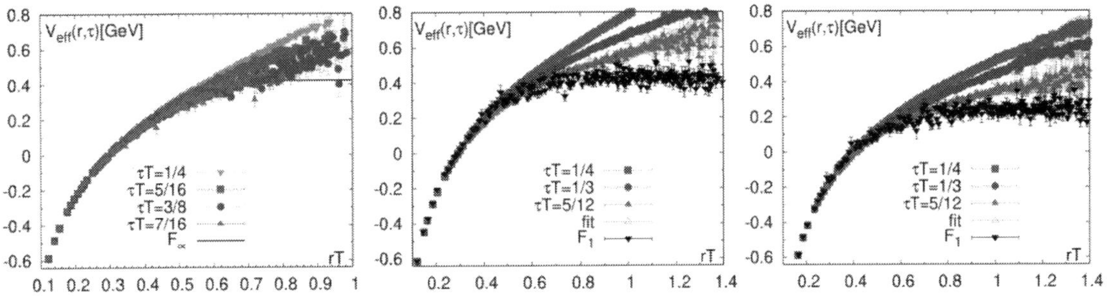

Figure 3. The effective potential $V_{eff}(r,\tau)$ as function of rT calculated for $48^3 \times 16$ lattice and $\beta = 7.5$ (left), $48^3 \times 12$ lattice and $\beta = 7.28$ (middle), and $48^3 \times 12$ lattice and $\beta = 7.5$ (right). The left, middle and right panels correspond to temperatures of 225 MeV, 249 MeV and 300 MeV respectively.

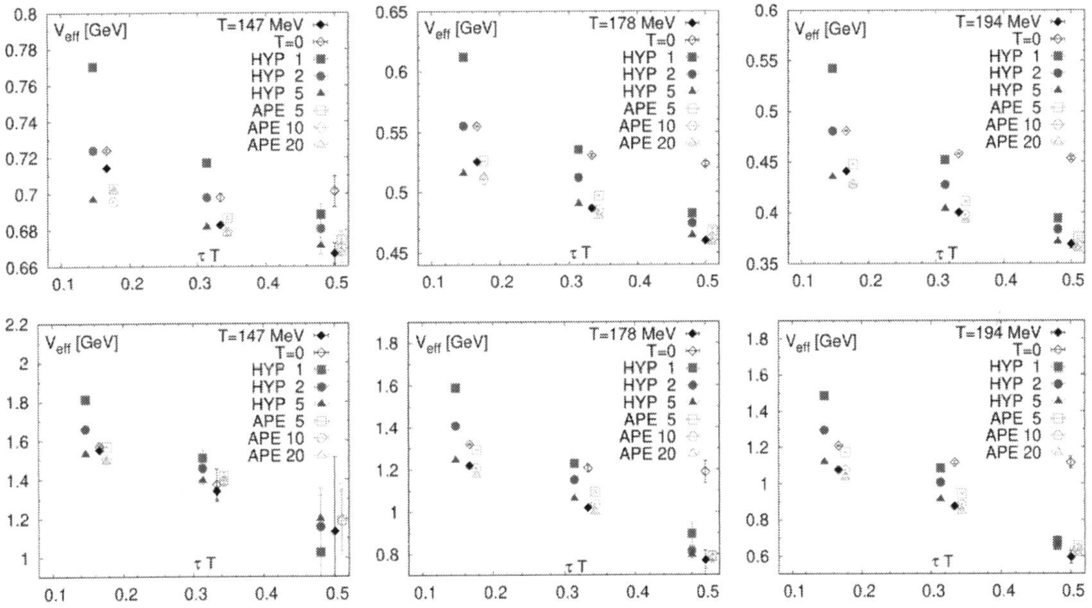

Figure 4. The effective potential $V_{eff}(r,\tau)$ as function of τT calculated on $24^3 \times 6$ lattices for three different temperatures. The upper panels correspond to $rT = 1/2$ while the lower panels correspond to $rT = 1$. Filled and open diamonds correspond to the Coulomb gauge results at finite and zero temperature respectively.

$F_1(r,T) - F_\infty(T)$ and the leading order relation $(F_1(r,T) - F_\infty(T))/(F_8(r,T) - F_\infty(T)) = -8$. The corresponding results are shown in the right panel of Fig. 2. As one can see from the figure the numerical data for $F(r,T)$ are in reasonable agreement with the ones reconstructed from this procedure. The reconstruction works better with increasing temperature. Thus the expected cancellation of the singlet and octet contributions to the free energy of static $Q\bar{Q}$ pair seems to be confirmed by lattice calculations.

5. Wilson loops at non-zero temperature

In the previous section we considered correlation function of static $Q\bar{Q}$ pair evaluated at Euclidean time $t = 1/T$. These correlators are related to the free energy of static $Q\bar{Q}$ pair.

One can consider Wilson loops for Euclidean times $t < 1/T$ which have no obvious relation to the free energy of a static $Q\bar{Q}$ pair. Wilson loops at non-zero temperature have been first studied in Refs. [23, 24] in connection with heavy quark potential at non-zero temperature and a spectral decomposition has been conjectured for the Wilson loops

$$W(r,\tau) = \int_0^\infty \sigma(\omega, r, T) e^{-\omega\tau}. \tag{21}$$

At zero temperature the spectral function is proportional to sum of delta functions $\sigma(r,\omega) = \sum_n c_n \delta(E_n(r) - \omega)$ and thus the spectral decomposition is just the generalization of Eq. (14). At high temperatures the spectral function will be proportional to sum of smeared delta functions and the position and the width of the lowest peak are related to real and imaginary part of the potential, respectively [23, 24]. Motivated by this we calculated Wilson loops on finite temperature lattices in 2+1 flavor QCD using the HISQ action with physical strange quark mass and light quark masses $m_l = m_s/20$. We performed calculations using $48^3 \times 16$ and $48^3 \times 12$ at $\beta = 7.5$ as well as $48^3 \times 12$ lattices at $\beta = 7.28$ ($\beta = 10/g^2$). These correspond to temperatures 225 MeV, 300 MeV and 249 MeV respectively. In addition we performed calculations using $24^3 \times 6$ lattices for the lattice parameters discussed in the previous section. One of the problems in extracting physical information from the Wilson loops on the lattice is large noise associated with them. To reduce the noise smeared gauge fields are used in the spatial gauge transporters $U(\vec{x}, \vec{y}; \tau)$ that enter the Wilson loops. Alternatively one can fix the Coulomb gauge and omit the spatial gauge connections, i.e calculate correlation function of Wilson lines of extent $t < 1/T$. This method was used by the MILC collaboration [25] as well as by the HotQCD collaboration [20] to calculate the static potential at zero temperature. We used both approaches. If the Wilson loop is dominated by the ground state for some value of τ we may try to extract the static energies at non-zero temperature from single exponential fits or from the ratio of the Wilson loops at two neighboring time-slices separated by single lattice spacing a

$$aV_{eff}(r,\tau) = \ln W(r, \tau/a)/W(r, \tau/a + 1). \tag{22}$$

At zero temperature for sufficiently large τ the effective potential $V_{eff}(r,\tau)$ should reach a plateau. For non-zero temperature the situation is more complicated due to the backward propagating contribution. Lattice calculations of the Wilson loops at non-zero temperature in SU(3) gauge theory show exponential decay in τ but at distance around $\tau T = 1$ the Wilson loops increase again [23, 24]. Similar behavior was observed in 2+1 flavor QCD [26]. While no temporal boundary conditions are imposed on static quarks the gluon fields are periodic in time and this may give rise to a contribution that propagates backward in time. Such backward propagating contribution was also observed in the study of bottomonium spectral function in NRQCD at non-zero temperature [27]. In Fig. 3 we show the effective potential calculated on $N_\tau = 12$ and 16 lattices as function of the distance r for different τ. For $N_\tau = 16$ lattice that corresponds to the temperature of 225 MeV plateau seems to be reached for $\tau T \le 1/2$. For these values of τ the backward propagating contribution is expected to be small. However, the statistical errors are very large for $rT > 1$. For $N_\tau = 12$ the effective potential does not reach a plateau for $\tau T \le 1/2$. We do not consider larger values of τ because of the backward propagation contribution. We attempted to extract the static energy by removing the backward propagating contribution and fitting the remainder by single exponential. The results are shown in Fig. 3 as open symbols and agree quite well with $V_{eff}(r, \tau = 5/(12T))$. Thus it is reasonable to assume that the static energy is well approximated by $V_{eff}(r, \tau = 5/(12T))$ at these temperatures. For $rT > 1$ the static energy is larger than the free energy. These findings are in agreement with earlier findings based on $24^3 \times 6$ lattices [26].

The above analysis as well as the analysis performed in Ref. [26] is based on using the Coulomb gauge. It is important to check how the results depend on the choice of the static meson operator. In addition it is interesting to study the onset of medium effects as function of τ. We calculated rectangular Wilson loops using smeared gauge fields in the spatial gauge transporters. To reduce the noise we used several iterations of APE [28] and HYP [29] smearings. Namely, we used 5, 10 and 20 steps of APE smearing and 1, 2 and 5 steps of HYP smearing. The numerical results are shown in Fig. 4. As expected the HYP smearing is more efficient than APE smearing but for the coarse lattices used in our study the difference is not that large. One needs 5 steps of HYP smearing or 10 steps of APE smearing to get results comparable to the Coulomb gauge results. Fig. 4 shows that except for the lowest temperature and distances $rT \leq 1/2$ the static potential is affected by the medium. While V_{eff} seems to reach a plateau at zero temperature no plateau is observed at finite temperature. Overall, the behavior of the Wilson loops with smeared spatial links is similar to the Coulomb gauge result if sufficient number of smearing steps is used.

6. Quarkonium spectral functions

Heavy meson correlation functions in Euclidean time $G(\tau, T)$ are related to the meson spectral functions $\sigma(\omega, T)$

$$G(\tau, T) = \int_0^\infty d\omega \sigma(\omega, T) \frac{\cosh(\omega(\tau - 1/(2T)))}{\sinh(\omega/(2T))}. \quad (23)$$

Attempts to reconstruct quarkonium spectral function from lattice QCD using the above equation and Maximum Entropy Method have been presented in Ref. [30, 31, 32]. In these studies it was concluded that charmonium ground state may survive in the deconfined medium up to the temperature 1.6 times the transition temperature or maybe even higher contrary to the expectations based on color screening. However, the reconstruction of meson spectral functions from Euclidean correlator is very difficult [33], and the spectral functions are also strongly modified by cutoff effects [34]. The suggested survival of quarkonium states in the deconfined medium is closely related to the weak temperature dependence of the Euclidean time quarkonium correlators [35]. Using potential models, quarkonium spectral functions have been calculated and it was shown that Euclidean correlation functions do not show significant temperature dependence even if bound states are dissolved due to the limited Euclidean time extent [36]. This seems to be confirmed by the study of spatial charmonium correlators which indicate the dissolution of the ground state for $T > 300$ MeV [37] as well as by the study of P-wave bottomonium correlators using lattice NRQCD [27], where larger values of the Euclidean time can be used.

Potential models can be related to pNRQCD. As the temperature increases the binding energy becomes smaller and eventual,ly will be the smallest scale in the problem: $E_{bin} \ll \Lambda_{QCD} \ll T,\ m_D,\ 1/r$, and all the other scales in the problem can be integrated out [38]. In this case the potential will be equal to the static energy. The real part of the static energy can be estimated using lattice QCD. In Ref. [38] a phenomenological form based on lattice QCD calculations of the singlet free energy was used for the real part of the potential while for the imaginary part of the potential the hard thermal loop result [10] was used. The spectral functions calculated in this approach show that most quarkonia states melt at temperatures $T > 250$MeV, while ground state bottomonium melts at temperature $T > 450$ MeV [38]. The estimates of the static energy obtained from Wilson loops and discussed in the previous section turn out to be very close to the phenomenological potential used in Ref. [38]. Therefore the above estimates of the maximal temperatures that permit the existence of quarkonium states still hold.

7. Conclusion

Color electric screening in high-temperature QCD can be studied using correlation functions of static mesons that go around the periodic Euclidean time direction. These are related to the free energy of static quark anti-quark pair. Determination of quarkonium spectral functions from the meson correlation functions calculated on the lattice is very difficult. The study of Wilson loops at non-zero temperature offers the possibility to extract the potential that can be used in potential model calculations to extract the quarkonium spectral functions.

Acknowledgements
This work was supported by U.S. Department of Energy under Contract No. DE-AC02-98CH10886. Computations have been performed on BlueGene/L computers of the New York Center for Computational Sciences (NYCCS) at Brookhaven National Laboratory and on clusters of the USQCD collaboration in JLab and FNAL.

References

[1] Petreczky P 2012 *J.Phys.* **G39** 093002 (*Preprint* 1203.5320)
[2] Matsui T and Satz H 1986 *Phys. Lett.* **B178** 416
[3] Mocsy A 2009 *Eur.Phys.J.* **C61** 705–710 (*Preprint* 0811.0337)
[4] Bazavov A, Petreczky P and Velytsky A 2009 (*Preprint* 0904.1748)
[5] Brambilla N, Ghiglieri J, Vairo A and Petreczky P 2008 *Phys. Rev.* **D78** 014017 (*Preprint* 0804.0993)
[6] Brambilla N, Pineda A, Soto J and Vairo A 2005 *Rev.Mod.Phys.* **77** 1423 (*Preprint* hep-ph/0410047)
[7] Brambilla N, Eidelman S, Heltsley B, Vogt R, Bodwin G *et al.* 2011 *Eur.Phys.J.* **C71** 1534 (*Preprint* 1010.5827)
[8] Caswell W and Lepage G 1986 *Phys.Lett.* **B167** 437
[9] Petreczky P 2005 *Eur. Phys. J.* **C43** 51–57 (*Preprint* hep-lat/0502008)
[10] Laine M, Philipsen O, Romatschke P and Tassler M 2007 *JHEP* **0703** 054 (*Preprint* hep-ph/0611300)
[11] Pisarski R 1993 *Phys.Rev.* **D47** 5589–5600
[12] Beraudo A, Blaizot J P and Ratti C 2008 *Nucl.Phys.* **A806** 312–338 (*Preprint* 0712.4394)
[13] McLerran L D and Svetitsky B 1981 *Phys. Rev.* **D24** 450
[14] Jahn O and Philipsen O 2004 *Phys. Rev.* **D70** 074504 (*Preprint* hep-lat/0407042)
[15] Gross D J, Pisarski R D and Yaffe L G 1981 *Rev.Mod.Phys.* **53** 43
[16] Nadkarni S 1986 *Phys. Rev.* **D33** 3738
[17] Nadkarni S 1986 *Phys. Rev.* **D34** 3904
[18] Brambilla N, Ghiglieri J, Petreczky P and Vairo A 2010 *Phys.Rev.* **D82** 074019 (*Preprint* 1007.5172)
[19] Burnier Y, Laine M and Vepsalainen M 2010 *JHEP* **1001** 054 (*Preprint* 0911.3480)
[20] Bazavov A, Bhattacharya T, Cheng M, DeTar C, Ding H *et al.* 2012 *Phys. Rev.* **D85** 054503 (*Preprint* 1111.1710)
[21] Digal S, Fortunato S and Petreczky P 2003 *Phys. Rev.* **D68** 034008 (*Preprint* hep-lat/0304017)
[22] Kaczmarek O, Karsch F, Petreczky P and Zantow F 2002 *Phys. Lett.* **B543** 41–47 (*Preprint* hep-lat/0207002)
[23] Rothkopf A, Hatsuda T and Sasaki S 2009 *PoS* **LAT2009** 162 (*Preprint* 0910.2321)
[24] Rothkopf A, Hatsuda T and Sasaki S 2012 *Phys.Rev.Lett.* **108** 162001 (*Preprint* 1108.1579)
[25] Aubin C, Bernard C, DeTar C, Osborn J, Gottlieb S *et al.* 2004 *Phys. Rev.* **D70** 094505 (*Preprint* hep-lat/0402030)
[26] Bazavov A and Petreczky P 2012 (*Preprint* 1210.6314)
[27] Aarts G, Kim S, Lombardo M, Oktay M, Ryan S *et al.* 2011 *Phys.Rev.Lett.* **106** 061602 (*Preprint* 1010.3725)
[28] Albanese M *et al.* (APE) 1987 *Phys. Lett.* **B192** 163
[29] Hasenfratz A and Knechtli F 2001 *Phys. Rev.* **D64** 034504 (*Preprint* hep-lat/0103029)
[30] Umeda T, Nomura K and Matsufuru H 2005 *Eur. Phys. J.* **C39S1** 9–26 (*Preprint* hep-lat/0211003)
[31] Asakawa M and Hatsuda T 2004 *Phys. Rev. Lett.* **92** 012001 (*Preprint* hep-lat/0308034)
[32] Datta S, Karsch F, Petreczky P and Wetzorke I 2004 *Phys. Rev.* **D69** 094507 (*Preprint* hep-lat/0312037)
[33] Wetzorke I, Karsch F, Laermann E, Petreczky P and Stickan S 2002 *Nucl.Phys.Proc.Suppl.* **106** 510–512 (*Preprint* hep-lat/0110132)
[34] Karsch F, Laermann E, Petreczky P and Stickan S 2003 *Phys. Rev.* **D68** 014504 (*Preprint* hep-lat/0303017)
[35] Petreczky P 2009 *Eur.Phys.J.* **C62** 85–93 (*Preprint* 0810.0258)
[36] Mocsy A and Petreczky P 2008 *Phys. Rev.* **D77** 014501 (*Preprint* 0705.2559)
[37] Karsch F, Laermann E, Mukherjee S and Petreczky P 2012 *Phys.Rev.* **D85** 114501 (*Preprint* 1203.3770)
[38] Petreczky P, Miao C and Mocsy A 2011 *Nucl.Phys.* **A855** 125–132 (*Preprint* 1012.4433)

Extreme QCD 2012 (xQCD) IOP Publishing
Journal of Physics: Conference Series **432** (2013) 012004 doi:10.1088/1742-6596/432/1/012004

Lattice QCD in Background Fields

Massimo D'Elia

Dipartimento di Fisica dell'Università di Pisa and INFN - Sezione di Pisa, Largo Pontecorvo 3, I-56127 Pisa, Italy

E-mail: delia@df.unipi.it

Abstract. We review some recent results about the exploration of strong interactions in presence of external background fields by means of lattice QCD simulations. We discuss in particular studies concerning the influence of the external fields on the non-perturbative properties of the QCD vacuum and of the QCD phase diagram.

1. Introduction

The influence of external background fields on Quantum Chromodynamics has attracted growing interest in the recent years. At a purely theoretical level, external fields represent useful probes to investigate the non-perturbative properties of strong interactions. Moreover, they are relevant to many phenomenological contexts. Large magnetic or chromomagnetic background fields, of the order of 10^{16} Tesla, i.e. $\sqrt{|e|B} \sim 1.5$ GeV, may have been produced at the time of the cosmological electroweak phase transition [1]. Large magnetic fields are generated in non-central heavy ion collisions (up to $\sim 10^{15}$ Tesla at LHC [2, 3]), where they may give rise, in presence of non-trivial topological charge fluctuations, to CP-odd effects like the *chiral magnetic effect* [4, 2, 5]: the net unbalance of chirality induced by the topological background would lead, in presence of a magnetic field strong enough to align the magnetic moments of quarks, to a net separation of electric charge along the field direction. Large magnetic (or even chromomagnetic [6]) fields are also expected in compact astrophysical objects like magnetars [7], where they may reach up to 10^{10} Tesla.

At the level of the QCD vacuum, one of the major effects predicted in presence of an external magnetic field is known as magnetic catalysis [8, 9, 10, 11, 12, 13, 14, 15, 16, 17, 18, 19, 20, 21, 22, 23, 24, 25, 26, 27, 28, 29, 30, 31, 32, 33, 34, 35, 36, 37, 38]. It consists in an enhancement of chiral symmetry breaking or spontaneous mass generation, it is common also to other systems characterized by chiral fermion fields and can be related to the dimensional reduction of particle dynamics in a strong magnetic field. While such effect seems naively related to fermion properties only, it has been found that the interrelation with gluon fields, which interact with electromagnetic fields only indirectly through quark loops, may play indeed a relevant role [39]: actually this is not a big surprise for a non-perturbative, strongly interacting theory. Unexpected phenomena have been found, like an inverse magnetic catalysis at high enough temperature [40, 41], which may be intimately related to the non-perturbative properties of strong interactions and therefore claim for a deeper investigation.

The phenomenology regarding the influence of electromagnetic fields on the properties of QCD vacuum is not limited to magnetic catalysis. One may indeed consider the vacuum as any other medium and study its reaction to various kinds of external stimulation, involving for instance its magnetic susceptibility [42, 43, 44, 45], its electrical conductivity [46] or even its susceptibility to CP-odd electromagnetic background fields, giving rise to an effective topological θ parameter [47].

Extending the interest beyond the properties of the vacuum state alone, one can consider the introduction of a background field, either magnetic of chromomagnetic, on the same grounds as any other external parameter, like e.g. a finite baryon chemical potential, capable of modifying the phase diagram of the theory. The interest in such issue is both theoretical and phenomenological, since strong background fields may be relevant to the cosmological QCD transition, to relativistic heavy ion collisions and to the physics of compact astrophysical objects. The main questions regarding the QCD transition are the following: 1) is a strong enough magnetic field capable of clearly separating the temperatures at which deconfinement and chiral symmetry restoration take place? 2) what is the dependence of the temperature and of the nature of the transition on the background field strength? 3) does any new, exotic phase emerge for strong enough background fields, like e.g. an anisotropic superconductive phase [48, 49, 50, 51]? Such questions are of utmost importance and they have indeed stimulated an intense activity in the recent years, based mostly on model computations [52, 53, 54, 55, 56, 57, 58, 59, 60, 61, 62, 63, 64, 65, 66, 67, 68, 69, 70, 71].

Lattice QCD simulations offer the possibility of a first principle investigation of such exciting issues, and indeed many studies have been performed in the last few years. Lattice studies in presence of electromagnetic fields have been done since long, originally with the purpose of studying the electromagnetic properties of hadrons [72, 73, 74, 75, 76]. Similarly to the continuum theory, electromagnetic fields affect quark propagation by a modification of the covariant derivative $D_\mu = \partial_\mu + i\,g A_\mu^a T^a + i\,q a_\mu$ which can be easily discretized on the lattice in terms of $SU(3)$ and $U(1)$ link variables, e.g. for the simplest symmetric discretization on an isotropic cubic lattice of spacing a:

$$D_\mu \psi \to \frac{1}{2a}\left(U_\mu(n) u_\mu(n)\psi(n+\hat{\mu}) - U_\mu^\dagger(n-\hat{\mu}) u_\mu^*(n-\hat{\mu})\psi(n-\hat{\mu})\right) \tag{1}$$

where $U_\mu \in SU(3)$ and $u_\mu = \exp(i\,q \int_{an}^{a(n+\hat{\mu})} dx_\mu a_\mu) \simeq \exp(i\,a\,q\,a_\mu(n)) \in U(1)$. In presence of periodic boundary conditions, which is the usual choice in lattice simulations in order to minimize finite size effects, one can show that a constant and uniform electromagnetic field can be imposed only if the components $F_{\mu\nu}$ are integer multiples of the elementary quanta $2\pi/(qa^2 L_\mu L_\nu)$, where L_μ is the lattice extension in direction μ in dimensionless lattice units [77, 78, 79, 80]. The finite lattice spacing (ultraviolet cutoff) places further limitations, setting a limiting value π/a^2 for $|qF_{\mu\nu}|$, but constraining in fact $|qF_{\mu\nu}| \ll \pi/a^2$ if one wants to keep away from unphysical saturation effects.

The fermion determinant is real and positive in the case of a magnetic field, allowing for a probabilistic interpretation of the path integral measure and thus for the application of standard Monte Carlo methods. However, it is easy to realize that real components $F_{0i} \neq 0$ in Euclidean space corresponds, when continued to Minkowski space, to purely imaginary electric fields [81, 76]. In order to have a real electric field, one needs imaginary components for the gauge potential in Euclidean space, which make the quark determinant complex: such sign problem hinders numerical simulations, similarly to what happens in presence of a baryon chemical potential. A possible approach, followed e.g. by lattice studies of the electric polarizabilities of hadrons, is to perform numerical simulations in presence of an imaginary electric field, then exploiting analytic continuation.

The introduction of a chromomagnetic field requires a different approach, since in this case the background variables are strictly related to the quantum gluon degrees of freedom: a standard procedure is that defined in the framework of the lattice Schrödinger functional [82, 83, 84,

85, 86, 87, 88, 89, 90]. One considers functional integration over Euclidean space-time, with frozen spatial field configurations at the temporal boundaries, $A_i^{ext1}(\vec{x}, \tau_1)$ and $A_i^{ext2}(\vec{x}, \tau_2)$: the quantum amplitude of passing from the field eigenstate $|A_i^{ext1}\rangle$ to the field eigenstate $|A_i^{ext2}\rangle$ is dominated, in the classical limit, by the configuration having the minimal action among those respecting the given boundary conditions; functional integration is therefore performed around such classical background field.

In the following we will discuss in detail only few of the many interesting topics mentioned above. In particular, in Section 2 we will discuss about the properties of the QCD vacuum and of the QCD phase diagram in presence of strong magnetic fields. Section 3 is devoted to lattice studies in presence of chromomagnetic background fields. Finally, in Section 4, we will discuss about the breaking of CP symmetry in the strong sector induced by CP-odd electromagnetic backgrounds.

2. QCD in strong magnetic fields

Exploratory lattice studies of the QCD vacuum in strong magnetic fields have considered magnetic catalysis in the quenched approximation for $SU(2)$ [91] and $SU(3)$ [43] gauge theories, while later studies [92, 39, 40, 93, 41] have considered effects deriving from the inclusion of dynamical fermions coupled to the external field. A useful quantity, which can be used to quantify magnetic catalysis, is the relative increment of the chiral condensate, defined as

$$r(B) = \frac{\langle\bar{\psi}\psi\rangle(B) - \langle\bar{\psi}\psi\rangle(B=0)}{\langle\bar{\psi}\psi\rangle(B=0)} \,. \tag{2}$$

The mass dependent additive renormalization of the chiral condensate cancels out in the numerator of Eq. (2); a residual additive renormalization remains in the denominator, leading to an overall, B independent multiplicative renormalization for $r(B)$. Chiral perturbation theory predicts a linear, non-analytic dependence of $r(B)$ on $|B|$ in the chiral limit [18], while for non-zero pion masses an analytic dependence is recovered, which is quadratic for small fields [20, 21]

$$r(B) \simeq \frac{(|e|B)^2}{96\pi^2 F_\pi^2 m_\pi^2} \,. \tag{3}$$

In Ref. [39], where a standard staggered fermion action and a plaquette gauge action have been adopted to investigate $N_f = 2$ QCD, an attempt has been made to determine the contribution to magnetic catalysis deriving from the indirect effect of the magnetic field, via dynamical quark loops, on the gluon field distribution. To that aim, the authors have measured $r(B)$ on gauge configurations sampled in presence of the magnetic field, however without including the magnetic field in the observable, i.e. in the Dirac operator which is inverted to determine the condensate; the relative increment of the condensate measured in this way is named "dynamical", $r^{dyn}(B)$. The complementary quantity is the "valence" contribution, $r^{val}(B)$, which is determined by including the magnetic field in the observable but not in the gluon field distribution. One can show that such separation can be done consistently in the limit of small external fields [39], in the sense that $r(B) \simeq r^{dyn}(B) + r^{val}(B)$, so that one can determine the contribution to catalysis stemming from indirect effects of the magnetic background on gluon fields. Numerical data, which are reported in Fig. 1 for a pion mass $m_\pi \sim 200$ MeV and for zero temperature, show that the separation is well defined in an extended range of magnetic field strengths, and that the dynamical contribution can be as large as 40 % of the total signal.

An improved investigation of magnetic catalysis in the QCD vacuum has appeared recently [41], making use of (stout) rooted staggered discretization of $N_f = 2+1$ QCD: the authors have adopted physical quark masses and performed the extrapolation to the continuum limit. As in Ref. [39], also in this case one observes, for the relative increment $r(B)$ (see Fig. 1

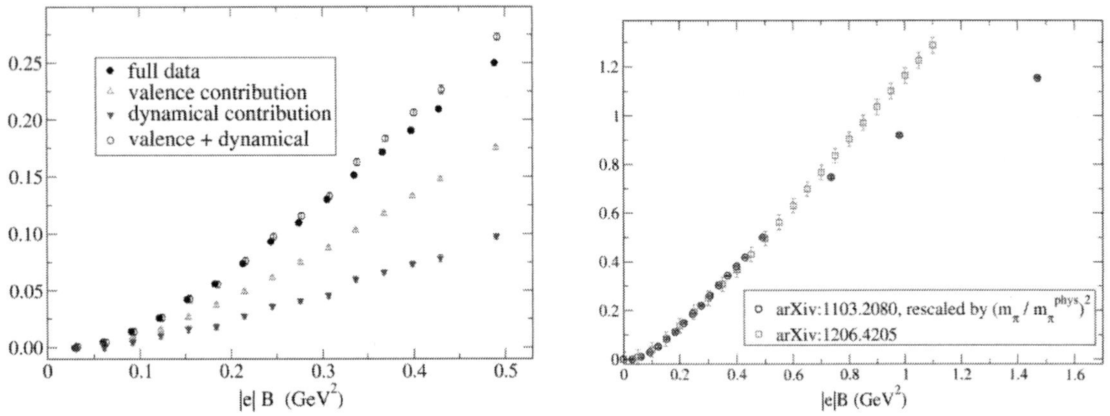

Figure 1. Left: relative increment of the average of the u and d quark condensates as a function of the magnetic field. $r(B), r^{val}(B), r^{dyn}(B)$ and $r^{val}(B) + r^{dyn}(B)$ are reported separately. From [39]. Right: relative increment of the average condensate, comparison of the data from Ref. [39] with those from Ref. [40], after rescaling the former to a physical pion mass according to Eq. (3).

of Ref. [41]), a quadratic dependence of $r(B)$ on $|e|B$ for small magnetic fields, followed by an almost linear dependence as $|eB| \gg m_\pi^2$.

It is interesting to observe that, in the regime of small fields, the results of Ref. [41] are compatible with those of Ref. [39] also quantitatively, if they are rescaled by an appropriate factor given by the ratio of the squared pion masses adopted in the two studies, as suggested by Eq. (3). A direct comparison is presented in Fig. 1 (right) [1]: a nice agreement is observed for small B, while the discrepancies visible at larger values of B can be ascribed to the finite cutoff (saturation) effects affecting the results at large B of Ref. [39] (where $a^{-1} \sim 650$ MeV), while those of Ref. [41] are already extrapolated to the continuum limit.

The issue of magnetic catalysis becomes even more interesting as one approaches the high temperature, deconfined phase: results from Refs. [40, 41] have shown that the growth of the chiral condensate with the magnetic field stops and turns into an inverse magnetic catalysis at high enough temperatures, in particular close and above the deconfinement transition. Such effect, which has not been observed in studies with unimproved actions or unphysical quark masses [92, 93] has different possible interpretations, like those based on the effects of a strong magnetic field on gluodynamics [19, 37] or that pointing to dimensional reduction of neutral pion dynamics induced by the magnetic field [94]; however, in fact the effect has not yet been fully understood.

A topic which is strictly related to inverse magnetic catalysis regards the fate of chiral symmetry restoration in a strong magnetic background. Present studies [92, 40, 93] agree on two facts: 1) there is no clear splitting of deconfinement and chiral symmetry restoration for $|e|B$ up to 1 GeV^2; 2) the magnetic field leads to a strengthening of the transition. However, while studies with unimproved actions and unphysical quark masses observe an increase in the pseudo-critical temperature T_c as the magnetic field is switched on [92, 93] (even if the observed increase is quite modest in Ref. [92] and of the order of 2% at $eB \sim 1$ GeV^2), the improved studies observing inverse magnetic catalysis also observe a decrease in T_c, which is of the order of 10-20% at $|e|B \sim 1$ GeV^2. It is not surprising that various improvements may turn the slowly

[1] The author thanks G. Endrodi for giving him access to the continuum extrapolated data of Ref. [41].

increasing function $T_c(B)$ of Ref. [92] into a decreasing function, however we believe that it would be of great importance to understand which aspect is more directly related to the appearance of inverse catalysis and to the different behavior observed for $T_c(B)$: is it just a problem of approach to the continuum limit, or can the difference be traced back to the different number of flavors and quark mass spectra? A clear answer to those questions may be given by considering each single aspect separately.

We will not discuss in due detail many other interesting aspects of the influence of strong magnetic fields on strong interactions. Among the studies regarding chiral dynamics, we mention those addressing the determination of the magnetic susceptibility of the chiral condensate, both in the quenched approximation [42, 43] and in full QCD [45]: such quantity is part of the total contribution to the magnetic susceptibility of the vacuum and recent unquenched results [45] show that it is of diamagnetic nature. Other studies have investigated the issue of the electrical conductivity of the QCD vacuum by measuring Euclidean vector current correlators, and pointed out that the vacuum may turn from an insulator to an anisotropic conductor for strong enough magnetic fields [46]. Also the possible emergence of an anisotropic superconductive phase for strong enough magnetic fields, due to charged ρ meson condensation, is currently under investigation [48, 49, 95, 50, 51].

Finally, let us briefly mention lattice studies aimed at the verification or quantification of the chiral magnetic effect. Early studies have looked directly at electric charge separation for gauge configurations containing a non-trivial topological background and in presence of the magnetic field [96], or at the increased fluctuations of the electric current along the direction of the magnetic field [97]; lattice studies about the modifications of the Dirac spectrum in presence of combined topological (instanton) and magnetic background [96] have been compared with analytic studies in Ref. [98]. Later lattice studies [99] have considered instead the introduction of an axial chemical potential μ_5, in order to create a net unbalance between left-handed and right-handed quarks, together with a background magnetic field, then checking for the presence of an electric current directed along the magnetic field direction and proportional to both B and μ_5, as predicted in Ref. [5].

3. QCD in strong chromomagnetic fields

A chromomagnetic background field influences directly gluodynamics, for that reason its effects on QCD dynamics have been studied both in pure Yang-Mills theories and in full QCD and for various kinds of backgrounds, going from those corresponding to an uniform magnetic field to those produced by magnetic monopoles [84, 85, 86, 87, 88, 89, 90]; in the following we shall focus on the case of a constant background field [86, 87, 88].

Also a chromomagnetic background field leads to an equal shift of chiral symmetry restoration and of deconfinement. Moreover, present lattice results indicate that, both for pure gauge and full QCD, the transition temperature decreases, as shown in Fig. 2 (left) [88], where the (pseudo)critical temperature T_c (expressed in units of the parameter $\Lambda \sim 6$ MeV), is plotted as a function of gB for the pure gauge theory and for $N_f = 2$ QCD; T_c is determined in various consistent ways, like looking at the peaks in the susceptibilities of the chiral condensate or of the Polyakov loop. The change in the critical temperature is, in general, quite larger than what happens in presence of a magnetic background: a possible interpretation may be based of the fact that a colored background affects directly gluodynamics, while the influence of magnetic fields on gluons is only indirect. An extrapolation of lattice results would hint, as shown in Fig. 2, at the presence of a deconfining transition at zero temperature for $\sqrt{gB}$ of the order of 1 GeV: such possibility should be further investigated in the future, for its possible cosmological and astrophysical implications. In particular, one should repeat the study of Ref. [88] in presence of physical quark masses (with the values adopted in [88], m_π is of the order of 400-500 MeV) and approaching the continuum limit: that would also permit to reach higher values of the external

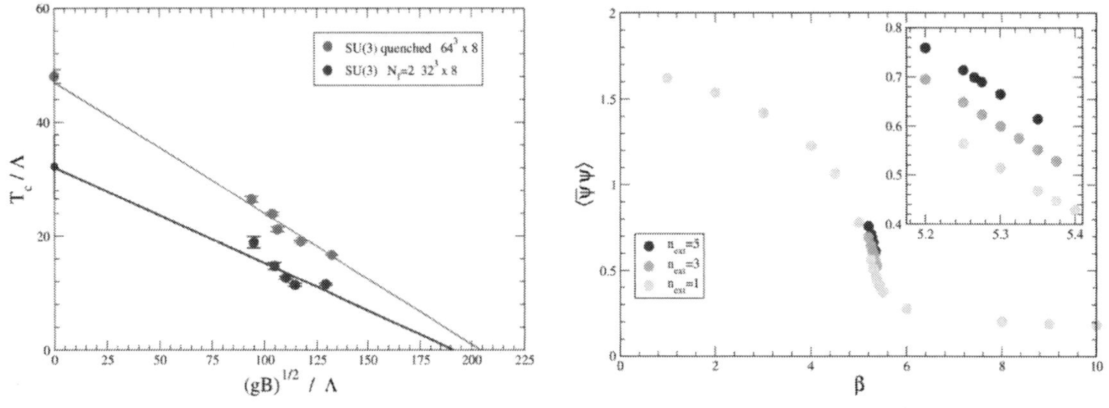

Figure 2. Left: dependence of the transition temperature on the chromomagnetic background field for pure gauge and for $N_f = 2$ QCD; quantities are expressed in terms of the parameter $\Lambda \sim 6$ MeV. Right: chiral condensate on a 32^3 lattice as a function of the inverse gauge coupling β (hence of the temperature) for different values of the chromomagnetic background field $gB = \pi n_{\rm ext}/(16a^2)$ and $N_f = 2$ QCD. The inset is a zoom on the transition region. Figure taken from [88].

field.

Finally, we notice that Ref. [88] also reports evidence about magnetic catalysis induced by the chromomagnetic background field. In Fig. 2 (right), in particular, we report the behavior of the chiral condensate as a function of the temperature (actually of the inverse gauge coupling) for a few values of gB: at least around the transition, where data for all chromomagnetic fields are available, the chiral condensate is an increasing function of gB. In this case, however, the transition temperature decreases anyway, meaning that inverse catalysis is not a necessary condition for a decreasing T_c, at least in presence of finite quark masses.

4. QCD in CP-odd electromagnetic fields

An aspect which clearly emerges from lattice simulations and from model studies [37, 19, 100] is that an electromagnetic background field may have a significant influence also on the gluonic sector, even if it is directly coupled only to quark fields. A possible way to further clarify such issue is to investigate how the explicit breaking of an exact symmetry by the background field propagates to the gluon sector. The case of charge conjugation symmetry has been discussed in Ref. [101], while the case of CP symmetry, which is discussed in the following, has been investigated in Ref. [47].

Let us consider a constant background electromagnetic field such that $F_{\mu\nu}\tilde{F}_{\mu\nu} \propto \vec{E}\cdot\vec{B} \neq 0$: such background induces an effective CP-violating θ term in the gluon sector, $\theta_{\rm eff}\, q(x) = \theta_{\rm eff}\frac{g^2}{64\pi^2} G^a_{\mu\nu}(x)\tilde{G}^a_{\mu\nu}(x)$ where $q(x)$ is the topological charge density. $\theta_{\rm eff}$ must be odd in $\vec{E}\cdot\vec{B}$ and, at the lowest order, one can write

$$\theta_{\rm eff} \simeq \chi_{CP}\, e^2 \vec{E}\cdot\vec{B} + O((\vec{E}\cdot\vec{B})^3) \tag{4}$$

where χ_{CP} is defined as the susceptibility of the QCD vacuum to CP-breaking electromagnetic fields. Such effect is complementary to the chiral magnetic effect, where a CP-violating non-Abelian background gives rise to charge separation, hence to an electric field parallel to a background magnetic field, thus propagating CP violation from the gluonic to the

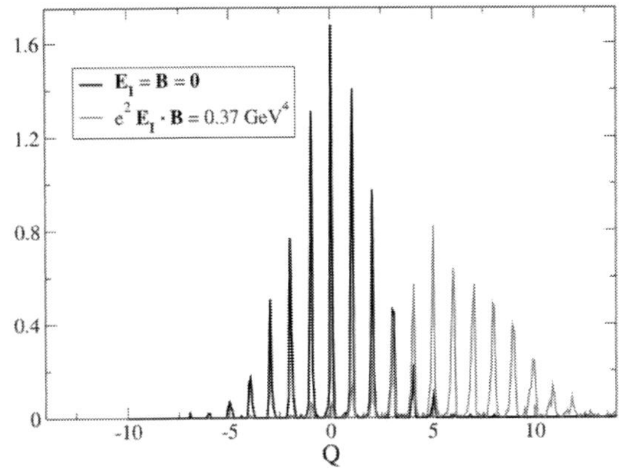

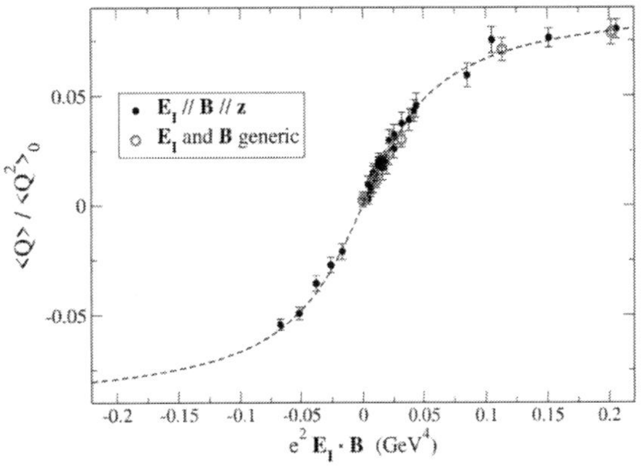

Figure 3. Left: topological charge distribution for two different backgrounds fields; data obtained on a 16^4 lattice for $m_\pi \simeq 480$ MeV and $a \simeq 0.12$ fm. Right: $\langle Q\rangle(\vec{E}_I, \vec{B})/\langle Q^2\rangle_0$ for various $(\vec{E}_I, \vec{B})$ on a 16^4 lattice for $m_\pi \simeq 480$ MeV and $a \simeq 0.3$ fm, the dashed line is a fit according to an inverse tangent dependence. Figures taken from Ref. [47].

electromagnetic sector. χ_{CP} is related to the strength of the effective pseudoscalar QED-QCD interaction, $\mathcal{L}^{eff}_{QED-QCD} = \chi_{CP}\, q(x)\, e^2\, \vec{E} \cdot \vec{B}$ [102, 103, 104, 105, 106].

As we have emphasized earlier, the introduction of electric background fields leads to a sign problem. A possible way out is to take the electric background field purely imaginary, $\vec{E} = i\,\vec{E}_I$. As a consequence, when $\vec{E}_I \cdot \vec{B} \neq 0$, one expects to produce a purely imaginary effective parameter $\theta_{\rm eff} = i\,\theta_{I\rm eff}$. On the other hand, taking θ as purely imaginary is also a possible way to avoid a similar sign problem in the study of θ dependence [107, 108, 109, 110, 111]. An imaginary θ_I adds a factor $\exp(\theta_I Q)$ to the probability distribution of gauge configurations, where $Q = \int d^4x q(x)$ is the topological charge. That shifts the distribution of Q, leading, at the lowest order in θ_I, to

$$\langle Q\rangle_{\theta_I} \simeq V\,\chi\,\theta_I = \langle Q^2\rangle_{\theta=0}\,\theta_I \tag{5}$$

where χ is the topological susceptibility at $\theta_I = 0$ and V is the spacetime volume. Therefore one can determine $\theta_{I\rm eff}$ as follows

$$\theta_{I\rm eff} \simeq \frac{\langle Q\rangle(\vec{E}_I, \vec{B})}{\langle Q^2\rangle_0} + O((\vec{E}_I \cdot \vec{B})^3) \tag{6}$$

where $\langle\cdot\rangle_0$ is the average taken at zero background field.

In Fig. 3 we report some results taken from Ref. [47], where QCD with two standard rooted staggered flavors has been investigated: the distribution of the topological charge, determined by a standard gluonic operator after cooling, loses its symmetry around zero, thus signalling CP symmetry breaking, when $\vec{E}_I \cdot \vec{B} \neq 0$, as expected. Moreover, the average value $\langle Q\rangle(\vec{E}_I, \vec{B})$, from which $\theta_{I\rm eff}$ can be extracted according to Eq. (6), seems to be dependent on $\vec{E}_I \cdot \vec{B}$ alone, thus permitting to extract χ_{CP} unambiguously from the small field behaviour of $\theta_{I\rm eff}$. After checking for finite lattice spacing and finite volume effects, and after analytic continuation to real electric fields and real θ, one obtains [47] $\chi_{CP} = (7 \pm 1)$ GeV^{-4} for $m_\pi = 480$ MeV and, preliminarly, $\chi_{CP} \simeq 10(1)$ GeV^{-4} for $m_\pi \simeq 280$ MeV, suggesting that χ_{CP} tends to increase when approaching the chiral limit.

5. Conclusions

The last few years have been characterized by a considerably activity regarding lattice QCD simulations in presence of external background fields. We have reviewed part of this activity, with more emphasis on issues regarding modifications of the QCD vacuum and of the QCD phase diagram in presence of the external fields.

Together with the verification of well established facts, like the presence of magnetic catalysis for the QCD vacuum, new phenomena have been found, like the inversion of catalysis at high enough temperatures, which still await for a clear explanation. This and other issues may be strictly related to one of the aspects that emerges more clearly from present studies, i.e. the fact that the gluon field distribution is strongly affected by the electromagnetic background field, even though indirectly, by means of quark loops: such aspect should be better investigated by future studies.

Acknowledgments

The author is grateful to P. Cea, L. Cosmai, M. Mariti, S. Mukherjee, F. Negro and F. Sanfilippo for collaboration on some of the topics discussed in this review. He also acknowledges communication with G. Endrodi and many useful discussions with C. Bonati, M. Chernodub, E. Fraga, K. Fukushima, V. Miransky and M. Ruggieri.

References

[1] T. Vachaspati, Phys. Lett. B **265**, 258 (1991).
[2] D. E. Kharzeev, L. D. McLerran and H. J. Warringa, Nucl. Phys. A **803**, 227 (2008).
[3] V. Skokov, A. Y. Illarionov and V. Toneev, Int. J. Mod. Phys. A **24**, 5925 (2009).
[4] A. Vilenkin, Phys. Rev. D **22**, 3080 (1980).
[5] K. Fukushima, D. E. Kharzeev and H. J. Warringa, Phys. Rev. D **78**, 074033 (2008).
[6] P. Cea, Int. J. Mod. Phys. D **13**, 1917 (2004). [astro-ph/0301578].
[7] R. C. Duncan and C. Thompson, Astrophys. J. **392**, L9 (1992).
[8] A. Salam and J. A. Strathdee, Nucl. Phys. B **90**, 203 (1975).
[9] A. D. Linde, Phys. Lett. B **62**, 435 (1976).
[10] S. Kawati, G. Konisi, H. Miyata, Phys. Rev. **D28**, 1537-1541 (1983).
[11] S. P. Klevansky and R. H. Lemmer, Phys. Rev. D **39**, 3478 (1989).
[12] H. Suganuma and T. Tatsumi, Annals Phys. **208**, 470 (1991).
[13] K. G. Klimenko, Z. Phys. C **54**, 323 (1992); Theor. Math. Phys. **94**, 393 (1993).
[14] S. Schramm, B. Muller, A. J. Schramm, Mod. Phys. Lett. **A7**, 973-982 (1992).
[15] K. G. Klimenko, B. V. Magnitsky, A. S. Vshivtsev, Nuovo Cim. **A107**, 439-452 (1994).
[16] V. P. Gusynin, V. A. Miransky and I. A. Shovkovy, Phys. Rev. Lett. **73**, 3499 (1994) [Erratum-ibid. **76**, 1005 (1996)].
[17] V. P. Gusynin, V. A. Miransky and I. A. Shovkovy, Phys. Lett. B **349**, 477 (1995).
[18] I. A. Shushpanov and A. V. Smilga, Phys. Lett. B **402**, 351 (1997).
[19] V. A. Miransky and I. A. Shovkovy, Phys. Rev. D **66**, 045006 (2002).
[20] N. O. Agasian, Phys. Atom. Nucl. **64**, 554 (2001) [Yad. Fiz. **64**, 608 (2001)] [hep-ph/0112341].
[21] T. D. Cohen, D. A. McGady and E. S. Werbos, Phys. Rev. C **76**, 055201 (2007).
[22] A. Y. .Babansky, E. V. Gorbar, G. V. Shchepanyuk, Phys. Lett. **B419**, 272-278 (1998).
[23] D. Ebert, K. G. Klimenko, M. A. Vdovichenko and A. S. Vshivtsev, Phys. Rev. D **61**, 025005 (2000).
[24] A. Goyal, M. Dahiya, Phys. Rev. D **D62**, 025022 (2000).
[25] N. O. Agasian and I. A. Shushpanov, Phys. Lett. B **472**, 143 (2000).
[26] D. Ebert, V. V. Khudyakov, V. C. Zhukovsky, K. G. Klimenko, Phys. Rev. **D65**, 054024 (2002).
[27] D. N. Kabat, K. -M. Lee, E. J. Weinberg, Phys. Rev. **D66**, 014004 (2002).
[28] V. G. Filev, C. V. Johnson, R. C. Rashkov and K. S. Viswanathan, JHEP **0710**, 019 (2007) [hep-th/0701001].
[29] T. Albash, V. G. Filev, C. V. Johnson and A. Kundu, JHEP **0807**, 080 (2008) [arXiv:0709.1547 [hep-th]].
[30] C. V. Johnson and A. Kundu, JHEP **0812**, 053 (2008).
[31] E. Rojas, A. Ayala, A. Bashir, A. Raya, Phys. Rev. **D77**, 093004 (2008).
[32] O. Bergman, G. Lifschytz and M. Lippert, Phys. Rev. D **79**, 105024 (2009).
[33] A. V. Zayakin, JHEP **0807**, 116 (2008).
[34] K. Fukushima and J. M. Pawlowski, Phys. Rev. D **86**, 076013 (2012) [arXiv:1203.4330 [hep-ph]].

[35] M. S. Alam, V. S. Kaplunovsky and A. Kundu, JHEP **1204**, 111 (2012) [arXiv:1202.3488 [hep-th]].
[36] A. Ayala, M. Loewe, J. C. Rojas and C. Villavicencio, Phys. Rev. D **86**, 076006 (2012).
[37] I. A. Shovkovy, arXiv:1207.5081 [hep-ph].
[38] F. Preis, A. Rebhan and A. Schmitt JHEP **1103**, 033 (2011); J. Phys. G G **39**, 054006 (2012); arXiv:1209.4468 [hep-ph].
[39] M. D'Elia and F. Negro, Phys. Rev. D **83**, 114028 (2011).
[40] G. S. Bali, F. Bruckmann, G. Endrodi, Z. Fodor, S. D. Katz, S. Krieg, A. Schafer and K. K. Szabo, JHEP **1202**, 044 (2012).
[41] G. S. Bali, F. Bruckmann, G. Endrodi, Z. Fodor, S. D. Katz and A. Schafer, Phys. Rev. D **86**, 071502 (2012) [arXiv:1206.4205 [hep-lat]].
[42] P. V. Buividovich, M. N. Chernodub, E. V. Luschevskaya and M. I. Polikarpov, Nucl. Phys. B **826**, 313 (2010) [arXiv:0906.0488 [hep-lat]].
[43] V. V. Braguta, P. V. Buividovich, T. Kalaydzhyan, S. V. Kuznetsov, M. I. Polikarpov, PoS **LATTICE2010**, 190 (2010). [arXiv:1011.3795 [hep-lat]].
[44] M. Frasca and M. Ruggieri, Phys. Rev. D **83**, 094024 (2011).
[45] G. S. Bali, F. Bruckmann, M. Constantinou, M. Costa, G. Endrodi, S. D. Katz, H. Panagopoulos and A. Schafer, Phys. Rev. D **86**, 094512 (2012) [arXiv:1209.6015 [hep-lat]].
[46] P. V. Buividovich, M. N. Chernodub, D. E. Kharzeev, T. Kalaydzhyan, E. V. Luschevskaya and M. I. Polikarpov, Phys. Rev. Lett. **105**, 132001 (2010).
[47] M. D'Elia, M. Mariti and F. Negro, arXiv:1209.0722 [hep-lat].
[48] M. N. Chernodub, Phys. Rev. D **82**, 085011 (2010); Phys. Rev. Lett. **106**, 142003 (2011).
[49] V. V. Braguta, P. V. Buividovich, M. N. Chernodub and M. I. Polikarpov, Phys. Lett. B **718**, 667 (2012) [arXiv:1104.3767 [hep-lat]].
[50] Y. Hidaka and A. Yamamoto, arXiv:1209.0007 [hep-ph].
[51] M. N. Chernodub, Phys. Rev. D **86**, 107703 (2012) [arXiv:1209.3587 [hep-ph]].
[52] N. O. Agasian and S. M. Fedorov, Phys. Lett. B **663**, 445 (2008).
[53] E. S. Fraga and A. J. Mizher, Phys. Rev. D **78**, 025016 (2008).
[54] J. K. Boomsma and D. Boer, Phys. Rev. D **81**, 074005 (2010).
[55] K. Fukushima, M. Ruggieri and R. Gatto, Phys. Rev. D **81**, 114031 (2010).
[56] A. J. Mizher, M. N. Chernodub and E. S. Fraga, Phys. Rev. D **82**, 105016 (2010).
[57] R. Gatto and M. Ruggieri, Phys. Rev. D **82**, 054027 (2010).
[58] S. -i. Nam, C. -W. Kao, Phys. Rev. D **83**, 096009 (2011).
[59] J. O. Andersen and R. Khan, Phys. Rev. D **85**, 065026 (2012).
[60] S. .Fayazbakhsh and N. Sadooghi, Phys. Rev. D **83**, 025026 (2011).
[61] N. Evans, T. Kalaydzhyan, K. -y. Kim and I. Kirsch, JHEP **1101**, 050 (2011).
[62] R. Gatto and M. Ruggieri, Phys. Rev. D **83**, 034016 (2011).
[63] A. Rabhi and C. Providencia, Phys. Rev. C **83**, 055801 (2011).
[64] K. Kashiwa, Phys. Rev. D **83**, 117901 (2011).
[65] B. Chatterjee, H. Mishra and A. Mishra, Phys. Rev. D **84**, 014016 (2011).
[66] V. Skokov, Phys. Rev. D **85**, 034026 (2012).
[67] E. S. Fraga and L. F. Palhares, Phys. Rev. D **86**, 016008 (2012) [arXiv:1201.5881 [hep-ph]].
[68] S. S. Avancini, D. P. Menezes, M. B. Pinto and C. Providencia, Phys. Rev. D **85**, 091901 (2012).
[69] J. O. Andersen and A. Tranberg, JHEP **1208**, 002 (2012) [arXiv:1204.3360 [hep-ph]].
[70] G. N. Ferrari, A. F. Garcia and M. B. Pinto, Phys. Rev. D **86**, 096005 (2012) [arXiv:1207.3714 [hep-ph]].
[71] E. S. Fraga, J. Noronha and L. F. Palhares, arXiv:1207.7094 [hep-ph].
[72] G. Martinelli, G. Parisi, R. Petronzio and F. Rapuano, Phys. Lett. B **116**, 434 (1982).
[73] C. W. Bernard, T. Draper, K. Olynyk and M. Rushton, Phys. Rev. Lett. **49**, 1076 (1982).
[74] W. Detmold, B. C. Tiburzi and A. Walker-Loud, Phys. Rev. D **73**, 114505 (2006) [hep-lat/0603026]; Phys. Rev. D **79**, 094505 (2009) [arXiv:0904.1586 [hep-lat]]; Phys. Rev. D **81**, 054502 (2010) [arXiv:1001.1131 [hep-lat]].
[75] B. C. Tiburzi, PoS LATTICE **2011**, 020 (2011) [arXiv:1110.6842 [hep-lat]]; Nucl. Phys. A **814**, 74 (2008) [arXiv:0808.3965 [hep-ph]]; Phys. Lett. B **674**, 336 (2009) [arXiv:0809.1886 [hep-lat]].
[76] A. Alexandru and F. X. Lee, PoS LATTICE **2008**, 145 (2008) [arXiv:0810.2833 [hep-lat]]; PoS LAT **2009**, 144 (2009) [arXiv:0911.2520 [hep-lat]].
[77] G. 't Hooft, Nucl. Phys. B **153**, 141 (1979).
[78] J. Smit and J. C. Vink, Nucl. Phys. B **286**, 485 (1987).
[79] P. H. Damgaard and U. M. Heller, Nucl. Phys. B **309**, 625 (1988).
[80] M. H. Al-Hashimi and U. J. Wiese, Annals Phys. **324**, 343 (2009).
[81] E. Shintani, S. Aoki, N. Ishizuka, K. Kanaya, Y. Kikukawa, Y. Kuramashi, M. Okawa and A. Ukawa *et al.*,

Phys. Rev. D **75**, 034507 (2007).
[82] M. Luscher, R. Narayanan, P. Weisz and U. Wolff, Nucl. Phys. B **384**, 168 (1992).
[83] M. Luscher and P. Weisz, Nucl. Phys. B **452**, 213 (1995).
[84] P. Cea and L. Cosmai, Phys. Lett. B **264**, 415 (1991).
[85] P. Cea, L. Cosmai and A. D. Polosa, Phys. Lett. B **392**, 177 (1997).
[86] P. Cea and L. Cosmai, JHEP **0302**, 031 (2003).
[87] P. Cea and L. Cosmai, JHEP **0508**, 079 (2005).
[88] P. Cea, L. Cosmai and M. D'Elia, JHEP **0712**, 097 (2007); PoS LAT **2006**, 062 (2006) [hep-lat/0610014]; PoS LAT **2007**, 295 (2007) [arXiv:0710.1449 [hep-lat]].
[89] S. Antropov, M. Bordag, V. Demchik and V. Skalozub, Int. J. Mod. Phys. A **26**, 4831 (2011).
[90] P. Cea, L. Cosmai and M. D'Elia, JHEP **0402**, 018 (2004).
[91] P. V. Buividovich, M. N. Chernodub, E. V. Luschevskaya and M. I. Polikarpov, Phys. Lett. B **682**, 484 (2010).
[92] M. D'Elia, S. Mukherjee, F. Sanfilippo, Phys. Rev. D **82**, 051501 (2010).
[93] E. -M. Ilgenfritz, M. Kalinowski, M. Muller-Preussker, B. Petersson and A. Schreiber, Phys. Rev. D **85**, 114504 (2012).
[94] K. Fukushima and Y. Hidaka, Phys. Rev. Lett. **110**, 031601 (2013) [arXiv:1209.1319 [hep-ph]].
[95] E. V. Luschevskaya and O. V. Larina, arXiv:1203.5699 [hep-lat].
[96] M. Abramczyk, T. Blum, G. Petropoulos and R. Zhou, PoS LAT **2009**, 181 (2009).
[97] P. V. Buividovich, M. N. Chernodub, E. V. Luschevskaya and M. I. Polikarpov, Phys. Rev. D **80**, 054503 (2009).
[98] G. Basar, G. V. Dunne and D. E. Kharzeev, Phys. Rev. D **85**, 045026 (2012) [arXiv:1112.0532 [hep-th]].
[99] A. Yamamoto, Phys. Rev. Lett. **107**, 031601 (2011).
[100] B. V. Galilo and S. N. Nedelko, Phys. Rev. D **84**, 094017 (2011).
[101] M. D'Elia, arXiv:1209.0374 [hep-lat].
[102] M. M. Musakhanov and F. C. Khanna, hep-ph/9605232.
[103] H. T. .Elze and J. Rafelski, In *Sandansky 1998, Frontier tests of QED and physics of the vacuum* 425-439 [hep-ph/9806389].
[104] H. T. .Elze, B. Muller and J. Rafelski, hep-ph/9811372.
[105] M. Asakawa, A. Majumder and B. Muller, Phys. Rev. C **81**, 064912 (2010).
[106] B. Muller and A. Schafer, Phys. Rev. C **82**, 057902 (2010) [arXiv:1009.1053 [hep-ph]].
[107] V. Azcoiti, G. Di Carlo, A. Galante and V. Laliena, Phys. Rev. Lett. **89**, 141601 (2002).
[108] B. Alles and A. Papa, Phys. Rev. D **77**, 056008 (2008).
[109] S. Aoki, R. Horsley, T. Izubuchi, Y. Nakamura, D. Pleiter, P. E. L. Rakow, G. Schierholz and J. Zanotti, arXiv:0808.1428 [hep-lat].
[110] H. Panagopoulos and E. Vicari, JHEP **1111**, 119 (2011).
[111] M. D'Elia and F. Negro, Phys. Rev. Lett. **109**, 072001 (2012).

Deconfinement to Quark Matter in Magnetars

Veronica Dexheimer
UFSC, Florianopolis, BR
Gettysburg College, Gettysburg, USA
E-mail: vantoche@gettysburg.edu

Rodrigo Negreiros
Instituto de Fisica, UFF, Niteroi, BR

Stefan Schramm
FIAS - Johann Wolfgang Goethe University, Frankfurt, DE

Abstract. We model magnetars as hybrid stars, which have a core of quark matter surrounded by hadronic matter. For this purpose, we use an extended version of the SU(3) non-linear realization of the sigma model in which the degrees of freedom change naturally from hadrons to quarks as the temperature/density increases. The presence of a variable magnetic field allows us to study in detail the influence of Landau quantization and the anomalous magnetic moment on the particle population of the star, more precisely on particles with different spin projections. This allows us to calculate the polarization of the system throughout different phases of the star, hadronic, quark and also a mixed phase.

1. Introduction and Model Description

Magnetars are compact stars that have extremely high magnetic fields. The highest magnetic field observed on the surface of a star is on the order of 10^{15} G. The highest possible magnetic field in the center of stars, on the other hand, can only be estimated through models, even when applying the virial theorem. Some results indicate limiting magnetic fields ranging between $B = 10^{18} - 10^{20}$ G [1–8]. Here we assume that magnetars are not necessarily composed of hadronic matter and describe them using a model that contains both hadronic and quark degrees of freedom.

There is no first principles solution for QCD in both the hadronic and quark phases at finite temperature and density. In order to model the deconfinement transition, we use a model that agrees with nuclear saturation properties and reproduces reasonable hyperon optical potentials in the low energy limit and incorporates known QCD properties in the high energy limit. In order to constrain the model at intermediate energies, we compare our model predictions with lattice results such as: a first order phase transition and a pressure functional P(T) similar to Ref. [9] at $\mu = 0$ for pure gauge theory, a crossover at vanishing chemical potential with a transition temperature determined as the peak of the change of the chiral condensate and the Polyakov loop, and the location of the critical end-point [10, 11].

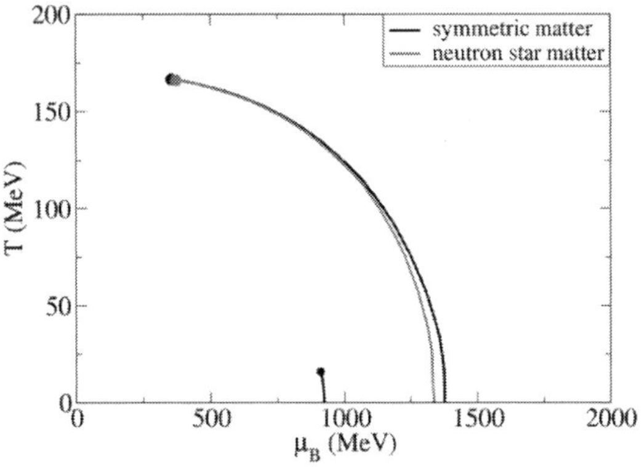

Figure 1. (Color online) QCD Phase Diagram - Temperature as a function of baryon chemical potential showing first-order phase transitions. The dots represent critical points.

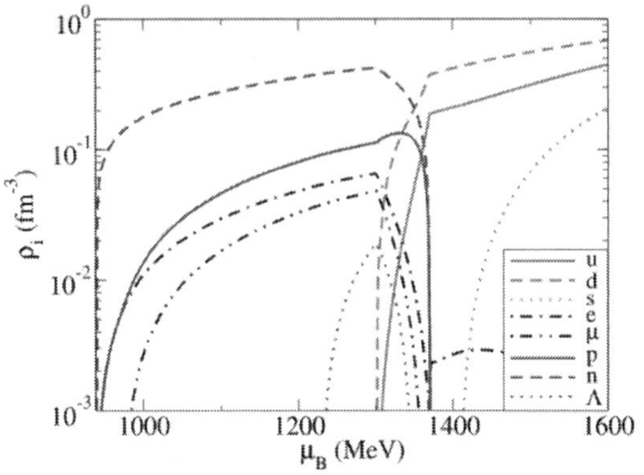

Figure 2. (Color online) Particle densities as a function of baryon chemical potential assuming global charge neutrality and chemical equilibrium at T=0. Quark densities are divided by 3.

The Lagrangian density of the non-linear sigma model (shown in Ref. [12]) represents the interactions between baryons (and quarks) and vector and scalar mesons, the self interactions of scalar and vector mesons and includes an explicit chiral symmetry breaking term which is responsible for producing the masses of the pseudo-scalar mesons. The mesons are treated as classical fields within the mean-field approximation. Finite-temperature calculations include a heat bath of hadronic and quark quasiparticles within the grand canonical ensemble.

Within our approach, the hadrons (whole baryon octet) are replaced by quarks (up, down, strange) at high densities and/or temperatures. This happens as the effective masses of the hadrons increase and the effective masses of the quarks decrease within these limits. The aforementioned model (see Ref. [12] for details) is an extended version of the SU(3) non-linear realization of the sigma model. Changes in the order parameters of the model σ and Φ signal chiral symmetry restoration and quark deconfinement, respectively. The potential for Φ is an extension of the Polyakov loop potential [9] modified to also depend on baryon chemical potential. In this way our model is able to describe the entire QCD phase diagram, even at zero temperature. Fig. 1 shows that the model is in good agreement with lattice QCD constraints and that it reproduces the liquid-gas phase transition for symmetric matter. In this figure we also show results for neutron star matter, which is charge neutral and in chemical equilibrium. The phase transitions at low temperatures and high densities are of first order, whereas at high temperatures and low densities the model exhibits smooth crossovers.

The SU(3) non-linear realization of the sigma model and its extension (that also contains quarks) have been successful in reproducing nuclear matter properties [13], heavy ion collision data [14], compact star and proto-neutron star properties [15–17]. As compact stars have temperatures of the order of 1 MeV, we can safely set their temperature to zero. As already mentioned, for star calculations we have to take into account charge neutrality and chemical equilibrium. Here we assume that the surface tension between the hadronic and quark phases is small [18] and allow charge neutrality to be global (only the combination of both phases sum up to zero charge). As a consequence, a mixed phase appears in the star. This can be seen in Fig. 2, which also shows that hyperons are almost completely suppressed by the appearance of the quarks.

We include in the model a magnetic field in the z-direction that has varying magnitude. This is a more realistic approach than considering a constant magnetic field throughout the star and

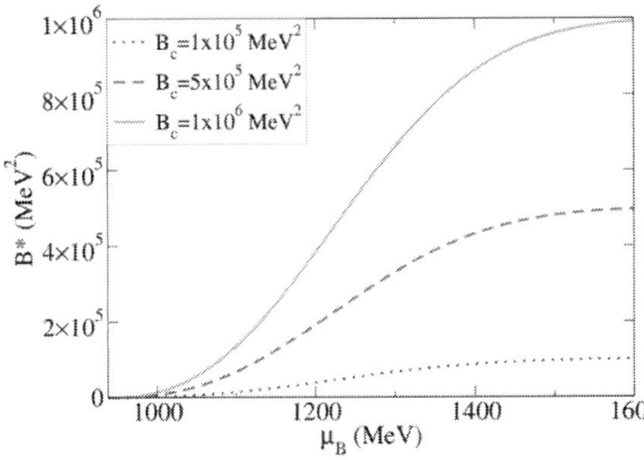

Figure 3. (Color online) Effective magnetic field as a function of baryon chemical potential shown for different central magnetic fields. Recall that using Gaussian natural units 1 $\text{MeV}^2 = 1.44 \times 10^{13}$ G.

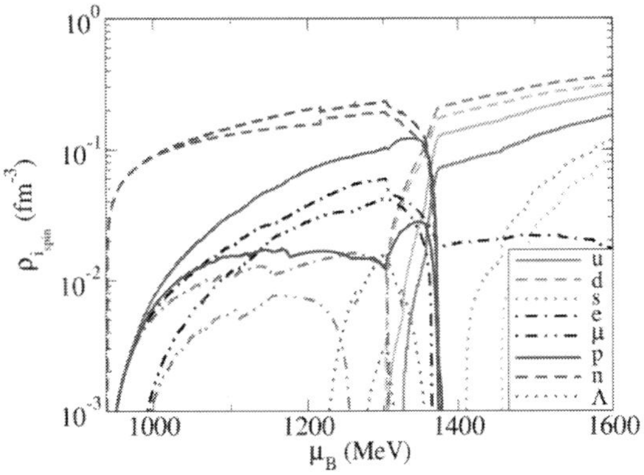

Figure 4. (Color online) Particle densities as a function of baryon chemical potential for $B_c = 5 \times 10^5$ MeV^2 $= 7.22 \times 10^{18}$ G including AMM. Black, blue and red stand for negative, while green, purple and orange stand for positive spin projections. Quark densities are divided by 3.

can prevent the creation of hydrodynamical instabilities due to pressure anisotropy [19–23]. This happens because, in our approach, the magnetic field only becomes extremely high in the center of the star, where the matter pressure is also high (see Ref. [24] for more details). More precisely, we assume an effective magnetic field B^* that increases with chemical potential, running from a surface value $B_{surf} = 69.25$ MeV$^2 = 10^{15}$ G (when $\mu_B = 938$ MeV) to different central values B_c at large values of baryon chemical potential following [17]

$$B^*(\mu_B) = B_{surf} + B_c[1 - e^{b\frac{(\mu_B - 938)^a}{938}}], \tag{1}$$

with $a = 2.5$, $b = -4.08 \times 10^{-4}$ and μ_B given in MeV. As can be seen in Fig. 3, the values of the effective magnetic field only approach B_c at very high baryon chemical potentials and, in practice, only about 70% of B_c can be reached inside stars. The use of an explicit dependence of B on the baryon chemical potential instead of on density was chosen to prevent discontinuities in the magnetic field at the phase transition, where the baryon density is discontinuous. The constants a and b and the form of the B^* expression were chosen to reproduce (in the absence of quarks) the effective magnetic field curve as a function of density from Refs. [5, 25, 26].

2. Results and Conclusions

The magnetic field in the z-direction forces the eigenstates in the x and y directions of charged particles to be quantized into Landau levels. The energy levels of all baryons are further split due to the alignment/anti-alignment of their spins with the magnetic field (anomalous magnetic moment effect, AMM). But even when the AMM is not taken into account, like in the quark phase in our model, only one of the spin projections contributes to the zeroth Landau level, creating a spin projection asymmetry in the system. In this work, we focus on the analysis of magnetic field effects on the chemical composition of the neutron star, the total spin polarization and the magnetization of the system. Studies of magnetic field effects on compact star observables can be found in Refs. [17, 27–29].

The particle population is shown in Fig. 4 when a central magnetic field $B_c = 5 \times 10^5$ MeV$^2 = 7.22 \times 10^{18}$ G with AMM is considered. The "wiggles" in the charged particle densities

Journal of Physics: Conference Series **432** (2013) 012005 doi:10.1088/1742-6596/432/1/012005

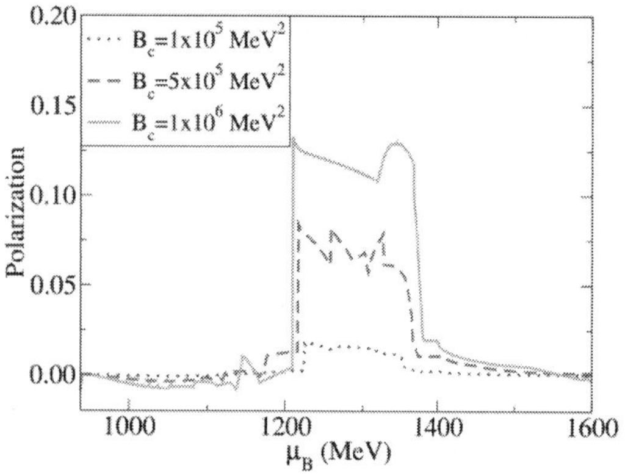

Figure 5. (Color online) Total spin polarization as a function of baryon chemical potential for different central magnetic fields (with 1 MeV^2 = 1.44×10^{13} G) including AMM.

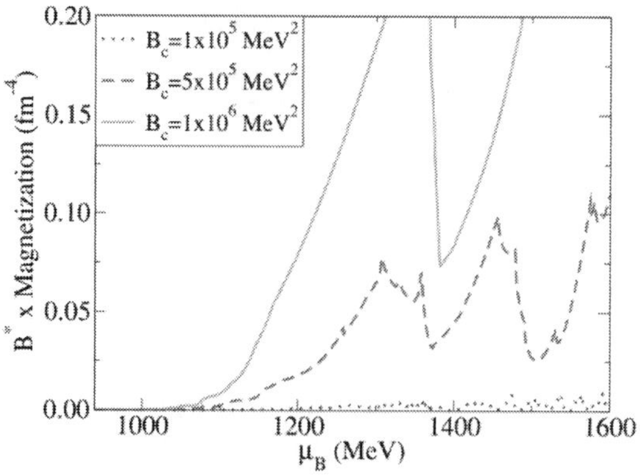

Figure 6. (Color online) Effective magnetic field times magnetization as a function of baryon chemical potential for different central magnetic fields (with 1 MeV^2 = 1.44×10^{13} G) including AMM.

mark the μ_B values, for which their Fermi energies cross the discrete threshold of a Landau level. The charged particle population is enhanced due to B, as their chemical potentials increase. Although the AMM is known to make the EOS stiffer, it does not have a very significant effect in the particle population. This fact can be easily understood in terms of polarization, when, instead of looking at the total particle density (sum of spin up and down particle densities) for each species, we look at the spin up/spin down particle densities separately. In this case some of these particles are enhanced while others are suppressed (see Ref. [30] for details).

The total polarization of the system is defined as

$$\frac{\sum_i Q_{Bi}(\rho_{+i} - \rho_{-i})}{\sum_i Q_{Bi}(\rho_{+i} + \rho_{-i})}, \tag{2}$$

where Q_{Bi} stands for the baryon number of each species (3 for baryons, 1 for quarks and 0 for leptons). We can see in Fig. 5 that the total polarization of the system (taking into account hadrons and quarks) is larger for larger magnetic fields and increases with chemical potential until the quarks appear. After this point the polarization decreases. This happens because the AMM for the quarks is not included in our calculations, as it is not fully understood for these particles. In this case, the spin asymmetry comes exclusively from the spin asymmetric contribution to the zeroth Landau level. For an example of a complete study of the polarization without the AMM see Ref. [31]. The "wiggles" in Fig. 5 mark again when the Fermi energies of the charged particles cross the discrete threshold of a Landau level.

Finally, we turn our attention to the calculation of the magnetization of the system. It is important to notice that in the case with the AMM, not only the magnetization of the charged particles has to be included, but also the magnetization of the uncharged particles. For details on the calculation of the magnetization including the AMM see Ref. [32]. Fig. 6 shows that the magnetization is larger for larger magnetic fields and it increases with baryon chemical potential. As in Figs. 4 and 5, the "wiggles" mark when the Fermi energies of the charged particles cross the discrete threshold of a Landau level. The magnetization of the system is a very important quantity as it relates with the pressure anisotropy of the system [19–23].

We have shown in this work some possible effects of strong magnetic fields in hybrid stars. The presence of different hadronic and quark degrees of freedom makes this quark-hadron sigma model an ideal tool for such an analysis in the different possible phases of the star. More specifically, we analyzed the effects of strong chemical potential dependent magnetic fields on particles with different spin projections, their total polarization and magnetization.

Acknowledgments

V. D. acknowledges support from CNPq (National Counsel of Technological and Scientific Development - Brazil).

References

[1] Bocquet M, Bonazzola S, Gourgoulhon E and Novak J 1995 *Astron.Astrophys.* **301** 757 (*Preprint* gr-qc/9503044)
[2] Cardall C Y, Prakash M and Lattimer J M 2001 *Astrophys.J.* **554** 322–339 (*Preprint* astro-ph/0011148)
[3] Lai D and Shapiro S L 1991 *Astrophys. J.* **383** 745–751
[4] Chakrabarty S, Bandyopadhyay D and Pal S 1997 *Phys.Rev.Lett.* **78** 2898–2901 (*Preprint* astro-ph/9703034)
[5] Bandyopadhyay D, Chakrabarty S and Pal S 1997 *Phys. Rev. Lett.* **79** 2176–2179 (*Preprint* astro-ph/9703066)
[6] Broderick A E, Prakash M and Lattimer J M 2002 *Phys. Lett.* **B531** 167–174 (*Preprint* astro-ph/0111516)
[7] Ferrer E J, de la Incera V, Keith J P, Portillo I and Springsteen P P 2010 *Phys. Rev.* **C82** 065802 (*Preprint* 1009.3521)
[8] Malheiro M, Ray S, Mosquera Cuesta H J and Dey J 2007 *Int.J.Mod.Phys.* **D16** 489–499 (*Preprint* astro-ph/0411675)
[9] Roessner S, Ratti C and Weise W 2007 *Phys.Rev.* **D75** 034007 (*Preprint* hep-ph/0609281)
[10] Fodor Z and Katz S 2004 *JHEP* **0404** 050 (*Preprint* hep-lat/0402006)
[11] Aoki Y, Endrodi G, Fodor Z, Katz S and Szabo K 2006 *Nature* **443** 675–678 (*Preprint* hep-lat/0611014)
[12] Dexheimer V A and Schramm S 2010 *Phys. Rev.* **C81** 045201 (*Preprint* 0901.1748)
[13] Papazoglou P *et al.* 1999 *Phys. Rev.* **C59** 411–427 (*Preprint* nucl-th/9806087)
[14] Steinheimer J, Dexheimer V, Petersen H, Bleicher M, Schramm S *et al.* 2010 *Phys.Rev.* **C81** 044913 (*Preprint* 0905.3099)
[15] Dexheimer V and Schramm S 2008 *Astrophys. J.* **683** 943–948 (*Preprint* 0802.1999)
[16] Negreiros R, Dexheimer V and Schramm S 2010 *Phys.Rev.* **C82** 035803 (*Preprint* 1006.0380)
[17] Dexheimer V, Negreiros R and Schramm S 2012 *Eur.Phys.J.* **A48** 189 (*Preprint* 1108.4479)
[18] Pinto M B, Koch V and Randrup J 2012 *Phys.Rev.* **C86** 025203 (*Preprint* 1207.5186)
[19] Chaichian M, Masood S, Montonen C, Perez Martinez A and Perez Rojas H 2000 *Phys.Rev.Lett.* **84** 5261–5264 (*Preprint* hep-ph/9911218)
[20] Perez Martinez A, Perez Rojas H, Mosquera Cuesta H J, Boligan M and Orsaria M G 2005 *Int. J. Mod. Phys.* **D14** 1959 (*Preprint* astro-ph/0506256)
[21] Perez Martinez A, Perez Rojas H and Mosquera Cuesta H 2008 *Int. J. Mod. Phys.* **D17** 2107–2123 (*Preprint* 0711.0975)
[22] Huang X G, Huang M, Rischke D H and Sedrakian A 2010 *Phys.Rev.* **D81** 045015 (*Preprint* 0910.3633)
[23] Paulucci L, Ferrer E J, de la Incera V and Horvath J E 2011 *Phys. Rev.* **D83** 043009 (*Preprint* 1010.3041)
[24] Dexheimer V, Menezes D and Strickland M 2012 (*Preprint* 1210.4526)
[25] Mao G J, Iwamoto A and Li Z X 2003 *Chin. J. Astron. Astrophys.* **3** 359–374 (*Preprint* astro-ph/0109221)
[26] Rabhi A, Pais H, Panda P K and Providencia C 2009 *J. Phys.* **G36** 115204 (*Preprint* 0909.1114)
[27] Menezes D, Benghi Pinto M, Avancini S and Providencia C 2009 *Phys.Rev.* **C80** 065805 (*Preprint* 0907.2607)
[28] Menezes D, Benghi Pinto M, Avancini S, Perez Martinez A and Providencia C 2009 *Phys.Rev.* **C79** 035807 (*Preprint* 0811.3361)
[29] Avancini S S, Menezes D P and Providencia C 2011 *"Phys. Rev. C"* **83** 065805
[30] Dexheimer V, Negreiros R, Schramm S and Hempel M 2012 (*Preprint* 1208.1320)
[31] Isayev A and Yang J 2013 *J.Phys.* **G40** 035105 (*Preprint* 1210.3322)
[32] Strickland M, Dexheimer V and Menezes D 2012 *Phys.Rev.* **D86** 125032 (*Preprint* 1209.3276)

Effective matrix models for deconfinement in SU(N) and G(2) gauge theories

Adrian Dumitru

RIKEN BNL Research Center, Brookhaven National Laboratory, Upton, NY 11973, USA
Department of Natural Sciences, Baruch College, CUNY, 17 Lexington Avenue, New York, NY 10010, USA

E-mail: adrian.dumitru@baruch.cuny.edu

Abstract. I present a simple matrix model for the deconfined phase of $SU(N)$ theories at temperatures just above T_c. The model is designed to reproduce the anomaly, in particular the flatness of $(e-3p)/T^2$ above $\sim 1.2T_c$ observed by lattice simulations with up to six colors. Furthermore, it predicts the existence of an adjoint Higgs phase where the masses of diagonal and off-diagonal gluons split and that the phase transition of $SU(N)$ with three or more colors, and for exceptional groups with a trivial center such as $G(2)$, is generically of first order. For $G(2)$ gauge group, a small value of the Polyakov loop at T_c^- can be obtained with a non-perturbative potential summed over the roots of $SU(7)$ rather than $G(2)$, which implements the principle of maximal eigenvalue repulsion.

1. Introduction

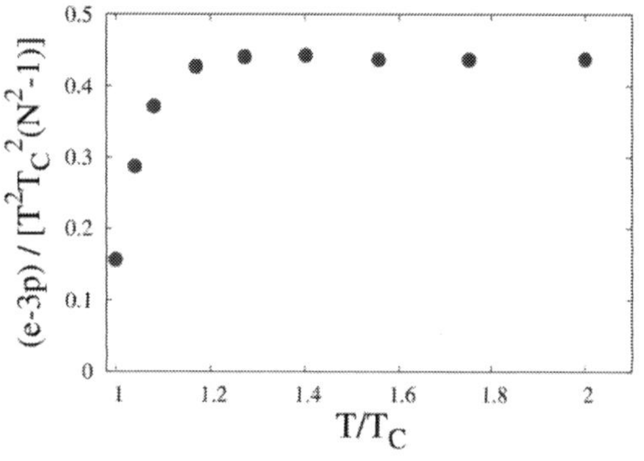

Figure 1. $SU(3)$ trace anomaly $e-3p$ divided by $8\,T^2T_c^2$ [1].

Lattice gauge theory simulations for $N = 2$, 3, 4, and 6 colors find that the conformal anomaly $e-3p$ scaled by T^2 is approximately constant from $\simeq (1.2-4)T_c$, where T_c is the critical temperature for deconfinement. Data from the WHOT collaboration for $N = 3$ is shown in fig. 1. If a bag constant were the dominant correction to perturbation theory in this range of temperature, this quantity would fall off as $\sim 1/T^2$. Rather, in this range, the pressure can be approximated as

$$p(T) \approx c_1 \left(T^4 - c_2\, T_c^2\, T^2\right) . \tag{1}$$

Thus, any non-perturbative terms which we introduce are assumed to be proportional to $\sim T^2$, $\sim T^0$, *etc.* [2, 3, 4, 5]. To describe the deconfined phase just above T_c we use a *matrix model*, where the basic variables are the eigenvalues of the thermal Wilson line. The vacuum is found by minimizing an effective potential with respect to these N matrix variables. Clearly the simplest way of obtaining a term $\sim T^2$ in the pressure is to introduce a similar term in the potential for the eigenvalues. We stress that this is merely an educated guess and that we have not performed an exhaustive search in the space of possible non-perturbative contributions.

With only the one-loop perturbative potential plus our model for non-perturbative contributions one can obtain decent fits of the pressure, energy density (incl. latent heat for $N \geq 3$) and the interaction measure of $SU(N)$ gauge theories up to a few times T_c with only two parameters [4, 5]. One can also predict various interface tensions and the renormalized Polyakov loop; the latter approaches unity much more rapidly than indicated by lattice data as in our current model the condensate for A_0 evaporates by $\simeq 1.2T_c$. It will be interesting to understand the origin of this discrepancy and to compute perturbative corrections due to quantum fluctuations about the classical background field.

The model also provides other basic insight. As a matrix model, it generically predicts that deconfining phase transitions are first order except when forbidden by symmetry such as $Z(2)$ center symmetry in case of $SU(2)$ Yang-Mills [6]. In particular, this is true even for gauge groups with a trivial center, such as $G(2)$, where deconfinement does not correspond to spontaneous breaking of the global center symmetry. To obtain a (nearly) vanishing Polyakov loop at T_c^-, we find that we are required to include terms which lead to maximal eigenvalue repulsion. Finally, with a condensate which is a linear combination of elements of the Cartan subalgebra, the matrix model predicts that masses of diagonal and off-diagonal gluons split (as the latter do not commute with the background field) until the condensate evaporates at high T.

2. Model for $SU(2)$ and $SU(3)$

For two colors, we introduce a classical condensate [5]

$$A_0 = \frac{\pi T}{g}\, q\, \sigma_3 \ , \quad \mathbf{L}(\vec{x}) = \mathbf{P}\exp\left(ig\int_0^{1/T} A_0(\vec{x},\tau)\, d\tau\right) = \begin{pmatrix} e^{i\pi q} & 0 \\ 0 & e^{-i\pi q}\end{pmatrix} . \tag{2}$$

The perturbative vacuum is $q=0$, $\mathbf{L}=\mathbf{1}$. The $Z(2)$ transform is $q=1$, $\mathbf{L}=-\mathbf{1}$. As an angular variable, q can be restricted to $[0,1]$ so that a $Z(2)$ transformation corresponds to $q \to 1-q$. In this model, $q_c = 1/2$ corresponds to maximal eigenvalue repulsion: $\mathbf{L}_c = \mathrm{diag}(i,-i)$, tr $\mathbf{L}_c = 0$, which mimics the confined vacuum.

The free energy in the presence of the background field (2) at one loop order is

$$V_{\rm pt}(q) = \pi^2 T^4 \left(-\frac{1}{15} + \frac{4}{3}\, q^2(1-q)^2\right) . \tag{3}$$

For $q=0,1$ this is the free energy of an ideal gas of three, massless gluons. The confined state at $q_c = 1/2$ is a maximum. To describe the transition to a confined phase we add non-perturbative terms to force the vacuum to go from the perturbative vacua, at $q=0$ and 1, to $q_c = 1/2$. Given the behavior of the interaction measure, we assume that any such term is proportional to T^2. It must also be $Z(2)$ symmetric, $q \to 1-q$, and so should be constructed in terms of $q(1-q)$. The simplest possibility, which also avoids unwanted additional phase transitions above T_c, is $V_{\rm npt} \sim -T^2 T_c^2\, c_1\, q(1-q)$. To fit to the lattice data, we find it necessary to add two more terms,

$$V_{\rm npt} = -\frac{4\pi^2}{3}\, T^2 T_c^2 \left(\frac{1}{5} c_1\, q(1-q) \,+\, c_2\, q^2(1-q)^2 - c_3(t)\right) , \quad (t \equiv \frac{T}{T_c}) . \tag{4}$$

Note that we need to constrain the parameters so that the transition occurs at T_c, and we need to fix the pressure in the confined phase to zero. Thus c_1 and c_2 are not free. To generate terms $\sim T^0$ we let $c_3(t) = c_3(\infty) + [c_3(1) - c_3(\infty)]/t^2$. The term $\sim c_3(1)$ in fact corresponds to a positive bag constant and is needed to fit the latent heat for $N \geq 3$. In all, this is then a model with two free parameters.

Viewing the A_0 as an adjoint Higgs condensate we observe that for $\langle q \rangle \neq 0$ there is a splitting of masses. Since $A_0 \sim \sigma_3$, the off diagonal components $\sim \sigma_1$, σ_2 develop a mass $\sim q$, while the diagonal ones do not [4]. Away from T_c all components develop an equal mass from Debye screening. In our model the theory is in an adjoint Higgs phase for all $T > T_c$ though in practice, for the parameters of the model, we find that the condensate evaporates above $\sim 1.2\, T_c$.

For three colors the background field formed by the Cartan subalgebra is

$$A_0 = \frac{\pi T}{3\, g}\, (q_3\, \lambda_3 + q_8\, \lambda_8) \ \ ; \ \ \lambda_3 = \mathrm{diag}(1, -1, 0) \ \ ; \ \ \lambda_8 = \mathrm{diag}(1, 1, -2) \ . \tag{5}$$

Moving along the λ_8 direction generates $Z(3)$ transformations: $\mathbf{L} = \mathbf{1}$ when $q = 0$, and $\mathbf{L} = \exp(2\pi i/3)\mathbf{1}$ when $q = 1$. The λ_3 direction takes one to the ground state $\mathbf{L}_c = \mathrm{diag}\left(\mathrm{e}^{2\pi i/3}, \mathrm{e}^{-2\pi i/3}, 1\right)$ characterized by $\mathrm{tr}\, \mathbf{L}_c = 0$, like the confining vacuum. This path can be parametrized as $q_j(s) = (N - 2j + 1)/(2N)\, s$ and the potential is then minimized in s. In the limit of infinite N, this ansatz gives a uniform eigenvalue density to some maximum. We thus refer to it as the *uniform eigenvalue ansatz*.

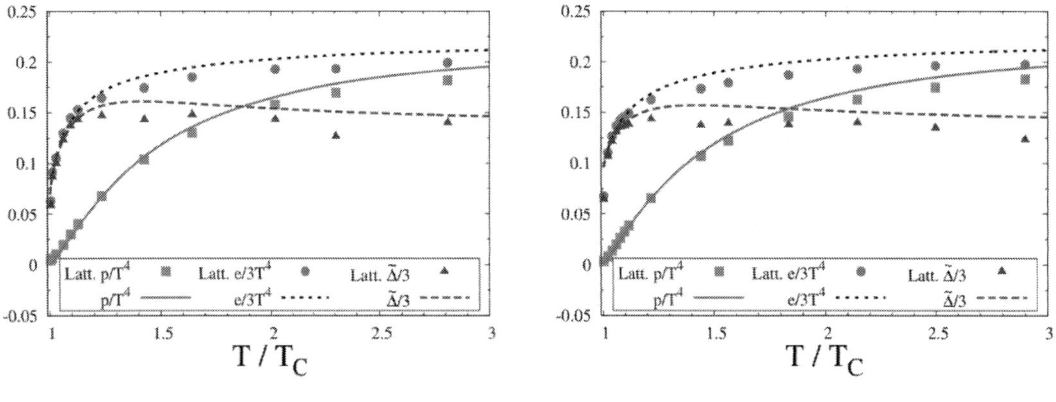

Figure 2. Thermodynamics of $SU(3)$ (left) and $SU(4)$ (right): pressure p/T^4, energy density $e/(3T^4)$, and the rescaled interaction measure $\widetilde{\Delta}$. All quantities are also scaled by $1/(N^2 - 1)$. Lattice data from ref. [7].

Numerical results for the thermodynamics of $SU(3)$ and $SU(4)$ are shown in fig. 2. Overall, the simple model with two free parameters performs quite well, especially near T_c. As already mentioned above, it will be interesting to see whether higher-loop perturbative corrections can be accommodated as well. The expectation value of the Polyakov loop approaches unity much more rapidly than on the lattice, see ref. [5], because in the current model the condensate is required to vanish quickly above T_c to reproduce the pressure and energy density.

3. $G(2)$ gauge group and the principle of maximal eigenvalue repulsion

For $SU(N)$ gauge groups the Polyakov loop vanishes below T_c which is a consequence of the global $Z(N)$ symmetry. There is maximal repulsion of the eigenvalues in the confined vacuum, as alluded to above. In fact, it may be more useful to think of confinement as arising not from the center symmetry *per se*, but from eigenvalue repulsion. To see this consider groups with a trivial center such as $G(2)$, which is also simply connected. Thus, for $G(2)$ the Polyakov loop in the fundamental representation can be screened dynamically as there is no $Z(N)$ center symmetry. It may therefore appear similar to QCD with dynamical light quarks where there is no deconfining phase transition but just a cross-over. Rather strikingly, lattice simulations find that there is a first order transition [8] from a "confined phase" where the expectation value of the Polyakov loop in the fundamental representation is very small to a deconfined phase. Thus we need to consider effective potentials which give confinement, and a deconfining phase transition, without relying on any center symmetry.

The simplest approach is to note that in the nonperturbative potential there is nothing to forbid a term where we sum over powers of the fundamental $G(2)$ loop, the **7**. If such a term is large, and of the right sign, one can drive the theory to a confined phase at low temperatures. Alternatively, we can construct a non-perturbative potential using both $SU(7)$ and $G(2)$ potentials since $G(2)$ is a subgroup of $SU(7)$; we must however restrict the $SU(7)$ potentials to the two dimensional Cartan space of $G(2)$. That way, it is possible to ensure that the stable vacuum is the $SU(7)$ confining vacuum at T_c^-. Confinement in $SU(7)$ then enforces the vanishing of the **7** loop in $G(2)$. In both cases, confinement is driven not by the center symmetry but through complete repulsion of eigenvalues.

The $SU(7)$ Wilson line is $\mathbf{L} = \exp(i2\pi\mathbf{q}_{SU7})$, where $\mathbf{q}_{SU7} = \mathrm{diag}(q_1, q_2, q_3, q_4, q_5, q_6, q_7)$ with $q_1+\cdots+q_7 = 0$ so that $\mathbf{q}_{SU7}$ is a traceless matrix. This constraint also defines the six dimensional Cartan space of $SU(7)$. $G(2)$ obeys the additional constraints $q_7 = q_1 + q_4 = q_2 + q_5 = 0$, which restricts to the Cartan space of $SO(7)$; and finally $q_1 + q_2 + q_3 = 0$, which leaves two degrees of freedom: q_1 and q_2.

Then, up to a factor of $4\pi^2T^4/3$ and a shift by $-14\pi^2T^4/45$, the one-loop perturbative $G(2)$ potential is given by

$$V_2^{G(2)}(\mathbf{q}_{G2}) = B_4(q_1) + B_4(q_2) + B_4(q_1+q_2) + B_4(q_1-q_2) + B_4(2q_1+q_2) + B_4(q_1+2q_2) \quad (6)$$

where $B_{2n}(q) \equiv q^n(1-q)^n$. Ref [5] considered a nonperturbative potential of the form

$$\begin{aligned} V_{\rm npt}(\mathbf{q}_{G2}) = & -\frac{4\pi^2}{3t^2}\Big(c_1^{G(2)}\,V_1^{G(2)}(\mathbf{q}_{G2}) + c_1^{SU(7)}\,V_1^{SU(7)}(\mathbf{q}_{G2}) \\ & + c_2^{G(2)}\,V_2^{G(2)}(\mathbf{q}_{G2}) + c_2^{SU(7)}\,V_2^{SU(7)}(\mathbf{q}_{G2}) + d_2^{G(2)}\,V_2^{\mathbf{7}}(\mathbf{q}_{G2}) + c_3\Big). \end{aligned} \quad (7)$$

where $V_1^{G(2)}$ is analogous to (6) with the replacement $B_4 \to B_2$. The $SU(7)$ potentials are

$$\begin{aligned} V_n^{SU(7)}(q_1,q_2) = & \; B_{2n}(2q_1) + B_{2n}(2q_2) + B_{2n}(2q_1+2q_2) \\ & + 2\left(B_{2n}(q_1-q_2) + B_{2n}(2q_1+q_2) + B_{2n}(q_1+2q_2)\right) \\ & + 4\left(B_{2n}(q_1) + B_{2n}(q_2) + B_{2n}(q_1+q_2)\right), \end{aligned} \quad (8)$$

while

$$V_2^{\mathbf{7}}(\mathbf{q}_{G2}) = B_4(q_1) + B_4(q_2) + B_4(q_1+q_2)\,. \quad (9)$$

The difference between (8) and (6) is that the former sums over the roots of $SU(7)$ rather than those of $G(2)$.

To restrict the number of parameters, one can consider a minimal $G(2)$ model,

$$c_1^{G(2)},\, c_3 \neq 0 \;\;;\;\; c_2^{G(2)} = c_1^{SU(7)} = c_2^{SU(7)} = d_2^{G(2)} = 0\,, \quad (10)$$

in analogy to the zero parameter $SU(N)$ model of ref. [2]. Alternatively, a model with a single fundamental loop,

$$d_2^{G(2)},\, c_1^{G(2)},\, c_2^{G(2)},\, c_3 \neq 0 \;\;;\;\; c_1^{SU(7)} = c_2^{SU(7)} = 0\,. \quad (11)$$

Finally, a $SU(7)$ type model,

$$c_2^{G(2)} = 1 \; ; \; c_1^{SU(7)}, c_2^{SU(7)}, c_3 \neq 0 \; ; \; c_1^{G(2)} = d_2^{G(2)} = 0 \,. \tag{12}$$

Notice that here $c_2^{G(2)} = 1$ so that the $G(2)$ part of the nonperturbative potential cancels the perturbative $G(2)$ potential at T_c fully. This ensures that the confining effects of the $SU(7)$ potential are maximized at T_c.

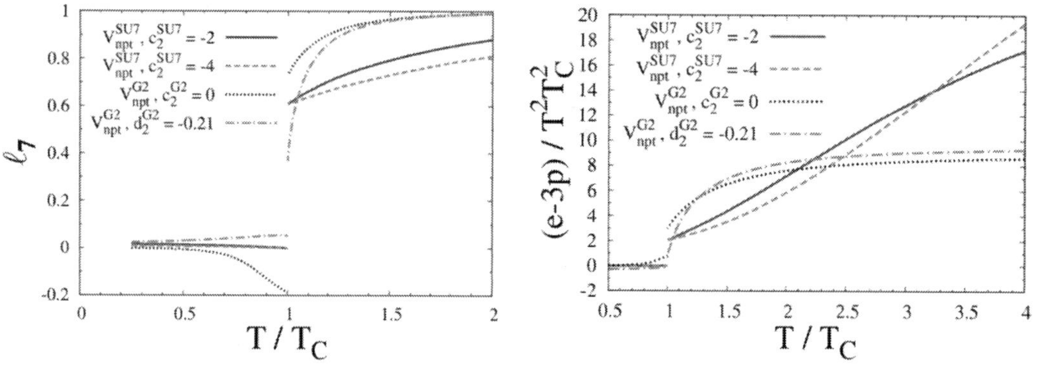

Figure 3. Left: Expectation value of the Polyakov loop in the fundamental representation of $G(2)$ for the minimal $G(2)$ model (10); the fundamental loop model (11); and two $SU(7)$ models (12). Right: Rescaled interaction measure for the four models.

Fig. 3 shows the fundamental Polyakov loop from the $G(2)$ models. The minimal $G(2)$ model appears to be excluded as it leads to a large (negative) value at T_c^-. The other models give fundamental loops which are small in the low temperature phase. In the fundamental loop model (11) this expectation value is positive at T_c^- and the model is near a critical end point, with a small discontinuity for the Polyakov loop. The lattice measurements from [8] indicate a small and perhaps negative value of the fundamental loop at T_c^- but correspond to bare expectation values. Lastly, the $SU(7)$ models automatically give a vanishing fundamental loop below T_c. The rescaled interaction measure for the minimal $G(2)$ and the fundamental loop models are flat and resemble those of $SU(N)$, while that for the $SU(7)$ model is not. Additional lattice simulations [9] for $G(2)$ could help understand the similarities and differences of its phase transition to that of $SU(N)$ and test the principle of eigenvalue repulsion.

Acknowledgments

I thank the organizers of XQCD2012 for the opportunity to present this work; my collaborators Y. Guo, C. Korthals Altes, Y. Hidaka and R. Pisarski; the DOE Office of Nuclear Physics for support through Grant No. DE-FG02-09ER41620 and The City University of New York through the PSC-CUNY Research Award Program, grant 65041-0043.

References

[1] Umeda T, Ejiri S, Aoki S, Hatsuda T and Kanaya K, 2009 *Phys. Rev.* D **79** 051501
[2] Meisinger P N, Miller T R and Ogilvie M C 2002 *Phys. Rev.* D **65** 034009
[3] Pisarski R D 2006 *Phys. Rev.* D **74** 121703
[4] Dumitru A, Guo Y, Hidaka Y, Korthals-Altes C P and Pisarski R D 2011 *Phys. Rev.* D **83** 034022
[5] Dumitru A, Guo Y, Hidaka Y, Korthals-Altes C P and Pisarski R D 2012 *Phys. Rev.* D **86** 105017
[6] Svetitsky B and Yaffe L G 1982 *Nucl. Phys* B **210** 423
[7] Datta S and Gupta S (2010) *Phys. Rev.* D **82** 114505
[8] Wellegehausen B H, Wipf A and Wozar C 2011 *Phys. Rev* D **83** 114502
[9] Panero M, *et al.*, in preparation

Thermodynamic Lattice Study for Preconformal Dynamics in Strongly Flavored Gauge Theory

Kohtaroh Miura

INFN - Laboratori Nazionali di Frascati Via Enrico Fermi, n.40 Frascati Rome Italy

E-mail: `Kohtaroh.Miura@lnf.infn.it`

Abstract. By using the lattice Monte-Carlo simulation, we investigate the finite temperature chiral phase transition in color SU(3) gauge theories with various species of fundamental fermions, and discuss the signals of the (pre-)conformality at large N_f (number of flavors) via their comparisons. With increasing N_f, we confirm stronger fermion screening which results from a larger fermion multiplicity. We investigate a finite T step-scaling which is attributed to the uniqueness of the critical temperature (T_c) at each N_f, then the vanishing step-scaling signals the emergence of the conformality around $N_f^* \sim 10 - 12$. Further, motivated by the recent functional renormalization group analyses, we examine the N_f dependence of T_c, whose vanishing behavior indicates that the conformal phase sets in around $N_f^* \sim 9 - 10$.

1. Introduction

Conformal invariance is anticipated to emerge in asymptotically free non-Abelian gauge theories when the number of fermion species, N_f, exceeds a critical value $N_f = N_f{}^*$ [1,2]. The approach to the conformality from below is in principle associated with a pre-conformal (walking) behavior of the running coupling, which has been advocated as a basis for strongly interacting mechanisms of electroweak symmetry breaking [3].

Recent lattice studies [4] focused on the computation of $N_f{}^*$ and the analysis of the conformal window itself, either with fundamental fermions or other representations. Among the many interesting results with fundamental fermions, we single out the observation that the color SU(3) gauge theory with $N_f = 8$ is still in the hadronic phase [5,6], while $N_f = 12$ seems to be close to the critical number of flavors, with some groups favoring conformality [6–9], and others chiral symmetry breaking [10].

In order to attack the walking and the conformal dynamics, it is more informative beyond a fixed N_f to investigate the vanishing or reducing chiral dynamics with increasing N_f. To this end, we investigate the N_f dependence of the chiral phase transition at finite temperature (T) based on our recent work [5,11]: The vanishing (reducing) finite T step-scaling which attributes to the uniqueness of T_c at each N_f signals the emergence of the (pre-)conformality. Further, motivated by the recent functional renormalization group (FRG) studies [12], we examine the N_f dependence of T_c, whose vanishing (decreasing) behavior indicates the (pre-)conformality. This *thermodynamic* lattice study for the large N_f non-Abelian gauge theory has played a crucial role to extract a notion of more strongly interacting non-Abelian plasma [13], and it is expected to provide a new connection between the lattice and the Gauge/Gravity duality [14].

2. Simulation setups

Simulations have been performed by utilizing the publicly available MILC code [15]. We use an improved version of the staggered action, the Asqtad action, with a one-loop Symanzik [16] and tadpole [17] improved gauge action. The tadpole factor u_0 is determined by performing zero temperature simulations on the 12^4 lattice, and used as an input for finite temperature simulations. To generate configurations with mass degenerate dynamical flavors, we have used the rational hybrid Monte Carlo algorithm (RHMC) [18].

Our observables are the chiral condensate and the Polykov loop

$$a^3\langle\bar{\psi}\psi\rangle = \frac{N_f}{4N_s^3N_t}\Big\langle \mathrm{Tr}[\mathrm{M}^{-1}]\Big\rangle\,, \quad L = \frac{1}{N_cN_s^3}\sum_{\mathbf{x}}\mathrm{Re}\Big\langle \mathrm{tr}_c\prod_{t=1}^{N_t}U_{4,t\mathbf{x}}\Big\rangle\,, \tag{1}$$

where N_s (N_t) represents the number of lattice sites in the spatial (temporal) direction and $U_{4,t\mathbf{x}}$ is the temporal link variable. The other important observable is the ratio of a scalar and a pseudo-scalar susceptibility [19],

$$R_\pi \equiv \frac{\chi_\sigma}{\chi_\pi} = \frac{\partial\langle\bar{\psi}\psi\rangle/\partial m}{\langle\bar{\psi}\psi\rangle/m} = \frac{\chi_{\mathrm{conn}}+\chi_{\mathrm{disc}}}{\langle\bar{\psi}\psi\rangle/m}\,, \tag{2}$$

where $a^2\chi_{\mathrm{conn}} = -N_f\langle\mathrm{Tr}[(MM)^{-1}]\rangle/(4N_s^3N_t)$ and $a^2\chi_{\mathrm{disc}} = N_f^2[\langle\mathrm{Tr}[M^{-1}]^2\rangle - \langle\mathrm{Tr}M^{-1}\rangle^2]/(16N_s^3N_t)$. Here, $R_\pi \sim \mathcal{O}(1)$ indicates the scalar and pseudo-scalar degeneracy attributed to the approximate chiral restoration, while $R_\pi \ll 1$ indicates the chiral symmetry breaking [5, 19].

3. Results

We evaluate the thermalized ensemble averages of the chiral condensate (PBP) and Polyakov loop (PLOOP) for various lattice couplings β_{L}, lattice sizes with the finite T set up $N_s \gg N_t$, and the number of flavors N_f. All results have been obtained by using a single value for a lattice bare fermion mass $am = 0.02$. Then we locate the lattice bare coupling ${\beta_{\mathrm{L}}}^{\mathrm{c}}$ associated with the chiral crossover which is signaled by the drastic decrease (increase) of PBP (PLOOP) as a function of β_{L}. In practice, the ratio of the scalar and pseudo-scalar susceptibility R_π gives a stronger signal of the chiral crossover, owing to its renormalization invariant property. In table 1, we summarize the obtained critical lattice couplings as a function of (N_f, N_t). We have confirmed the (approximate) asymptotic scaling for the normalized critical temperature T_c/Λ_{L} varying N_t at each N_f, where Λ_{L} is so-called lattice Lambda. This indicates that our ${\beta_{\mathrm{L}}}^{\mathrm{c}}$ have been determined near the continuum limit [20].

Table 1. Summary of the (pseudo) critical lattice couplings ${\beta_{\mathrm{L}}}^{\mathrm{c}}$ for the theories with $N_f = 0,\ 4,\ 6,\ 8$, $am = 0.02$ and varying $N_t = 4,\ 6,\ 8,\ 12$. The result has partially been extracted from our recent paper [5, 11]. All results are obtained by using the action with the same level of improvements.

$N_f\backslash N_t$	4	6	8	12
0	7.35 ± 0.1	7.88 ± 0.05	8.20 ± 0.1	–
4	5.60 ± 0.1	5.89 ± 0.05	6.10 ± 0.1	–
6	4.65 ± 0.05	5.05 ± 0.05	5.2 ± 0.05	5.55 ± 0.1
8	–	4.1125 ± 0.0125	4.20 ± 0.1	4.34 ± 0.04

We shall now discuss the emergence of the (pre-)conformality by using our critical lattice ${g_{\mathrm{L}}}^{\mathrm{c}} = \sqrt{10/{\beta_{\mathrm{L}}}^{\mathrm{c}}}$ collection. The uniqueness of a critical temperature at each N_f, $T_c^{-1} =$

$N_t\ a(\beta_{\rm L}{}^{\rm c}) = N_t'\ a(\beta_{\rm L}{}^{\rm c\prime})$ with $N_t \neq N_t'$ gives a *thermal* step-scaling $\Delta\beta_{\rm L}{}^{\rm c} = \beta_{\rm L}{}^{\rm c} - \beta_{\rm L}{}^{\rm c\prime}$. A vanishing (decreasing) $\Delta\beta_{\rm L}{}^{\rm c}$ can signal the (pre-)conformality.

Our thermal step-scaling is a function of N_f and two lattice temporal extensions, $\Delta\beta_{\rm L}{}^{\rm c}(N_f; N_t, N_t')$, and tends to be smaller with increasing N_f. We here estimate the number of flavors satisfying $\Delta\beta_{\rm L}{}^{\rm c}(N_f^*) = 0$ by extrapolating our $\beta_{\rm L}{}^{\rm c}$ collection into the larger flavor region. To this end, we plot the (pseudo) critical lattice coupling $g_{\rm L}{}^{\rm c} = \sqrt{10/\beta_{\rm L}{}^{\rm c}}$ as a function of N_f in Fig. 1, which gives an extension of Miransky-Yamawaki diagram [21] to finite T cases.

Let us first pick up the lattice critical couplings for $N_f = 6$ and 8, and consider a "constant N_t" line. As shown in the left panel of Fig. 1, $N_t = 6$ and 12 lines join at $(g_{\rm L}{}^{\rm c}, N_f^*) = (1.825\pm0.02,\ 11.57\pm0.17)$, and $N_t = 6$ and 12 lines at $(g_{\rm L}{}^{\rm c}, N_f^*) = (1.753\pm0.02,\ 10.715\pm0.17)$, indicating an infra-red fixed point with vanishing thermal step scalings. Next we shall investigate the critical lattice couplings at $N_t = 6$ and 8 for the whole range of $N_f = 0-8$. They can be well fitted by assuming the functional expression $N_f(g_{\rm L}{}^{\rm c}) = A \cdot \log\left[B \cdot \left(g_{\rm L}{}^{\rm c} - g_{\rm L}{}^{\rm c}|_{N_f=0}\right) + 1\right]$ giving $(g_{\rm L}{}^{\rm c}, N_f^*) = (1.88\pm0.09,\ 11.39\pm0.78)$ (right panel of Fig. 1). Thus, the thermal step-scaling with the use of our lattice critical couplings supports the emergence of conformal window near the 12-flavor system, whose (pre-)conformality is now under debate in the recent lattice studies [4].

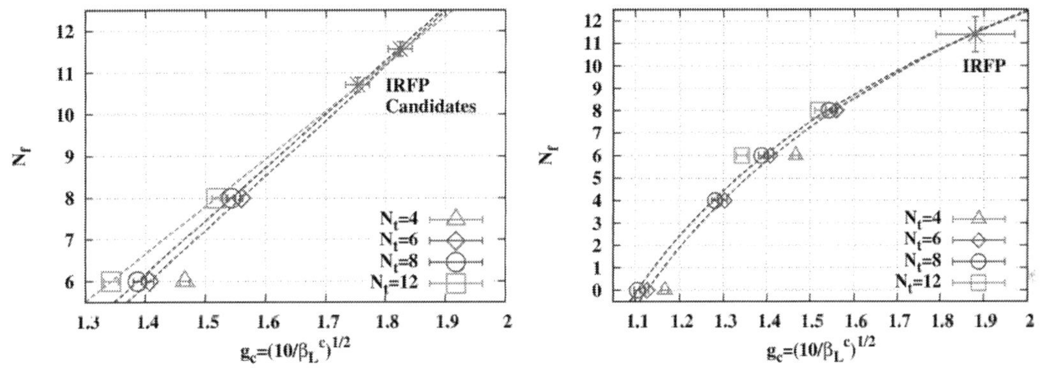

Figure 1. (Pseudo) critical values of the lattice coupling $g_{\rm L}{}^{\rm c} = \sqrt{10/\beta_{\rm L}{}^{\rm c}}$ for theories with $N_f = 0,\ 4,\ 6,\ 8$ and for several values of N_t in the Miransky-Yamawaki phase diagram. Left: We have picked up $g_{\rm L}{}^{\rm c} = \sqrt{10/\beta_{\rm L}{}^{\rm c}}$ for $N_f = 6$ and 8, and consider the "constant N_t" line with $N_t = 6,\ 8,\ 12$. Candidates for the IRFP have been estimated by the crossing points of $N_t = 12$ line with $N_f = 6$ or $N_f = 8$ line. Right: The dashed line is a fit with the ansatz $N_f(g_{\rm L}{}^{\rm c}) = A \cdot \log\left[B \cdot (g_{\rm L}{}^{\rm c} - g_{\rm L}{}^{\rm c}|_{N_f=0}) + 1\right]$ for $N_t = 6$ and $N_t = 8$ results.

As indicated by the constant N_t line in Fig. 1, the critical coupling is an increasing function of N_f for a fixed lattice temporal extension. This behavior is the direct consequence of enhanced fermion screening effects due to the larger number of fermion species. We have observed the thermal step (or equivalently the asymptotic) scalings for our $\beta_{\rm L}{}^{\rm c}$, thereby, the enhancement of the screening effects gets to a physical significance. The test of the asymptotic scaling has been the historical homework since the pioneering finite T study at large N_f by J. Kogut and collaborators [22] and now we have completed it.

We shall now investigate the N_f dependence of the critical temperature. In order to compare a physical quantity such as a critical temperature among different theories with a different N_f, it is necessary to introduce a N_f independent reference scale by hand. To set the reference scale, ideally speaking, we would like to measure Wilson loops of various sizes at various numbers of flavors in Monte-Carlo simulations to obtain the running couplings $\bar{g}$ in the wide range of scales at each number of flavors. Instead of performing such a massive simulation, we approximately construct the renormalization flow via the integral of the two-loop beta-function $\beta(g) = -g^3(b_0 + b_1 g^2)$

$$\frac{T_c}{M(g_L{}^{\rm ref})} = \frac{1}{N_t}\frac{a^{-1}(g_L{}^{\rm c})}{M(g_L{}^{\rm ref})} = \frac{1}{N_t}\int_{g_L{}^{\rm ref}}^{g_L{}^{\rm c}} \frac{dg}{-g^3(b_0 + b_1 g^2)} \ . \tag{3}$$

To specify the reference scale $M(g_L{}^{\rm ref})$, we utilize our plaquette (tad-pole factor u_0) data shown in the left panel of Fig. 2. Note that the plaquettes can be regarded as a kind of renormalized couplings. Let us consider a constant u_0 without N_f dependence, for instance $u_0 = 0.9$ in figure, and read-off the corresponding bare lattice couplings at each N_f. The obtained $g_L(N_f) = \sqrt{10/\beta_L(N_f)}$ is used as a reference coupling $g_L{}^{\rm ref}$ in Eq. (3). This procedure imitates the scale setting in the potential scheme renormalization, and the use of N_f independent u_0 is motivated by the FRG scale setting method [12].

To be analogous to the FRG study, we should choose u_0 so as to get a UV $M(g_L)$ free from the chiral dynamics. The middle panel of Fig. 2 displays the N_f dependence of $T_c/M(g_L{}^{\rm ref})$ defined by Eq. (3) with $u_0 = 0.9$. Fitting $T_c/M(g_L{}^{\rm ref})$ with the FRG motivated ansatz $T_c = K|N_f^* - N_f|^{(-2b_0^2/b_1)(N_f^*)}$, we now read-off the lower edge of conformal window $N_f^* \sim 9.47\pm0.02$, which is somewhat smaller value comparing to those obtained by the vanishing thermal scale settings. In the middle panel of Fig. 2, we find $T_c/M(g_L{}^{\rm ref}) \ll 1$, indicating the UV nature of the reference scale $M(g_L{}^{\rm ref})$.

To get more transparent view for the UV reference scale, we here consider the particular reference coupling $g_L{}^{\rm ref}$ - the thermal critical coupling $g_T^{\rm c}$ which makes the reference scale $M(g_L{}^{\rm ref} = g_T^{\rm c})$ equivalent to T_c in Eq. (3): $T_c/M(g_T^{\rm c}) = 1$, giving a typical interaction strength at T_c. As shown in the right panel of Fig. 2, the increasing nature of $g_T^{\rm c}$ indicates a realization of more strongly interacting non-Abelian plasma at larger N_f as discussed in Ref. [13]. The criterion to set the UV reference scale $M(g_L{}^{\rm ref})$ at every N_f would be given by the condition $g_L{}^{\rm ref} \ll g_T^{\rm c}(N_f)$ for all N_f. We find that $u_0 \geq 0.84$ meets the requirement, while the use of too large u_0 suffers from the strong discritization errors. In practice, we find that the number of flavors giving the vanishing $T_c/M(g_L{}^{\rm ref})$ is relatively stable within the range $0.84 \leq u_0 \leq 0.94$ which results in $9.85 \geq N_f^* \geq 9.17$.

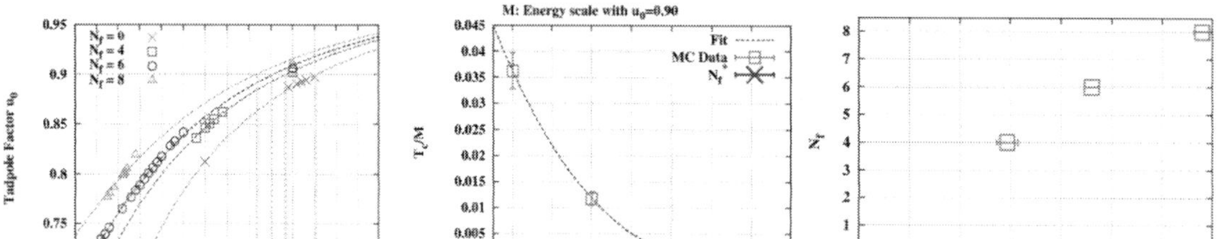

Figure 2. Left: The β_L dependences of the tadpole factor u_0 at zero temperature with the use of 12^4 lattice. Specifying a constant u_0 (*e.g.* $u_0 = 0.9$ in figure), we read off the corresponding lattice couplings β_L which are used to define the scale M at each theory with N_f. Middle: The N_f dependence of T_c/M where M is the UV scale with $u_0 = 0.9$ at each theory with N_f. The dashed line represents the fit for data by using the FRG motivated ansatz $T_c = K|N_f^* - N_f|^{(-2b_0^2/b_1)(N_f^*)}$ Right: The thermal critical coupling with increasing N_f.

4. Summary

We have investigated the (pre-)conformal dynamics in color SU(3) gauge theories with multi-species of fundamental fermions by using the lattice Monte-Carlo simulation. In order to study the conformality beyond the fixed number of flavors N_f, we have focused on a reducing chiral dynamics at finite T as a function of increasing N_f. We have observed stronger fermion screenings resulting from a larger fermion multiplicity at larger N_f. We have investigated a finite T step-scaling which follows from the uniqueness of critical temperature (T_c) at each N_f, then the vanishing step-scaling signals the conformal dynamics at $N_f^* \sim 10-12$. Further, motivated by the recent FRG based studies [12], we have examined the N_f dependence of T_c, by introducing a UV N_f independent reference scale $M({g_\mathrm{L}}^\mathrm{ref})$. We have used the thermal critical coupling g_T^c as a criterion to insure the UV nature of $M({g_\mathrm{L}}^\mathrm{ref})$ by imposing the condition ${g_\mathrm{L}}^\mathrm{ref} \ll g_\mathrm{T}^\mathrm{c}(N_f)$ for all N_f. We have found that the number of flavors giving a vanishing $T_c/M({g_\mathrm{L}}^\mathrm{ref})$ is relatively stable within the range $0.84 \leq u_0 \leq 0.94$, which results in $9.85 \geq N_f^* \geq 9.17$.

As a future perspective, we should measure Wilson loops of various sizes at various numbers of flavors, and perform more rigorous scale settings in the potential scheme. It is also mandatory to investigate the chiral limit and the thermodynamic limit at large N_f. This, together with a more extended set of flavor numbers, will allow a quantitative analysis of the critical behavior in the vicinity of the conformal IR fixed point.

Acknowledgments

The author thanks Maria Paola Lombardo and Elisabetta Pallante for continuous discussions. He thanks Edward Shuryak for fruitful discussion during the xQCD workshop. This work was in part based on the MILC Collaboration's public lattice gauge theory code [15]. The numerical calculations were carried out on the IBM-SP6 and BG/P at CINECA, Italian-Grid-Infrastructures in Italy, and the Hitachi SR-16000 at YITP, Kyoto University in Japan.

References

[1] Caswell W E 1974 *Phys. Rev. Lett.* **33** 244.
[2] Banks T and Zaks A 1982 *Nucl. Phys.* B **196** 189.
[3] For a recent review, see Sannino F 2009 *Acta Phys. Polon.* B **40** 3533.
[4] For recent reviews, see
Del Debbio L 2010 *PoS* **LAT2010** 004; Pallante E 2009 *PoS* **LAT2009** 015.
[5] Deuzeman A, Lombardo M P and Pallante E 2008 *Phys. Lett.* B **670** 41.
[6] Appelquist T, Fleming G T and Neil E T
2008 *Phys. Rev. Lett.* **100** 171607 [Erratum-ibid. 2009 **102** 149902], 2009 *Phys. Rev. D* **79** 076010.
[7] Deuzeman A, Lombardo M P and Pallante E 2010 *Phys. Rev.* D **82** 074503.
[8] Appelquist T, Fleming G T, Lin M F, Neil E T and Schaich D A 2011 *Phys. Rev.* D **84** 054501.
[9] Hasenfratz A 2010 *Phys. Rev.* D **82** 014506.
[10] Fodor Z, Holland K, Kuti J, Nogradi D and Schroeder C 2011 *Phys. Lett.* B **703** 348-58.
[11] Miura K, Lombardo M P and Pallante E 2012 *Phys. Lett.* B **710** 676.
[12] Braun J, Fisher C S and Gies H 2011 *Phys. Rev.* D **84** 034045;
J. Braun and H. Gies, 2010 *JHEP* **1005** 060; 2006 **0606** 024.
[13] Liao J, Shuryak E 2012 *Phys. Rev. Lett.* **109** 152001.
[14] For a recent development, see a review, Gursoy U, Kiritsis E, Mazzanti L, Michalogiorgakis G and Nitti F 2011 *Lect. Notes Phys.* **828** 79.
[15] MILC Collaboration, http://www.physics.indiana.edu/~sg/milc.html
[16] Bernard C *et al.* 2007 *Phys. Rev.* D **75** 094505; Luscher M and Weisz P 1985 *Phys. Lett.* B **158** 250; *Commun. Math. Phys.* **97** 59 [Erratum-ibid. **98** (1985) 433].
[17] Lepage G P and Mackenzie P B 1993 *Phys. Rev.* D **48** 2250.
[18] Clark M A 2006 *PoS* **LAT2006** 004.
[19] Kocic A, Kogut J B and Lombardo M P 1993 *Nucl. Phys.* B **398** 376.
[20] Gupta S 2001 *Phys. Rev.* D **64** 034507.
[21] Miransky V A and Yamawaki K 1997 *Phys. Rev.* D **55** 5051 [Erratum-ibid. 1997 D **56** 3768].
[22] Kogut J B, Polonyi J, Wyld H W and Sinclair D K 1985 *Phys. Rev. Lett.* **54** 1475.

String inspired solution to the sign problem and overlapping problem

Masanori Hanada, Yoshinori Mastuo

KEK Theory Center, High Energy Accelerator Research Organization (KEK), 1-1 Oho, Tsukuba, Ibaraki 305-0801, Japan

E-mail: hanada@post.kek.jp,ymatsuo@post.kek.jp

Naoki Yamamoto

Yukawa Institute for Theoretical Physics, Kyoto University, Kyoto 606-8502, Japan
Institute for Nuclear Theory, University of Washington, Seattle, Washington 98195-1550, USA
Maryland Center for Fundamental Physics, Department of Physics, University of Maryland, College Park, MD 20742-4111, USA

E-mail: nyama@umd.edu

Abstract. Because of the sign problem, it is difficult to study finite-density QCD. In order to circumvent the sign problem, we prove that the phase-quenched approximation is exact to $O(N_f/N_c)$ for any physical observables in a certain region of the phase diagram. We also find a quantitative evidence for the validity of the phase quenching from existing lattice QCD results at $N_c = 3$. Our results show that the phase-quenched approximation is good already at $N_c = 3$, and the $1/N_c$ correction can be incorporated by the phase reweighting method without suffering from a severe overlap problem. The same equivalence holds in effective models and holographic models. It gives us a theoretical understanding of the empirical facts (exactness of the phase quenching in the mean-field approximation) found in model calculations.

1. Introduction

QCD at a finite baryon chemical potential and/or finite temperature is an important subject. Although the lattice QCD simulations should play an important role for studying the strongly coupled parameter region of this theory, the notorious *sign problem* prevents us from a direct application of the simulation at finite baryon chemical potential. A possible option is the phase-quenched simulation. In principle, the effect of the phase can be taken into account by the phase reweighting. However whether it is practical or not is not clear a priori; when the number of the flavors N_f is two, the phase-quenched theory is the QCD with the isospin chemical potential[1], whose phase diagram is different from the original theory (especially the existence of the pion condensation), and hence a severe overlap problem can appear.

Actually the phase quenching and the phase reweighting are practically useful techniques. The first to emphasized this fact, albeit empirically, are probably Kogut and Sinclair [1]. They pointed out various model calculations give the same answer for certain observables in the full

[1] As we will see in Sec. 2.2, although this statement is correct for the partition function, there is a difference when one considers the physical observables.

and phase-quenched theories, as long as the pion condensation does not take place in the latter. They also pointed out that known lattice data give very similar results. Independently, Cohen [2] and Toublan [3] pointed out the similarity in the large-N_c limit. Recently these facts have been understood in theoretically unified manner [4, 5, 6]: there is an exact *equivalence* between the full and phase-quenched QCD at large-N_c, which provides a good approximation at $N_c = 3$.[2] The equivalence is a version of the large-N_c orbifold equivalence [7, 8], which was discovered through the study of the string theory. (For other interesting applications of the orbifold equivalence see e.g. [9].) Below we briefly summarize the equivalence, show the lattice data which verify that the equivalence can be seen already at $N_c = 3$, and point out the same equivalence holds for various effective models in the mean field approximation (MFA).

2. The (partial) equivalence between QCD$_B$ and QCD$_I$ in the large-N_c limit

2.1. The equivalence between QCD$_B$ and QCD$_I$

Let us start with the orbifold equivalence. First we choose the discrete symmetry P (subgroup of gauge, flavor, or spacetime symmetry) of the *parent* theory, which is the SO($2N_c$) or Sp($2N_c$) theory with the baryon chemical potential (SO$_B$ or Sp$_B$) in the present case. We then throw away all the degrees of freedom not invariant under P. This procedure is called the *orbifold projection*. After the projection, we obtain a new theory called the *daughter*. We consider two different projections, which give QCD with the baryon and isospin chemical potentials (QCD$_B$ and QCD$_I$) as daughters. The orbifold equivalence states that, in the large-N_c limit, correlation functions of operators $\mathcal{O}^{(p)}(A_\mu, \psi)$ invariant under P in the parent (called *neutral* operators) agree with those of the operators $\mathcal{O}^{(d)}(A_\mu^{\text{proj}}, \psi^{\text{proj}})$ that consist of projected fields in the daughter:

$$\langle \mathcal{O}_1^{(p)} \mathcal{O}_2^{(p)} \cdots \rangle_p = \langle \mathcal{O}_1^{(d)} \mathcal{O}_2^{(d)} \cdots \rangle_d. \tag{1}$$

Here the coupling constants are related as $g_{\text{SU}}^2 = g_{\text{SO}}^2 = g_{\text{Sp}}^2$, where the 't Hooft coupling $g_{\text{SU}}^2 N_c$ is kept finite. The field theoretic proof was given in [8] for a class of theories, which can be generalized to various cases. For QCD$_B$, QCD$_I$, SO$_B$ and Sp$_B$, evidence for nonperturbative equivalence were also provided by the weak-coupling analysis in QCD and QCD-like theories at high density limit [5], low-energy effective theories [10], chiral random matrix models [5] and holographic models [11].

For the projection from SO$_B$ to QCD$_B$, we use the $\mathbb{Z}_4$ discrete symmetries of SO$_B$ generated by $J_c = -i\sigma_2 \otimes 1_{N_c}$ (1_N is an $N \times N$ identity matrix) and $\omega = e^{i\pi/2} \in U(1)_B$. We require the gauge field $A^{\text{SO}}_{\mu,ab}$ and the fermion $\psi^{\text{SO}}_{\alpha,a}$ to be invariant under the following $\mathbb{Z}_2$ transformation embedded in the gauge and U(1)$_B$ transformation [4],

$$A^{\text{SO}}_{\mu,ab} = (J_c)_{aa'} A^{\text{SO}}_{\mu,a'b'} (J_c^{-1})_{b'b}, \tag{2}$$
$$\psi^{\text{SO}}_{\alpha,a} = \omega (J_c)_{aa'} \psi^{\text{SO}}_{\alpha,a'}. \tag{3}$$

The projection symmetry breaks down in the BEC/BCS crossover region (diquark condensation region) of SO$_B$, because the U(1)$_B$ symmetry is broken to $\mathbb{Z}_2$ there.

One can also construct the projection from SO$_B$ to QCD$_I$ for even N_f by choosing another $\mathbb{Z}_2$ symmetry [4, 5],

$$A^{\text{SO}}_{\mu,ab} = (J_c)_{aa'} A^{\text{SO}}_{\mu,a'b'} (J_c^{-1})_{b'b}, \tag{4}$$
$$\psi^{\text{SO}}_{\alpha,af} = (J_c)_{aa'} \psi^{\text{SO}}_{\alpha,a'f'} (J_i^{-1})_{f'f}, \tag{5}$$

[2] Previously we argued the equivalence is restricted to a class of observables. As we will see, however, the equivalence holds for any observables. We thank F. Karsch for a valuable critical comment, which made us revisit the issue and led to more precise statement.

where $J_i = -i\sigma_2 \otimes 1_{N_f/2}$ generates $\mathbb{Z}_4$ subgroup of SU(2) isospin symmetry and the projection condition for the gauge field is the same as (2). In this case, the isospin symmetry used for the projection is unbroken everywhere, and so the orbifold equivalence holds also in the BEC/BCS region of the phase diagram. Therefore, through the equivalence with SO_B, we obtain the equivalence between QCD_B and QCD_I outside the BEC/BCS region (pion condensation region) of the latter; the phase quenching is exact for neutral sectors in this region.

The $1/N_c$ corrections to the equivalence can be estimated as follows. In the 't Hooft large-N_c limit, expectation values of gluonic operators trivially agree because the fermions are not dynamical. Now consider finite-N_c, say $N_c = 3$ and $N_f = 2$. Then the largest correction to the 't Hooft limit comes from one-fermion-loop planar diagrams, which, as we have seen, do not distinguish μ_B and μ_I. Therefore the difference of expectation values of gluonic operators is at most $(N_f/N_c)^2$ (two-fermion-loop planar diagrams). In particular, the deconfinement temperatures, which are determined by the Polyakov loop, agree up to corrections of this order. A similar observation was made in [3].

2.2. The equivalence between QCD_B and phase-quenched QCD in the large-N_c limit

It is often said that QCD_I and the phase-quenched QCD are the same. However, although this statement is correct for the partition function, there is a difference when one studies the physical observables. In QCD_I, the propagators of up and down quarks are $D^{-1}(+\mu)$ and $D^{-1}(-\mu)$. On the other hand, in the phase-quenched QCD, both are taken as $D^{-1}(+\mu)$. (In the terms of the lattice QCD simulation, the configuration are generated by using QCD_I, while the same operators as QCD_B are used for the measurement.) Therefore the expectation values of the chiral condensate, the baryon density, and the isospin density are:

	QCD_B	QCD_I	phase-quenched QCD
chiral condensate	$2\langle \mathrm{Tr}D^{-1}(\mu)\rangle_B$	$\langle \mathrm{Tr}D^{-1}(\mu) + \mathrm{Tr}D^{-1}(-\mu)\rangle_I$	$2\langle \mathrm{Tr}D^{-1}(\mu)\rangle_I$
baryon density	$2\langle \mathrm{Tr}\gamma^0 D^{-1}(\mu)\rangle_B$	$\langle \mathrm{Tr}\gamma^0 D^{-1}(\mu) + \mathrm{Tr}\gamma^0 D^{-1}(-\mu)\rangle_I$	$2\langle \mathrm{Tr}\gamma^0 D^{-1}(\mu)\rangle_I$
isospin density	0	$\langle \mathrm{Tr}\gamma^0 D^{-1}(\mu) - \mathrm{Tr}\gamma^0 D^{-1}(-\mu)\rangle_I$	0

Here $\langle\ \rangle_B$ and $\langle\ \rangle_I$ are the expectation values with QCD_B and QCD_I ensembles, respectively. We can easily see the chiral condensate in QCD_I and the phase-quenched QCD take the same value because of the charge-conjugation invariance of the QCD_I ensemble. Therefore, the orbifold equivalence (the chiral condensate in QCD_B = the chiral condensate in QCD_I) tells us it is not affected by the phase quenching. For the baryon density, let us remind $\langle \mathrm{Tr}\gamma^0 D^{-1}(-\mu)\rangle_I = -\langle \mathrm{Tr}\gamma^0 D^{-1}(+\mu)\rangle_I$, again because of the charge-conjugation invariance of the QCD_I ensemble. Therefore, the isospin density in QCD_I and the baryon density in the phase-quenched QCD take the same value. (Also the baryon density in QCD_I becomes zero.) By combining it with the orbifold equivalence (the baryon density in QCD_B = the isospin density in QCD_I), we conclude that the phase quenching does not affect the expectation value of the baryon density. The same argument holds for other observables too, and the orbifold equivalence leads to the exactness of the phase quenching for any observable to $O(N_f/N_c)$.

3. Evidence from lattice simulations at $N_c = 3$

We have seen that the phase quenching is exact to $O(N_f/N_c)$. But the standard 't Hooft counting does not tell us the expansion coefficients. That motivates us to look at lattice data of $N_c = 3$ QCD. In the following we summarize lattice studies which compared QCD_B and the phase-quenched QCD.

- In [12], Nakamura et al. studied two-flavor QCD_B and phase-quenched QCD by using staggered fermions. QCD_B is obtained by the phase reweighting. For the chiral condensate and the Polyakov loop, they found a perfect agreement between QCD_B and the phase-quenched QCD within numerical errors.

- In the right panel of Fig. 1 and the left panel of Fig. 4 of [13], the free energy at various temperatures between $0.5T_c$ and $1.1T_c$ are plotted as functions of the baryon number, which show a nice agreement near the critical temperature.
- Fodor et al. [14] combined the phase reweighting and the density of states methods. In Fig. 4 of [14] they show the critical couplings at $a\mu = 0.3$ both in the phase-quenched and phase-reweighted cases, which take close values.
- The large-N_c equivalence holds for the imaginary baryon and isospin chemical potentials, $(\mu_u, \mu_d) = (i\mu_{\rm img}, i\mu_{\rm img})$ and $(\mu_u, \mu_d) = (i\mu_{\rm img}, -i\mu_{\rm img})$, without any modification. As a result, the chiral condensates $\langle\bar\psi\psi\rangle_B$ and $\langle\bar\psi\psi\rangle_I$ take the same value at finite imaginary potentials as long as the projection symmetries are unbroken.
  In [16], the chiral critical temperatures $T_c(\mu)$ in two-flavor QCD were exploited by the extrapolations from the imaginary chemical potential, by using a fitting ansatz $T_c(\mu)/T_c(0) = 1 + a_1\left(\frac{\mu}{\pi T}\right)^2$. They found [16] $a_1 = -0.470(13)$ for μ_I and $a_1 = -0.522(10)$ for μ_B, which provide a nice quantitative agreement already at $N_c = 3$.
- Let us consider the Taylor expansion method, in which the expectation value of an observable is expanded in powers of μ/T, $\langle\mathcal{O}\rangle_{B,I} = \sum_{n=0}^{\infty} c_n^{B,I}(\mu/T)^n$. Taylor coefficients c_n^B and c_n^I, which are functions of the temperature T, can be determined by the simulation at $\mu = 0$. The large-N_c equivalence tells that the coefficients agree in the large-N_c limit: $\lim_{N_c\to\infty} c_n^B = \lim_{N_c\to\infty} c_n^I$.
  In [15], the coefficient c_2^B and c_2^I for the chiral condensate and the pressure of the quark-gluon gas have been calculated[3] in two-flavor QCD. The agreement is good, especially for the chiral condensate.

4. The equivalence in the effective models

Let us we consider the Nambu–Jona-Lasinio (NJL) model as an example of the effective models. We concentrate on the chiral limit in order to simplify the discussion. The starting point is the Lagrangian with the U(N_c) color current interaction with N_f flavors,

$$\mathcal{L}_{\rm NJL} = \bar\psi_f\left(\gamma^\mu\partial_\mu + \mu_f\gamma^4\right)\psi_f + \frac{G}{N_c}J^{(\rm U)}_{\mu A}J^{(\rm U)}_{\mu A}, \tag{6}$$

where $J^{(\rm U)}_{\mu A} = \bar\psi_f\gamma_\mu T^A_{\rm U}\psi_f$ and $T^A_{\rm U}$ are the U(N_c) color generators and summation is taken over repeated indices. The coupling constant G is taken to be of order N_c^0. One rewrites it keeping only the interactions in the scalar and pseudoscalar channels after Fierz transformations:

$$\begin{aligned} \mathcal{L}_{\rm NJL} &= \bar\psi\left(\gamma^\mu\partial_\mu + \mu_f\gamma^4\right)\psi + \mathcal{L}_{\rm int}, \\ \mathcal{L}_{\rm int} &= \frac{G}{N_c}\left[(\bar\psi_f\psi_{f'})(\bar\psi_{f'}\psi_f) + (\bar\psi_f i\gamma^5\psi_{f'})(\bar\psi_{f'}i\gamma^5\psi_f)\right]. \end{aligned} \tag{7}$$

In the Lagrangians (6) and (7), the invariance under U(N_c) gauge symmetry and $\mathrm{U}(N_f)_L \times U(N_f)_R$ flavor symmetry are manifest. Here we ignore the effect of instantons or the $\mathrm{U}(1)_A$ anomaly which explicitly breaks the $\mathrm{U}(1)_A$ symmetry, because it is subleading in $1/N_c$. (From the viewpoint of the orbifold equivalence, there is no reason for the exactness of the phase quenching at the level of MFA if we take into account the $1/N_c$-suppressed instanton effects.

[3] For odd n, c_n^B and c_n^I vanish, and the first nontrivial μ-dependences appear in c_2^B and c_2^I. Although c_n^B $(n \geq 4)$ have been calculated, c_n^I $(n \geq 4)$ have not been calculated in [15]. (Note that, for $n \geq 4$, they use the same symbol c_n^I for another quantity.)

However, even if we incorporate them, the phase quenching for the chiral condensate turns out to be exact within the NJL model [6].) For SO($2N_c$) theory, we can construct the corresponding NJL model in the same manner, by using the SO($2N_c$) current. The proof of the large-N_c orbifold equivalence applies to the NJL model, by starting with the NJL model for SO_B and by using similar projection conditions as the previous section [6].

In order to consider the large-N_c limit, the correct $1/N_c$-counting scheme which reproduces the correct $1/N_c$-scaling in the large-N_c QCD is needed. The quark ψ has N_c colors so that a closed color loop gives a factor of N_c. The coupling constant of the four-fermi interaction should be taken as $O(N_c^{-1})$, and furthermore, the form of possible four-fermi interactions are restricted; in other words only the interactions which have origins in QCD are allowed. Then the right $1/N_c$-counting follows and we can use the same proof of the orbifold equivalence as the large-N_c QCD [6]. A rather standard argument relates this large-N_c equivalence to the equivalence at the level of MFA. Similar arguments hold also for various other theories, such as linear sigma model [6], Polyakov-Nambu-Jona-Lasinio model [6], Polyakov-quark-meson model [6], chiral random matrix model [5], Sakai-Sugimoto model [6] and D3/D7 model [11].

5. Conclusion

We have seen the exactness of the phase quenching to $O(N_f/N_c)$ outside the BEC/BCS crossover region. In other words, the effect of the phase is $1/N_c$-suppressed, and hence the reweighting method works without the overlap problem. Previous lattice studies confirm the effect of the phase is small already at $N_c = 3$. We have also shown the exactness of the phase quenching in effective models, which had been realized by explicit calculations and used to justify the reweighting method. Because the phase quenching and phase reweighting methods have been theoretically justified, it is important to study the QCD phase diagram by using them.

Acknowledgement

N.Y. is supported by JSPS Research Fellowships for Young Scientists.

References

[1] J. B. Kogut and D. K. Sinclair, Phys. Rev. D **77**, 114503 (2008).
[2] T. D. Cohen, Phys. Rev. D **70**, 116009 (2004).
[3] D. Toublan, Phys. Lett. B **621**, 145 (2005).
[4] A. Cherman, M. Hanada and D. Robles-Llana, Phys. Rev. Lett. **106**, 091603 (2011).
[5] M. Hanada and N. Yamamoto, JHEP **1202**, 138 (2012).
[6] M. Hanada, Y. Matsuo and N. Yamamoto, Phys. Rev. D **86**, 074510 (2012).
[7] S. Kachru and E. Silverstein, Phys. Rev. Lett. **80**, 4855 (1998); A. E. Lawrence, N. Nekrasov and C. Vafa, Nucl. Phys. **B533**, 199-209 (1998); M. Bershadsky, Z. Kakushadze and C. Vafa, Nucl. Phys. B **523**, 59 (1998).
[8] M. Bershadsky and A. Johansen, Nucl. Phys. B **536**, 141 (1998); P. Kovtun, M. Ünsal and L. G. Yaffe, JHEP **0507**, 008 (2005).
[9] M. Schmaltz, Phys. Rev. D **59**, 105018 (1999); M. J. Strassler, hep-th/0104032.
[10] A. Cherman and B. C. Tiburzi, JHEP **1106**, 034 (2011).
[11] M. Hanada, C. Hoyos, A. Karch and L. G. Yaffe, JHEP **1208**, 081 (2012).
[12] A. Nakamura, Y. Sasai and T. Takaishi, AIP Conf. Proc. **756**, 416 (2005); Y. Sasai, A. Nakamura and T. Takaishi, Nucl. Phys. Proc. Suppl. **129**, 539 (2004).
[13] P. de Forcrand, M. A. Stephanov and U. Wenger, PoS LAT **2007**, 237 (2007).
[14] Z. Fodor, S. D. Katz and C. Schmidt, JHEP **0703**, 121 (2007) [hep-lat/0701022].
[15] C. R. Allton, M. Doring, S. Ejiri, S. J. Hands, O. Kaczmarek, F. Karsch, E. Laermann and K. Redlich, Phys. Rev. D **71**, 054508 (2005) [hep-lat/0501030].
[16] P. Cea, L. Cosmai, M. D'Elia, A. Papa and F. Sanfilippo, Phys. Rev. D **85**, 094512 (2012).

Early thermalization of quark-gluon matter by elastic 3-to-3 scattering

Xiao-Ming Xu
Department of Physics, Shanghai University, Baoshan, Shanghai 200444, China
E-mail: xmxu@mail.shu.edu.cn

Abstract. The early thermalization is crucial to the quark-gluon plasma as a perfect liquid and results from many-body scattering. We calculate squared amplitudes for elastic parton-parton-parton scattering in perturbative QCD. Transport equations with the squared amplitudes are established and solved to obtain the thermalization time of initially produced quark-gluon matter and the initial temperature of quark-gluon plasma. We find that the thermalization times of quark matter and gluon matter are different.

1. Introduction

Quark-gluon matter initially produced in Au-Au collisions at RHIC energies is not in a thermal state and has not a temperature since gluon and quark distributions are anisotropic in momentum space. After a period of time of $t_{\rm therm}$, quark-gluon matter is in a thermal state and has a temperature while the distributions are isotropic, i.e. quark-gluon matter is a quark-gluon plasma. The transition from initially produced quark-gluon matter to the quark-gluon plasma is a thermalization process and $t_{\rm therm}$ is the thermalization time. It had not been expected that the thermalization time is small before the measured elliptic flow coefficient of hadrons was first reported. In order to explain the measured data, hydrodynamic calculations assume early thermalization and ideal relativistic fluid flow [1, 2, 3]. Early thermalization is that a thermal state is achieved with a time less than 1 fm/c from the moment when quark-gluon matter is initially created in heavy-ion collisions. Different thermalization times have been given in different hydrodynamic models [4]. Strongly interacting matter in thermal equilibrium can very easily flow in any direction without any breakup, i.e. this matter is a perfect liquid. Strong interaction of particles inside dense matter establishes a thermal state rapidly. This means that the quark-gluon plasma formed by the early thermalization is a perfect liquid. Therefore, the early thermalization is crucial to the perfect liquid the quark-gluon plasma.

The early thermalization is an assumption in hydrodynamic calculations. We need to understand the early thermalization. Then, two questions are raised. One is how initially produced quark-gluon matter establishes a thermal state, another is when initially produced quark-gluon matter establishes a thermal state. To answer the two questions, we examine the occurrence probabilities of two-parton scattering, three-parton scattering and so on in quark-gluon matter. This is motivated by the thought that thermalization is related to two-body scattering [5, 6, 7, 8]. One thousand and five hundred partons are generated from HIJING within $-0.3 < z < 0.3$ fm in the longitudinal direction and $r < 6.4$ fm in the transverse direction in central Au-Au collisions at $\sqrt{s_{NN}} = 200$ GeV. This volume of partons corresponds to a parton number density of 19.4 fm^{-3}. The maxima of the numbers of n-parton scattering with $n = 2, \cdots, 7$ are 750, 500, 375, 300, 250 and 214, respectively. The occurrence probability can be represented by the number of n-parton scattering divided by the sum of the maximum numbers of two-, three-, $\cdots$, and seven-parton scattering. The occurrence probability increases with increasing interaction range. At an interaction range of 0.62 fm the 2-parton scattering and 3-parton scattering have occurrence probabilities of 0.3 and 0.2, respectively [9]. Therefore, the three-parton scattering is important due to the high parton number density. One effect of elastic three-gluon scattering is the early thermalization of gluon matter initially produced in central Au-Au collisions [10]. The early thermalization is an effect of many-body interaction!

The occurrence probabilities for 4-parton, 5-parton, 6-parton, 7-parton, 8-parton, 9-parton, 10-parton and 11-parton scattering are 0.146, 0.11, 0.09, 0.075, 0.044, 0.023, 0.009 and 0.002, respectively. The occurrence probabilities for 12-parton scattering and 13-parton scattering are negligible. In a sphere with a radius the same as the interaction range the number of partons is at most 14. This means that at most 14-parton scattering is allowed. As a consequence, the occurrence probability of 15-parton scattering is zero and the one of the 14-parton scattering is very small. The sum of the occurrence probabilities from the 2-parton scattering equals 1 and the occurrence probabilities form a convergent series. From the occurrence probabilities the elastic 4-parton (5-parton) scattering is expected to give a smaller contribution to thermalization than the elastic 3-parton (4-parton) scattering and the contribution of the elastic n-parton scattering with $n \geq 6$ may be neglected.

Thermalization of quark-gluon matter at a high number density is governed by the two-parton scattering and the three-parton scattering in the present work. This is the answer to the first question. An answer to the second question is given in the next two sections.

2. Elastic three-parton scattering

Thermalization of water (air) is governed by elastic two-body scattering of water molecules (air molecules). Thermalization by elastic two-body scattering is conventional. Therefore, two-parton scattering is first taken into account to examine thermalization of quark-gluon matter. Elastic parton-parton scattering was studied by Cutler and Sivers in [11] and by Combridge, Kripfganz and Ranft in [12] in perturbative QCD. Spin- and color-averaged squared amplitudes of order α_s^2 were derived by hand and expressions for the squared amplitudes are presented in terms of the Mandelstam variables in [11, 12]. It is very convenient to use the expressions.

However, it is impractical to derive squared amplitudes for elastic parton-parton-parton scattering by hand and the squared amplitudes have to be derived from Fortran code. This is because many Feynman diagrams even at order α_s^4 are involved in the elastic 3-to-3 scattering. For example, there are 72 diagrams for elastic scattering of one gluon and two identical quarks and 76 diagrams for elastic gluon-quark-antiquark scattering [13]. Only one Feynman diagram has two four-gluon vertices and it is shown in figure 1. The squared amplitude for the diagram takes the shortest expression among all diagrams involved in elastic parton-parton-parton scattering at the tree level. Including the average over the spin and color states of the three initial gluons, the spin- and color-summed squared amplitude is [14]

$$\frac{118098\pi^4\alpha_s^4}{s^2}$$

where s is the square of the sum of the four-momenta of the three initial gluons. The expression has only one term, but the squared amplitudes for other diagrams have more terms with the reasons as follows. The other diagrams must have at least one triple-gluon vertex or the trace of a product of Dirac matrices. According to the Feynman rules the triple-gluon vertex consists of three terms. The trace of a product of even four Dirac matrices gives three terms. Then, the squared amplitude for each of the other Feynman diagrams or any interference term of different diagrams must have at least three terms.

According to the numbers of the triple-gluon vertex and the four-gluon vertex, Feynman diagrams are classified and their squared amplitudes are organized. For instance, the 72 diagrams for elastic gluon-quark-quark scattering are separated into 3 classes [13]. The first, second and third classes contain 40 diagrams with no triple-gluon vertex and no four-gluon vertex, 24 diagrams with one triple-gluon vertex and no four-gluon vertex and 8 diagrams with two triple-gluon vertices or one four-gluon vertex, respectively. The 76 diagrams for elastic gluon-quark-antiquark scattering are also divided into 3 classes even though quark-antiquark annihilation may occur [13]. The first class contains 40 diagrams with no triple-gluon vertex and no four-gluon vertex: 20 diagrams with no quark-antiquark annihilation and 20 diagrams with the annihilation. The second class contains 28 diagrams with one triple-gluon vertex and no four-gluon vertex: 12 with no quark-antiquark annihilation and 16 with the annihilation. The third class contains 8 diagrams with two triple-gluon vertices or one four-gluon vertex: 4 with no quark-antiquark annihilation and 4 with the annihilation.

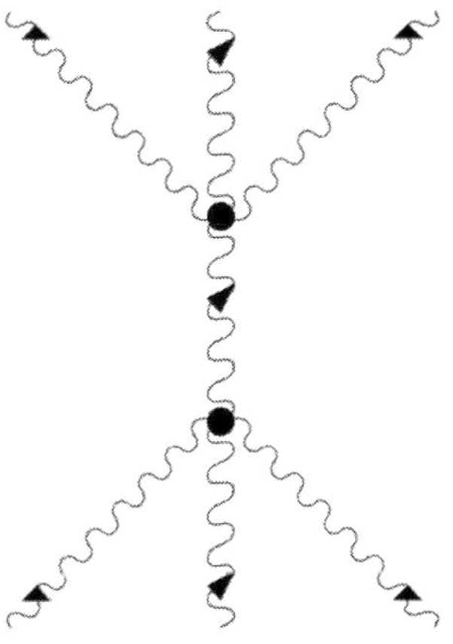

Figure 1. Scattering of three gluons. The wiggly lines stand for gluons.

3. Thermalization

We establish transport equations for quark-gluon matter which consists of gluons, quarks and antiquarks with up and down flavors. We assume that the quark distribution is symmetric in flavor and is identical with the antiquark distribution. The quark and gluon distribution functions are denoted by f_{qi} and f_{gi}, respectively, where i labels the ith quark or gluon. With the elastic parton-parton scattering and the elastic parton-parton-parton scattering, the transport equation for gluons is [13]

$$
\begin{aligned}
\frac{\partial f_{g1}}{\partial t} + \vec{v}_1 \cdot \vec{\nabla}_{\vec{r}} f_{g1} = & -\frac{1}{2E_1}\int \frac{d^3p_2}{(2\pi)^3 2E_2}\frac{d^3p_3}{(2\pi)^3 2E_3}\frac{d^3p_4}{(2\pi)^3 2E_4}(2\pi)^4\delta^4(p_1+p_2-p_3-p_4) \\
& \times\left\{\frac{g_G}{2} \mid \mathcal{M}_{gg\to gg}\mid^2 [f_{g1}f_{g2}(1+f_{g3})(1+f_{g4}) - f_{g3}f_{g4}(1+f_{g1})(1+f_{g2})]\right. \\
& +g_Q(\mid \mathcal{M}_{gu\to gu}\mid^2 + \mid \mathcal{M}_{gd\to gd}\mid^2 + \mid \mathcal{M}_{g\bar{u}\to g\bar{u}}\mid^2 + \mid \mathcal{M}_{g\bar{d}\to g\bar{d}}\mid^2) \\
& \left.\times[f_{g1}f_{q2}(1+f_{g3})(1-f_{q4}) - f_{g3}f_{q4}(1+f_{g1})(1-f_{q2})]\right\} \\
& -\frac{1}{2E_1}\int \frac{d^3p_2}{(2\pi)^3 2E_2}\frac{d^3p_3}{(2\pi)^3 2E_3}\frac{d^3p_4}{(2\pi)^3 2E_4}\frac{d^3p_5}{(2\pi)^3 2E_5}\frac{d^3p_6}{(2\pi)^3 2E_6}
\end{aligned}
$$

$$
\begin{aligned}
&\times(2\pi)^4\delta^4(p_1+p_2+p_3-p_4-p_5-p_6)\left\{\frac{g_G^2}{12}\mid \mathcal{M}_{ggg\to ggg}\mid^2\right.\\
&\times[f_{g1}f_{g2}f_{g3}(1+f_{g4})(1+f_{g5})(1+f_{g6})-f_{g4}f_{g5}f_{g6}(1+f_{g1})(1+f_{g2})(1+f_{g3})]\\
&+\frac{g_Gg_Q}{2}(\mid \mathcal{M}_{ggu\to ggu}\mid^2+\mid \mathcal{M}_{ggd\to ggd}\mid^2+\mid \mathcal{M}_{gg\bar{u}\to gg\bar{u}}\mid^2+\mid \mathcal{M}_{gg\bar{d}\to gg\bar{d}}\mid^2)\\
&\times[f_{g1}f_{g2}f_{q3}(1+f_{g4})(1+f_{g5})(1-f_{q6})-f_{g4}f_{g5}f_{q6}(1+f_{g1})(1+f_{g2})(1-f_{q3})]\\
&+g_Q^2[\frac{1}{4}\mid \mathcal{M}_{guu\to guu}\mid^2+\frac{1}{2}(\mid \mathcal{M}_{gud\to gud}\mid^2+\mid \mathcal{M}_{gdu\to gdu}\mid^2)+\frac{1}{4}\mid \mathcal{M}_{gdd\to gdd}\mid^2\\
&+\mid \mathcal{M}_{gu\bar{u}\to gu\bar{u}}\mid^2+\mid \mathcal{M}_{gu\bar{d}\to gu\bar{d}}\mid^2+\mid \mathcal{M}_{gd\bar{u}\to gd\bar{u}}\mid^2+\mid \mathcal{M}_{gd\bar{d}\to gd\bar{d}}\mid^2\\
&+\frac{1}{4}\mid \mathcal{M}_{g\bar{u}\bar{u}\to g\bar{u}\bar{u}}\mid^2+\frac{1}{2}(\mid \mathcal{M}_{g\bar{u}\bar{d}\to g\bar{u}\bar{d}}\mid^2+\mid \mathcal{M}_{g\bar{d}\bar{u}\to g\bar{d}\bar{u}}\mid^2)+\frac{1}{4}\mid \mathcal{M}_{g\bar{d}\bar{d}\to g\bar{d}\bar{d}}\mid^2]\\
&\left.\times[f_{g1}f_{q2}f_{q3}(1+f_{g4})(1-f_{q5})(1-f_{q6})-f_{g4}f_{q5}f_{q6}(1+f_{g1})(1-f_{q2})(1-f_{q3})]\right\}, \qquad (1)
\end{aligned}
$$

and the transport equation for up-quarks is

$$
\begin{aligned}
\frac{\partial f_{q1}}{\partial t} + \vec{v}_1\cdot\vec{\nabla}_{\vec{r}}f_{q1} = & -\frac{1}{2E_1}\int\frac{d^3p_2}{(2\pi)^32E_2}\frac{d^3p_3}{(2\pi)^32E_3}\frac{d^3p_4}{(2\pi)^32E_4}(2\pi)^4\delta^4(p_1+p_2-p_3-p_4)\\
&\times\left\{g_G\mid \mathcal{M}_{ug\to ug}\mid^2[f_{q1}f_{g2}(1-f_{q3})(1+f_{g4})-f_{q3}f_{g4}(1-f_{q1})(1+f_{g2})]\right.\\
&+g_Q(\frac{1}{2}\mid \mathcal{M}_{uu\to uu}\mid^2+\mid \mathcal{M}_{ud\to ud}\mid^2+\mid \mathcal{M}_{u\bar{u}\to u\bar{u}}\mid^2+\mid \mathcal{M}_{u\bar{d}\to u\bar{d}}\mid^2)\\
&\left.\times[f_{q1}f_{q2}(1-f_{q3})(1-f_{q4})-f_{q3}f_{q4}(1-f_{q1})(1-f_{q2})]\right\}\\
&-\frac{1}{2E_1}\int\frac{d^3p_2}{(2\pi)^32E_2}\frac{d^3p_3}{(2\pi)^32E_3}\frac{d^3p_4}{(2\pi)^32E_4}\frac{d^3p_5}{(2\pi)^32E_5}\frac{d^3p_6}{(2\pi)^32E_6}\\
&\times(2\pi)^4\delta^4(p_1+p_2+p_3-p_4-p_5-p_6)\left\{\frac{g_G^2}{4}\mid \mathcal{M}_{ugg\to ugg}\mid^2\right.\\
&\times[f_{q1}f_{g2}f_{g3}(1-f_{q4})(1+f_{g5})(1+f_{g6})-f_{q4}f_{g5}f_{g6}(1-f_{q1})(1+f_{g2})(1+f_{g3})]\\
&+g_Qg_G(\frac{1}{2}\mid \mathcal{M}_{uug\to uug}\mid^2+\mid \mathcal{M}_{udg\to udg}\mid^2+\mid \mathcal{M}_{u\bar{u}g\to u\bar{u}g}\mid^2+\mid \mathcal{M}_{u\bar{d}g\to u\bar{d}g}\mid^2)\\
&\times[f_{q1}f_{q2}f_{g3}(1-f_{q4})(1-f_{q5})(1+f_{g6})-f_{q4}f_{q5}f_{g6}(1-f_{q1})(1-f_{q2})(1+f_{g3})]\\
&+g_Q^2[\frac{1}{12}\mid \mathcal{M}_{uuu\to uuu}\mid^2+\frac{1}{4}(\mid \mathcal{M}_{uud\to uud}\mid^2+\mid \mathcal{M}_{udu\to udu}\mid^2)+\frac{1}{4}\mid \mathcal{M}_{udd\to udd}\mid^2\\
&+\frac{1}{2}\mid \mathcal{M}_{uu\bar{u}\to uu\bar{u}}\mid^2+\frac{1}{2}\mid \mathcal{M}_{uu\bar{d}\to uu\bar{d}}\mid^2+\mid \mathcal{M}_{ud\bar{u}\to ud\bar{u}}\mid^2+\mid \mathcal{M}_{ud\bar{d}\to ud\bar{d}}\mid^2\\
&+\frac{1}{4}\mid \mathcal{M}_{u\bar{u}\bar{u}\to u\bar{u}\bar{u}}\mid^2+\frac{1}{2}(\mid \mathcal{M}_{u\bar{u}\bar{d}\to u\bar{u}\bar{d}}\mid^2+\mid \mathcal{M}_{u\bar{d}\bar{u}\to u\bar{d}\bar{u}}\mid^2)+\frac{1}{4}\mid \mathcal{M}_{u\bar{d}\bar{d}\to u\bar{d}\bar{d}}\mid^2]\\
&\left.\times[f_{q1}f_{q2}f_{q3}(1-f_{q4})(1-f_{q5})(1-f_{q6})-f_{q4}f_{q5}f_{q6}(1-f_{q1})(1-f_{q2})(1-f_{q3})]\right\}, \qquad (2)
\end{aligned}
$$

where $\vec{v}_1$ is the parton velocity; g_G and g_Q are the color-spin degeneracy factors; $p_i(i=1,\cdots,6)$ denote the four-momenta of initial and final partons; E_i is the energy component of p_i. $\mid \mathcal{M}_{ab\to a'b'}\mid^2$ and $\mid \mathcal{M}_{abc\to a'b'c'}\mid^2$ are the squared amplitudes for $a+b\to a'+b'$ and $a+b+c\to a'+b'+c'$, respectively. The squared amplitude $\mid \mathcal{M}_{ggg\to ggg}\mid^2$ was obtained in [10]; $\mid \mathcal{M}_{qqq\to qqq}\mid^2$ in [15]; $\mid \mathcal{M}_{gqq\to gqq}\mid^2$, $\mid \mathcal{M}_{gq\bar{q}\to gq\bar{q}}\mid^2$ and $\mid \mathcal{M}_{g\bar{q}\bar{q}\to g\bar{q}\bar{q}}\mid^2$ in [13]. $\mid \mathcal{M}_{ggq\to ggq}\mid^2$ and $\mid \mathcal{M}_{gg\bar{q}\to gg\bar{q}}\mid^2$ have not been obtained and are temporarily set as zero. Similar equations for down-quarks, up-antiquarks and down-antiquarks can be established.

Parton-parton scattering affects only two partons' momenta while parton-parton-parton scattering changes three partons' momenta. If the 3-to-3 scattering has the same occurrence probability as the 2-to-2 scattering, the 3-to-3 scattering changes partons' momenta faster than the 2-to-2 scattering. Frequent changes of momenta establish a thermal state. Indeed, the elastic 3-to-3 scattering drives anisotropic matter described by the transport equations towards global thermal equilibrium [14], which is similar to the H-theorem. Solutions of the transport equations will show that a thermal state is established from a moment.

In the region $-0.3 < z < 0.3$ fm in the longitudinal direction and $r < 6.4$ fm in the transverse direction, 1500 gluons and 1000 fermions including up quarks, down quarks, up antiquarks and down antiquarks are generated from HIJING for central Au-Au collisions at $\sqrt{s_{NN}} = 200$ GeV. These gluons, quarks and antiquarks at $t = 0.2$ fm/c form initially produced quark-gluon matter and are anisotropically distributed in momentum space. Solving the transport equations, we found that at $t = 0.68$ fm/c gluon matter has an isotropic momentum distribution function, a thermal state with the temperature $T = 0.5$ GeV and a thermalization time of the order of 0.48 fm/c [13]. A thermal state of quark matter with $T = 0.3$ GeV is established at a later time of 1.56 fm/c. In other words, the thermalization time 1.36 fm/c of quark matter is longer than gluon matter [13]. The elastic scattering of gq, $g\bar{q}$, gqq, $gq\bar{q}$ and $g\bar{q}\bar{q}$ gives contributions to gluon matter and quark matter nearly identical in thermalization. The difference of the thermalization times of gluon matter and quark matter mainly comes from the difference that elastic ggg (gg) scattering has a larger squared amplitude than elastic qqq or $qq\bar{q}$ (qq or $q\bar{q}$) scattering and gluon matter is denser than quark matter.

4. Conclusions

Early thermalization is an intrinsic property of initially produced quark-gluon matter at the high number density and is an effect of many-body scattering. The elastic 3-to-3 scattering is important at such a number density. The elastic gluon-gluon scattering and the elastic gluon-gluon-gluon scattering lead to the early thermalization of gluon matter. The thermalization time of quark matter differs from that of gluon matter and we need to include the elastic scattering of both ggq and $gg\bar{q}$ in future work to explore whether quark matter thermalizes rapidly or not.

Acknowledgments

This work was supported by the National Natural Science Foundation of China under grant no 11175111.

References

[1] Kolb P F, Huovinen P, Heinz U and Heiselberg H 2001 *Phys. Lett.* B **500** 232
[2] Teaney D, Lauret J and Shuryak E V 2001 A hydrodynamic description of heavy ion collisions at the SPS and RHIC *Preprint* nucl-th/0110037
[3] Hirano T 2001 *Phys. Rev.* C **65** 011901
[4] Adcox K *et al* (PHENIX Collaboration) 2005 *Nucl. Phys.* A **757** 184
[5] Shuryak E 1992 *Phys. Rev. Lett.* **68** 3270
[6] Wong S M H 1996 *Phys. Rev.* C **54** 2588
[7] Nayak G C, Dumitru A, McLerran L and Greiner W 2001 *Nucl. Phys.* A **687** 457
[8] Shin G R and Müller B 2003 *J. Phys. G: Nucl. Part. Phys.* **29** 2485
[9] Xu X-M, Ma C-C, Chen A-Q and Weber H J 2007 *Phys. Lett.* B **645** 146
[10] Xu X-M, Sun Y, Chen A-Q and Zheng L 2004 *Nucl. Phys.* A **744** 347
[11] Cutler R and Sivers D 1978 *Phys. Rev.* D **17** 196
[12] Combridge B L, Kripfganz J and Ranft J 1977 *Phys. Lett.* B **70** 234
[13] Xu X-M and Xu L-S 2010 *J. Phys. G: Nucl. Part. Phys.* **37** 115003
[14] Xu X-M 2007 *High Energy Phys. Nucl. Phys.* **31** 880
[15] Xu X-M, Peng R and Weber H J 2005 *Phys. Lett.* B **629** 68

Some issues of chiral anomaly in quantum kinetic approach

Jian-hua Gao, Shi Pu, Qun Wang

Department of Modern Physics, University of Science and Technology of China, Hefei 230026, China

Abstract. We discuss some issues in our previous works in quantum kinetic approach to chiral anomaly, chiral magnetic and vortical effects. The Wigner function for massless fermions in a constant background electromagnetic field is solved in a power expansion. Chiral magnetic and vortical effects emerge naturally. The conservation equations with anomaly are derived in details.

Introduction. — Recently there have been some efforts to incorporate chiral anomaly in kinetic equations [1, 2, 3, 4]. The quantum kinetic theory with Wigner functions is a natural approach to describe chiral anomaly, chiral magnetic and vortical effects (CME and CVE) [5, 6, 7] which are all quantum effects. The CME and CVE have been widely studied in many respects, such as AdS/CFT duality [8, 9, 10, 11, 12], relativistic hydrodynamics [13, 14, 15, 16], and quantum field theory [17, 18, 19, 20, 21, 22, 23].

In one of our previous works [14], we have tried to describe chiral anomaly in a semi-classical Boltzmann equation. It seemed that non-trivial quantum effects have to be included in Boltzmann equation in a proper way. In our later work [1] we proposed a quantum kinetic approach with the gauge invariant Wigner functions [24, 25, 26] and showed that chiral anomaly, CME and CVE emerge naturally. In this paper we will discuss in detail some issues in our previous works. We will show in details how conservation equations with anomaly are derived. We use the metric convention $g^{\mu\nu} = \mathrm{diag}(1, -1, -1, -1)$ and the fluid velocity square becomes $u_\mu u^\mu = 1$.

Quantum kinetic approach. — In quantum kinetic theory, the particle distribution in a 8-dimensional phase space, four-coordinate x and four-momentum p, can be described by the Wigner function $W(x,p)$, defined as the ensemble average of the gauge invariant Wigner operator [24, 25, 26]. For spin-1/2 fermions, this operator is given by

$$\hat{W}_{\alpha\beta} = \int \frac{d^4y}{(2\pi)^4} e^{-ip\cdot y} \bar{\psi}_\beta(x_+) U(x_+, x_-) \psi_\alpha(x_-), \tag{1}$$

where $x_\pm \equiv x \pm \frac{1}{2}y$, and $U(x_+, x_-) \equiv \exp[-iQ\int_{x_-}^{x_+} dz^\mu A_\mu(z)]$ is the gauge link with Q the charge of particles. For simplicity, we only consider the massless fermions in a constant background field $F^{\mu\nu}$. In this case, the quantum kinetic equation for the Wigner function can be derived by the Dirac equation,

$$\gamma_\mu \left(p^\mu + \frac{1}{2} i \nabla^\mu \right) W(x,p) = 0, \tag{2}$$

where $\nabla^\mu \equiv \partial_x^\mu - QF^\mu{}_\nu \partial_p^\nu$. We can expand the Wigner function in terms of 16 generators of the Clifford algebra,

$$W(x,p) = \frac{1}{4}\left[\mathscr{F}(x,p) + i\gamma^5 \mathscr{P}(x,p) + \gamma^\mu \mathscr{V}_\mu(x,p) + \gamma^5\gamma^\mu \mathscr{A}_\mu(x,p) + \frac{1}{2}\sigma^{\mu\nu}\mathscr{S}_{\mu\nu}(x,p)\right], \quad (3)$$

where the coefficients can be obtained by $\mathscr{F} = \mathrm{Tr}[W]$, ,$\mathscr{P} = -i\mathrm{Tr}[\gamma^5 W]$, $\mathscr{V}_\mu = \mathrm{Tr}[\gamma^\mu W]$, $\mathscr{A}_\mu = \mathrm{Tr}[\gamma^\mu\gamma^5 W]$, and $\mathscr{S}_{\mu\nu} = \mathrm{Tr}[\sigma^{\mu\nu} W]$. The vector $\mathscr{V}^\mu$ and axial vector $\mathscr{A}^\mu$ are related to the $U(1)$ charge current j^μ and axial current j_5^μ, respectively,

$$\begin{aligned} j^\mu(x) &= <:\overline{\psi}(x)\gamma^\mu\psi(x):>= \int d^4p\gamma^\mu W(x,p) = \int d^4p\mathscr{V}^\mu, \\ j_5^\mu(x) &= <:\overline{\psi}(x)\gamma^\mu\gamma^5\psi(x):>= \int d^4p\gamma^\mu\gamma^5 W(x,p) = \int d^4p\mathscr{A}^\mu, \end{aligned} \quad (4)$$

where brackets indicate ensemble averaging. Therefore, we can get all hydrodynamics quantities from solving $\mathscr{V}^\mu$ and $\mathscr{A}^\mu$. Inserting the above decomposition into Eq. (2), we can write down a set of equations for the vector and axial vector component of the Wigner function, $\mathscr{V}^\mu$ and $\mathscr{A}^\mu$, as follows

$$p^\mu \mathscr{V}_\mu = 0, \quad p^\mu \mathscr{A}_\mu = 0, \quad (5)$$

$$\nabla^\mu \mathscr{V}_\mu = 0, \quad \nabla^\mu \mathscr{A}_\mu = 0, \quad (6)$$

$$\epsilon_{\mu\nu\rho\sigma}\nabla^\rho \mathscr{A}^\sigma = -2\left(p_\mu \mathscr{V}_\nu - p_\nu \mathscr{V}_\mu\right), \quad (7)$$

$$\epsilon_{\mu\nu\rho\sigma}\nabla^\rho \mathscr{V}^\sigma = -2\left(p_\mu \mathscr{A}_\nu - p_\nu \mathscr{A}_\mu\right). \quad (8)$$

Fortunately, $\mathscr{V}^\mu$ and $\mathscr{A}^\mu$ are decoupled with other components in Eq.(3). So we will only focus on these two vectors from now on.

One can expand hydrodynamical quantities in powers of Knudsen number K defined by the ratio of the mean free path to the macroscopic characteristic length scale. Since the Knudsen number is proportional to the space-time derivative ∂_μ^x, therefore this expansion is equivalent to the general gradient expansion. Similarly we can also expand $\mathscr{V}^\mu$ and $\mathscr{A}^\mu$ in terms of the space-time derivative ∂_μ^x and the field strength $F^{\mu\nu}$, i.e. $\mathscr{V}^\mu = \mathscr{V}_0^\mu + \mathscr{V}_1^\mu + O(K^2), \mathscr{A}^\mu = \mathscr{A}_0^\mu + \mathscr{A}_1^\mu + O(K^2)$. Then from Eqs. (7-8) which relate the lower to higher order contributions, we can solve $\mathscr{V}^\mu$ and $\mathscr{A}^\mu$ order by order.

In the zeroth order, we further assume that our system is very close to local equilibrium. Therefore, the vector and axial vector currents in the leading order should be identical to those in an ideal fluid. Meanwhile, $\mathscr{V}_\mu(x,p)$ and $\mathscr{A}_\mu(x,p)$ can be written as the functions of 4-velocity $u(x)$, temperature $T(x)$, chemical potential $\mu(x)$ and chiral chemical potential $\mu_5(x)$, i.e. the Wigner functions can be expressed in terms of $\{p, F^{\mu\nu}, u, T, \mu, \mu_5\}$ from the kinetic equations. Solving Eqs. (5, 7, 8) linear in the zeroth order, we get

$$\mathscr{V}_0^\mu = p^\mu \delta\left(p^2\right) V_0, \quad \mathscr{A}_0^\mu = p^\mu \delta\left(p^2\right) A_0, \quad (9)$$

with

$$\begin{aligned} [V_0, A_0] &= \sum_{s=\pm1} \theta(su\cdot p)\left[(f_{s,R} + f_{s,L}), (f_{s,R} - f_{s,L})\right], \\ f_{s,\chi} &= \frac{2}{(2\pi)^3}\frac{1}{e^{s(u\cdot p - \mu_\chi)/T} + 1}, (\chi = R, L), \end{aligned} \quad (10)$$

where $R(L)$ denotes the right (left)-handed fermions and $\mu_{R,L} = \mu \pm \mu_5$ [6] and $s = \pm 1$ for particles and anti-particles, respectively. The $\frac{(2\pi)^3}{2} f_{s,\chi}$ is the Fermi-Dirac distribution function.

Therefore, integration over d^4p, one will exactly get the currents at an equilibrium state. Since solutions (9) are only general solutions for Eq. (5, 7, 8), so we have to consider Eq. (6) as constraint conditions for the parameters set. Inserting the leading order solutions into Eq. (6) yields,

$$\begin{aligned}
&\Delta^{\sigma\alpha}\Delta^{\rho\beta}\left(\partial_\alpha u_\beta + \partial_\beta u_\alpha - \frac{2}{3}\Delta_{\alpha\beta}\Delta^{\rho\sigma}\partial_\rho u_\sigma\right) = 0,\\
&T\Delta^{\sigma\rho}\partial_\rho\frac{\mu}{T} + QE^\sigma = 0, \quad u\cdot\partial u^\sigma - \Delta^{\sigma\rho}\partial_\rho \ln T = 0,\\
&\partial_\sigma\frac{\mu_5}{T} = 0, \quad u^\sigma\partial_\sigma\frac{\mu}{T} = 0, \quad u\cdot\partial T + \frac{1}{3}T\Delta^{\rho\sigma}\partial_\rho u_\sigma = 0,
\end{aligned} \tag{11}$$

where we define the electric field $E^\mu = F^{\mu\nu}u_\nu$ in a moving frame. Those equations mean all the dissipative terms vanish in the zeroth order. These are also identical to the equations of motion for the ideal fluid [27]. For simplicity, without loss of generality, we consider only the static case with a constant temperature $T = const$. Those constraints become

$$T = \text{const}, \quad \mu_5 = \text{const}, \quad u\cdot\partial u^\sigma = 0, \ \partial_\sigma u^\sigma = 0, \ \partial_\sigma\mu = -QE_\sigma. \tag{12}$$

Solving Eq.(5-8) in the first order with these constraint conditions (11, 12), we obtained the following solution up to the first order,

$$\begin{aligned}
\mathscr{X}^\mu &= p^\mu\delta(p^2)Z_0 + \frac{1}{2}p_\nu[u^\mu\omega^\nu - u^\nu\omega^\mu]\frac{\partial\bar{Z}_0}{\partial(u\cdot p)}\delta(p^2)\\
&\quad -Qp_\nu[u^\mu B^\nu - u^\nu B^\mu]\bar{Z}_0\delta'(p^2) + Q\epsilon^{\mu\lambda\rho\sigma}u_\lambda p_\rho E_\sigma\bar{Z}_0\delta'(p^2),
\end{aligned} \tag{13}$$

where $\mathscr{X} = (\mathscr{V}, \mathscr{A})$, $Z_0 = (V_0, A_0)$, $\bar{Z}_0 = (A_0, V_0)$ Here we have defined $B^\mu = \epsilon^{\mu\nu\rho\sigma}u_\nu F_{\rho\sigma}$, and $\omega_\mu = \frac{1}{2}\epsilon_{\mu\nu\rho\sigma}u^\nu\partial^\rho u^\sigma$ are the magnetic field and the vorticity, respectively.

Derivation of conservation equations. — After integration over four-momenta we obtain the currents and energy-momentum tensor up to the first order,

$$j^\mu = \int d^4p\mathscr{V}^\mu = nu^\mu + \xi\omega^\mu + \xi_B B^\mu, \tag{14}$$

$$j_5^\mu = \int d^4p\mathscr{A}^\mu = n_5u^\mu + \xi_5\omega^\mu + \xi_{B5}B^\mu, \tag{15}$$

$$\begin{aligned}
T^{\mu\nu} &= \mathrm{Tr}\,\frac{1}{2}\int d^4(p^\mu\gamma^\nu + p^\nu\gamma^\mu)W(x,p)\\
&= \frac{1}{2}\int d^4p(p^\mu\mathscr{V}^\nu + p^\nu\mathscr{V}^\mu)\\
&= (\epsilon + P)u^\mu u^\nu - Pg^{\mu\nu} + n_5(u^\mu\omega^\nu + u^\nu\omega^\mu) + \frac{1}{2}Q\xi(u^\mu B^\nu + u^\nu B^\mu).
\end{aligned} \tag{16}$$

The charge n, n_5 and energy density ϵ are given by

$$\begin{aligned}
n &= \frac{1}{3}[\mu T^2 + \frac{1}{\pi^2}(\mu^3 + 3\mu\mu_5^2)],\\
n_5 &= \frac{1}{3}[\mu_5 T^2 + \frac{1}{\pi^2}(\mu_5^3 + 3\mu_5\mu^2)],\\
\epsilon &= \frac{1}{2\pi^2}\left[\frac{7\pi^4}{30}T^4 + \pi^2T^2(\mu^2 + \mu_5^2) + \frac{1}{2}(\mu^4 + \mu_5^4) + 3\mu^2\mu_5^2\right],
\end{aligned} \tag{17}$$

which are identical to the results from the classical kinetic theory in an equilibrium state (also see Appendix B.4 of Ref.[27]). The pressure is found to be $P = \epsilon/3$. The CME and CVE coefficients are

$$\begin{aligned}\xi &= \frac{1}{\pi^2}\mu\mu_5,\ \xi_B = \frac{Q}{2\pi^2}\mu_5,\\ \xi_5 &= \frac{1}{6}T^2 + \frac{1}{2\pi^2}\left(\mu^2+\mu_5^2\right),\ \xi_{B5} = \frac{Q}{2\pi^2}\mu.\end{aligned} \tag{18}$$

which are the functions of μ, μ_5, T. These results are consistent to those from the second law of thermodynamics in Ref. [13, 14, 15] and also independently in Ref. [20] by Kubo formula. With these non-zero coefficients, one will observe a current $j^\mu \propto B^\mu$ induced by a magnetic field. Since the $\mu_B\omega^\mu$, with μ_B the baryon chemical potential, can be considered as an effective magnetic field [7], the current induced by total angular momentum (or locally, vorticity) is expect to be observed in the non-central heavy ion collisions.

The axial vector current induced by vorticity can lead to the local polarization effect (LPE) [1], similar to the global polarization effect in heavy ion collisions proposed in Refs. [29, 30, 31, 32, 33, 34]. Because there are three quadratic terms in T, μ, μ_5 in ξ_5 as shown in Eq.(18), this effect should be present in high/low energy heavy ion collisions with high temperature and low chemical potential or low temperature and high chemical potential.

The space-time divergence of the axial current j_5^μ can be evaluated as follows

$$\partial_\mu j_5^\mu = -\frac{Q^2}{2\pi^2}E\cdot B, \tag{19}$$

where we have used $\partial^\sigma\mu = -QE^\sigma$, $\partial\cdot u = 0$, $u\cdot E = 0$, $\partial_\mu\omega^\mu = 0$ and $\partial_\mu B^\mu = 2(\omega\cdot E)$. We see that the triangle anomaly (19) is automatically satisfied by vector and axial vector components of the Wigner function in Eq. (13). This is in contrast to the normal field theory derivation of anomaly where the regularization is needed. We can take a closer look at the definition of the axial vector current with the Winger function. Inserting Eq.(1) into Eq. (15), we have

$$j_5^\mu(x) = \lim_{\epsilon\to 0}\overline{\psi}(x+\frac{1}{2}\epsilon)\gamma^\mu\gamma^5\mathcal{P}\exp\left[-i\epsilon^\mu Q\int_0^1 dsA_\mu(x-\frac{1}{2}\epsilon+s\epsilon)\right]\psi(x-\frac{1}{2}\epsilon), \tag{20}$$

which is defined as in the same way as in the textbook (e.g. Ref.[28]). In this way, we translate the Wigner function method to the language of the normal quantum field theory. We thus see that our derivation of axial anomaly in the quantum kinetic theory is very straightforward without explicit regularization. The reason is simple: the two fermion fields in the Wigner function are separated in space-time (nonlocal) and therefore free of singularities which appear in the contraction of two fields as their spatial distance is approaching zero. Similarly we can obtain the vector current consevation

$$\partial_\mu j^\mu = 0. \tag{21}$$

There is an ambiguity in defining a 3-dimensional vector current $\mathbf{j}$ through kinetic approach in a non-relativistic theory: for an arbitrary vector $\mathbf{a}$, we can shift $\mathbf{j}$ by $\nabla\times\mathbf{a}$, i.e. $\mathbf{j}\to\mathbf{j}+\nabla\times\mathbf{a}$, while keeping the vector current still conserved: $\nabla\cdot\mathbf{j} = 0$. To remove this ambiguity, one can enforce energy-momentum conservation. Therefore it is non-trivial to keep $\partial_\mu T^{\mu\nu} = QF^{\nu\mu}j_\mu$ satisfied as follows

$$\begin{aligned}\partial_\mu T^{\mu\nu} &= -\partial^\nu P + (u^\mu\omega^\nu + u^\nu\omega^\mu)\partial_\mu n_5 + n_5\partial_\mu(u^\mu\omega^\nu + u^\nu\omega^\mu)\\ &\quad +\frac{1}{2}Q(u^\mu B^\nu + u^\nu B^\mu)\partial_\mu\xi + \frac{1}{2}Q\xi\partial_\mu(u^\mu B^\nu + u^\nu B^\mu)\\ &= QF^{\nu\mu}nu_\mu + Q\xi F^{\nu\mu}\omega_\mu + Q\xi_B F^{\nu\mu}B_\mu + \frac{1}{2}Q\xi\Delta^{\nu\lambda}F_{\mu\lambda}\omega^\mu\\ &= QF^{\nu\mu}j_\mu.\end{aligned} \tag{22}$$

Here we have used $u^\mu\partial_\mu\omega^\nu = u^\mu\partial_\mu B^\nu = 0$, $\partial^\mu u^\nu = \epsilon^{\mu\nu\tau\lambda}u_\tau\omega_\lambda$, $\epsilon^{\nu\tau\lambda\mu}u_\tau\omega_\lambda B_\mu = \Delta^{\nu\lambda}F_{\mu\lambda}\omega^\mu$, $F^{\rho\sigma}\omega_\sigma = -(\omega\cdot E)u^\rho - \Delta^{\rho\lambda}F_{\mu\lambda}\omega^\mu$ and $F^{\rho\sigma}B_\sigma = -(E\cdot B)u^\rho$. Note that the residual term $(1/2)Q\xi\Delta^{\nu\lambda}F_{\mu\lambda}\omega^\mu$ in the above equation can be shown to be vanishing due to $\omega_\rho \parallel B_\rho$ from the static conditions. To prove this, we start from $\partial_\rho\mu = -QE_\rho$ and then get

$$\begin{aligned}\epsilon^{\mu\nu\rho\sigma}F_{\rho\lambda}\partial_\sigma u^\lambda &= -Q\epsilon^{\mu\nu\rho\sigma}F_{\rho\lambda}\partial_\sigma u^\lambda = \epsilon^{\mu\nu\rho\sigma}\partial_\sigma\partial_\rho\mu = 0\\ &= \epsilon^{\mu\nu\rho\sigma}\epsilon_{\sigma\lambda\xi\tau}u^\xi\omega^\tau u_\nu F_\rho^{\ \lambda} = \Delta^{\mu\lambda}F_{\rho\lambda}\omega^\rho. \end{aligned} \quad (23)$$

Here we have used $F_{\rho\lambda}$ is constant but E and B are not since they all depend on fluid velocity u^σ. So it means there is no such ambiguity in our approach as we mentioned above.

Acknowledgment. This work was supported by the National Science Foundation of China under the grant No. 11105137, 11125524, 10975092 11221504 and 11205150.

[1] J. -H. Gao, Z. -T. Liang, S. Pu, Q. Wang and X. -N. Wang, arXiv:1203.0725 [hep-ph].
[2] D. T. Son and N. Yamamoto, Phys. Rev. Lett. **109**, 181602 (2012).
[3] M. A. Stephanov and Y. Yin, Phys. Rev. Lett. **109**, 162001 (2012).
[4] J. -W. Chen, S. Pu, Q. Wang and X. -N. Wang, arXiv:1210.8312 [hep-th].
[5] D. E. Kharzeev, L. D. McLerran and H. J. Warringa, Nucl. Phys. A **803**, 227 (2008).
[6] K. Fukushima, D. E. Kharzeev and H. J. Warringa, Phys. Rev. D **78**, 074033 (2008).
[7] D. E. Kharzeev and D. T. Son, Phys. Rev. Lett. **106**, 062301 (2011).
[8] J. Erdmenger, M. Haack, M. Kaminski and A. Yarom, JHEP **0901**, 055 (2009).
[9] N. Banerjee *et al.*, JHEP **1101**, 094 (2011).
[10] M. Torabian and H. U. Yee, JHEP **0908**, 020 (2009).
[11] A. Rebhan, A. Schmitt and S. A. Stricker, JHEP **1001**, 026 (2010).
[12] T. Kalaydzhyan and I. Kirsch, Phys. Rev. Lett. **106**, 211601 (2011).
[13] D. T. Son and P. Surowka, Phys. Rev. Lett. **103**, 191601 (2009).
[14] S. Pu, J. h. Gao and Q. Wang, Phys. Rev. D **83**, 094017 (2011).
[15] A. V. Sadofyev and M. V. Isachenkov, Phys. Lett. B **697**, 404 (2011)
[16] D. E. Kharzeev and H. -U. Yee, Phys. Rev. D **84**, 045025 (2011).
[17] M. A. Metlitski and A. R. Zhitnitsky, Phys. Rev. D **72**, 045011 (2005).
[18] G. M. Newman and D. T. Son, Phys. Rev. D **73**, 045006 (2006).
[19] J. Charbonneau and A. Zhitnitsky, JCAP **1008**, 010 (2010).
[20] M. Lublinsky and I. Zahed, Phys. Lett. B **684**, 119 (2010).
[21] M. Asakawa, A. Majumder and B. Muller, Phys. Rev. C **81**, 064912 (2010).
[22] K. Landsteiner, E. Megias and F. Pena-Benitez, Phys. Rev. Lett. **107**, 021601 (2011).
[23] D. Hou, H. Liu and H. -c. Ren, JHEP **1105**, 046 (2011).
[24] H. T. Elze, M. Gyulassy and D. Vasak, Nucl. Phys. B **276**, 706 (1986).
[25] D. Vasak, M. Gyulassy and H. T. Elze, Annals Phys. **173**, 462 (1987).
[26] H. T. Elze and U. W. Heinz, Phys. Rept. **183**, 81 (1989).
[27] S. Pu, arXiv:1108.5828 [hep-ph].
[28] M. E. Peskin and D. V. Schroeder, *An introduction to Quantum Field Theory*, Addison-Wesley Publishing, 1995.
[29] Z. T. Liang, X. N. Wang, Phys. Rev. Lett. **94**, 102301 (2005) [Erratum-ibid. **96**, 039901 (2006)].
[30] B. Betz, M. Gyulassy and G. Torrieri, Phys. Rev. C **76**, 044901 (2007).
[31] F. Becattini, F. Piccinini and J. Rizzo, Phys. Rev. C **77**, 024906 (2008).
[32] J. H. Gao *et al.* , Phys. Rev. C **77**, 044902 (2008).
[33] J. H. Gao, High Energy Phys. Nucl. Phys. **31**, 1181 (2007).
[34] S. w. Chen, J. Deng, J. h. Gao and Q. Wang, Front. Phys. China **4**, 509 (2009).

Journal of Physics: Conference Series **432** (2013) 012011 doi:10.1088/1742-6596/432/1/012011

Dissipative hydrodynamic effects on the quark-gluon plasma at finite baryon density

Akihiko Monnai[1,2]

[1] Department of Physics, The University of Tokyo, Tokyo 113-0033, Japan
[2] Theoretical Research Division, Nishina Center, RIKEN, Wako 351-0198, Japan

E-mail: monnai@nt.phys.s.u-tokyo.ac.jp

Abstract. The quark-gluon plasma behaves as a relativistic viscous fluid in high-energy heavy ion collisions. I develop a causal dissipative hydrodynamic model at finite baryon density for RHIC and LHC to estimate the net baryon rapidity distribution. The net baryon number is found to be carried to forward rapidity by the flow, effectively enhancing the transparency of the collisions. This suggests that the energy available for the production of a hot medium could be larger than that naively implied from experimental data. Also the distribution can be sensitive to baryon dissipation as much as to shear and bulk viscosity.

1. Introduction

The quark-gluon plasma (QGP) [1] is a many-body system of quarks and gluons deconfined from hadrons at high energies. It is considered to be experimentally producible in high-energy heavy ion collisions at the Relativistic Heavy Ion Collider (RHIC) at Brookhaven National Laboratory and the Large Hadron Collider (LHC) at the European Organization for Nuclear Research, providing great opportunities and challenges for developing realistic theories of hot and dense QCD matter. One of the most distinctive characteristics of the QGP implied from the experimental data is that it behaves as a relativistic fluid with extremely small viscosity. This allows one to construct a hydrodynamic effective model of high-energy nuclear collisions. Modern hydrodynamic studies incorporate viscosity and fluctuation for precision physics [2].

The net baryon number, so far, has been neglected in viscous hydrodynamic studies with an evolving flow. This could be due to the complexity of dissipative processes in causal relativistic formalisms and the boost invariant ansatz that fails to take account of the net baryon number conserved at forward rapidity. It should be noted that finite-density effects are generally more prominent near the beam rapidity since the net baryon number originates from the remnant of the shattered nuclei. Introducing net baryon density to the model would enable one to understand the matter-to-antimatter ratios of the particle spectra as well as the transport properties of finite-density media. Analyses of the finite-density matter via dissipative hydrodynamics would also be of great help in exploring the QCD phase diagram for the search of the critical point, given that first principle approaches have difficulties for calculating systems with a finite baryon chemical potential. As the beam energy scans are being performed and planned at RHIC, GSI, NICA and J-PARC, it is important to develop a finite-density model to analyze the data and to examine the validity of the model for those experiments.

Journal of Physics: Conference Series **432** (2013) 012011 doi:10.1088/1742-6596/432/1/012011

The net baryon distribution is quantified with *baryon stopping* defined by the mean rapidity loss $\langle \delta y \rangle = y_p - \langle y \rangle$ [3, 4]. Here y_p is the rapidity of the incoming projectile and

$$\langle y \rangle = \int_0^{y_p} y \frac{dN_{B-\bar{B}}(y)}{dy} dy \Big/ \int_0^{y_p} \frac{dN_{B-\bar{B}}(y)}{dy} dy, \tag{1}$$

is the average rapidity of the net baryon density. When $\langle \delta y \rangle = 0$, the collision is completely transparent and no hot medium would be produced. On the other hand, $\langle \delta y \rangle = y_p$ suggests that the nuclei comes to a complete halt at the collision, losing all the kinetic energy for the medium. Thus this quantity is related to the magnitude of the energy loss of the colliding nuclei that is available for the QGP production. Here I would like to focus on dissipative hydrodynamic effects on the net baryon distribution and the mean rapidity loss [5].

2. Dissipative hydrodynamic model

The energy momentum tensor $T^{\mu\nu}$ and the net baryon current N_B^μ are the conserved quantities in the system of interest. Hydrodynamics introduces macroscopic flow u^μ which is defined here as the local energy flux. Then the tensor decomposition in terms of the flow yields

$$T^{\mu\nu} = (e_0 + P_0 + \Pi) u^\mu u^\nu - (P_0 + \Pi) g^{\mu\nu} + \pi^{\mu\nu}, \ N_B^\mu = n_{B0} u^\mu + V^\mu, \tag{2}$$

where e_0, P_0, n_{B0}, Π, $\pi^{\mu\nu}$ and V^μ are the energy density, the hydrostatic pressure, the net baryon density, the bulk pressure, the shear stress tensor and the baryon dissipation current, respectively. The system is described by energy-momentum conservation $\partial_\mu T^{\mu\nu} = 0$, baryon number conservation $\partial_\mu N_B^\mu = 0$ and the law of increasing entropy $\partial_\mu s^\mu \geq 0$. By introducing second order corrections to the entropy current s^μ in terms of the dissipative quantities, the semi-positive definite condition yields causal hydrodynamic equations of motion [6, 7]:

$$\begin{aligned}
\Pi &= -\zeta \nabla_\mu u^\mu - \zeta_{\Pi\delta e} D\frac{1}{T} + \zeta_{\Pi\delta n_B} D\frac{\mu_B}{T} - \tau_\Pi D\Pi \\
&+ \chi_{\Pi\Pi\Pi}^a \Pi D\frac{\mu_B}{T} + \chi_{\Pi\Pi\Pi}^b \Pi D\frac{1}{T} + \chi_{\Pi\Pi\Pi}^c \Pi \nabla_\mu u^\mu + \chi_{\Pi V}^a V_\mu \nabla^\mu \frac{\mu_B}{T} + \chi_{\Pi V}^b V_\mu \nabla^\mu \frac{1}{T} \\
&+ \chi_{\Pi V}^c V_\mu D u^\mu + \chi_{\Pi V}^d \nabla^\mu V_\mu + \chi_{\Pi\pi} \pi_{\mu\nu} \nabla^{\langle\mu} u^{\nu\rangle}, \qquad (3) \\
V^\mu &= \kappa_V \nabla^\mu \frac{\mu_B}{T} - \kappa_{VW}\left(\nabla^\mu \frac{1}{T} + \frac{1}{T} D u^\mu\right) - \tau_V \Delta^{\mu\nu} D V_\nu \\
&+ \chi_{VV}^a V^\mu D\frac{\mu_B}{T} + \chi_{VV}^b V^\mu D\frac{1}{T} + \chi_{VV}^c V^\mu \nabla_\nu u^\nu + \chi_{VV}^d V^\nu \nabla_\nu u^\mu + \chi_{VV}^e V^\nu \nabla^\mu u_\nu \\
&+ \chi_{V\pi}^a \pi^{\mu\nu} \nabla_\nu \frac{\mu_B}{T} + \chi_{V\pi}^b \pi^{\mu\nu} \nabla_\nu \frac{1}{T} + \chi_{V\pi}^c \pi^{\mu\nu} D u_\nu + \chi_{V\pi}^d \Delta^{\mu\nu} \nabla^\rho \pi_{\nu\rho} \\
&+ \chi_{V\Pi}^a \Pi \nabla^\mu \frac{\mu_B}{T} + \chi_{V\Pi}^b \Pi \nabla^\mu \frac{1}{T} + \chi_{V\Pi}^c \Pi D u^\mu + \chi_{V\Pi}^d \nabla^\mu \Pi, \qquad (4) \\
\pi^{\mu\nu} &= 2\eta \nabla^{\langle\mu} u^{\nu\rangle} - \tau_\pi D \pi^{\langle\mu\nu\rangle} \\
&+ \chi_{\pi\Pi} \Pi \nabla^{\langle\mu} u^{\nu\rangle} + \chi_{\pi\pi}^a \pi^{\mu\nu} D\frac{\mu_B}{T} + \chi_{\pi\pi}^b \pi^{\mu\nu} D\frac{1}{T} + \chi_{\pi\pi}^c \pi^{\mu\nu} \nabla_\rho u^\rho + \chi_{\pi\pi}^d \pi^{\rho\langle\mu} \nabla_\rho u^{\nu\rangle} \\
&+ \chi_{\pi V}^a V^{\langle\mu} \nabla^{\nu\rangle} \frac{\mu_B}{T} + \chi_{\pi V}^b V^{\langle\mu} \nabla^{\nu\rangle} \frac{1}{T} + \chi_{\pi V}^c V^{\langle\mu} D u^{\nu\rangle} + \chi_{\pi V}^d \nabla^{\langle\mu} V^{\nu\rangle}. \qquad (5)
\end{aligned}$$

Here $D = u^\mu \partial_\mu$ and $\nabla_\mu = \partial_\mu - u_\mu D$. The angular bracket on the indices denotes traceless symmetrization. T and μ_B are the temperature and the chemical potential, respectively. ζ, κ_V and η are the bulk viscosity, the baryon charge conductivity and the shear viscosity, which characterize scalar, vector and tensor dissipative processes in a medium at the linear order. $\zeta_{\Pi\delta e}$

and $\zeta_{\Pi\delta n_B}$ are the bulk-energy and the bulk-baryon cross viscosities and κ_{VW} is the baryon-heat cross conductivity, which should satisfy the Onsager reciprocal relations. It is note-worthy that they can be negative. τ_Π, τ_V and τ_π are the relaxation times, which are required for causality and stability in relativistic systems. χ are second order transport coefficients.

One has to introduce the equation of state, the transport coefficients and the initial conditions to construct a model for high-energy heavy ion collisions. The equation of state is obtained via Taylor expansion with the results from latest lattice QCD calculations [8, 9]. The transport coefficients are obtained from the AdS/CFT approaches [10, 11], the non-equilibrium statistical operator method [12] and phenomenological analyses based on the dimensionality, the matter-antimatter symmetry and the semi-positive definiteness of the transport coefficient matrices [5]. It should be noted that the coefficients are introduced to investigate qualitative responses of the system. The initial conditions at $\tau_0 = 1$ fm are based on the color glass theory; the energy density is from the Nara Monte-Carlo version [13, 14] of the Kharzeev-Levin-Nardi model [15, 16] and the net baryon density from the distribution of the valence quarks [17, 18]. The list of input parameters can be found in Table 1, where $\chi_B^{(2)}$ is the quadratic baryon fluctuation, s is the entropy density and $c_s = [(\partial e_0/\partial P_0)_{s/n_{B0}}]^{1/2}$ is the sound velocity:

Table 1. Input to dissipative hydrodynamic model

Equation of state	Description
EoS at finite μ_B	$P_0(e_0, n_{B0}) = P_0(e_0, 0) + (\mu_B^2 T^2/2)\chi_B^{(2)}(e_0, 0)$
Transport coefficients	**Description**
Shear viscosity	$\eta = s/4\pi$
Effective bulk viscosity*	$\zeta_{\text{eff}} = (5/2)[(1/3) - c_s^2]\eta$
Baryon charge conductivity	$\kappa_V = c_V(\partial\mu_B/\partial n_B)_T^{-1}/2\pi$
Baryon heat conductivity	$\kappa_{VW} = c_{VW}[n_{B0}/(e_0 + P_0)](5\eta T\kappa_V)^{1/2}$
Initial conditions	**Description**
Energy density	$e_0(\tau_0, \eta_s) = (1/S_{\text{area}})dE_T/\tau_0 d\eta_s$
Net baryon density	$n_{B0}(\tau_0, \eta_s) = (1/S_{\text{area}})dN_{B-\bar{B}}/\tau_0 d\eta_s$

*The contributions of the scalar cross coefficients are merged into the effective bulk viscosity.

In the present analyses, I discuss dissipative hydrodynamic expansion in the longitudinal direction by integrating out the transverse geometry for the numerical analyses. It is note-worthy that the dependence of the net baryon yields on transverse geometry is shown to be very small experimentally in the analyses of centrality dependence [19].

3. Results

Fig. 1 shows the net baryon rapidity distributions for the Au-Au collisions at RHIC and the Pb-Pb collisions at LHC for the most central 0-5 % events. The parameters for the conductivities are $c_V = 1$ and $c_{VW} = 0$. The contribution of hadron resonances up to 2 GeV is considered at the freeze-out temperature $T_f = 0.16$ GeV. The experimental data points are scaled results from the corresponding net proton distribution by the BRAHMS experiments [20]. One can see that the net baryon number is carried to the forward rapidity region in hydrodynamic evolution. This is mainly due to the pressure gradient, to which the net baryon density is coupled via the flow. The effect is slightly reduced for the shear and bulk viscous case because the longitudinal pressure is effectively reduced by the viscosities. Baryon diffusion further steepens the distribution as

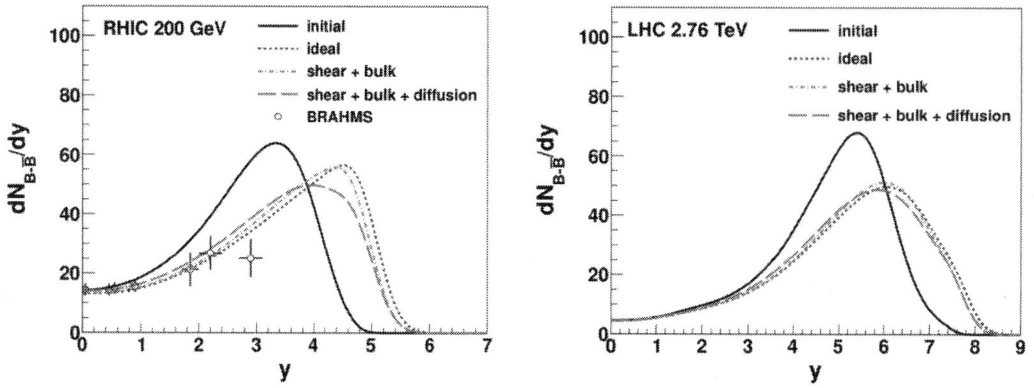

Figure 1. The net baryon distributions of the initial condition and ideal, viscous and viscous+baryon diffusive hydrodynamic results at RHIC (left) and at LHC (right) are shown as solid, dotted, dash-dotted and dashed curves, respectively. The experimental data are from the BRAHMS Collaboration [20].

it is induced by the chemical gradient into the mid-rapidity region. For the current parameter settings, the hydrodynamic results for RHIC roughly agree with the experimental data. The off-equilibrium corrections are more visible at RHIC than at LHC.

The hydrodynamic evolution reduces the mean rapidity loss from $\langle \delta y \rangle_{\text{initial}} = 2.67$ to $\langle \delta y \rangle_{\text{hydro}} = 2.09, 2.16$ and 2.26 for the ideal, the viscous and the dissipative systems at RHIC. Here the dissipative results include effects of baryon diffusion in addition to those of shear and bulk viscosities. It shows that the transparency of the collision is sizably enhanced as a medium effect. This could explain the abrupt deviation of the RHIC data from the simple extrapolation of the AGS and SPS results (Fig. 2) since the hot medium is created in heavy ion collisions with sufficiently high energies. The kinetic energy losses after the hydrodynamic evolutions are also about 10-15% less than those of the initial conditions, suggesting that more energy is available for the QGP production at the time of collision than naively implied from the experimental data. At LHC, when $\langle \delta y \rangle_{\text{initial}} = 3.36$, the rapidity loss is reduced to $\langle \delta y \rangle_{\text{hydro}} = 2.92$ for the dissipative hydrodynamic case. It should be emphasized that the result is dependent on the choice of the initial conditions and the transport coefficients.

Finally, the net baryon distributions in the presence of the baryon-heat cross-coupling term is shown in Fig. 3. The parameter for the cross coefficient is chosen as $c_{WV} = 5, 0$ and -5. The term induces Soret effect, a chemical diffusion process by the thermal gradient. The numerical results indicate that the effect is rather small. This is due to the fact that the cross conductivity has to stay small when the chemical potential is not large as $\kappa_{VW}(-\mu_B) = -\kappa_{VW}(\mu_B)$ follows from the matter-antimatter symmetry condition $V^\mu(-\mu_B) = -V^\mu(\mu_B)$ on Eq. (4).

4. Summary

Non-equilibrium hydrodynamic analyses are extended for the quark-gluon systems with finite baryon density. The finite-density equation of state is constructed in the Taylor expansion method with state-of-art data of lattice QCD calculations and the initial conditions in the color glass-based picture. The baryon stopping at RHIC and LHC show clear effects of hydrodynamic convection where both viscous and dissipative modifications are qualitatively visible at RHIC. The widening of the net baryon distribution suggests that the kinetic energy loss is effectively reduced during the hydrodynamic evolution. This implies that the energy available for the production of a hot medium is larger initially, and part of the energy is transferred back to

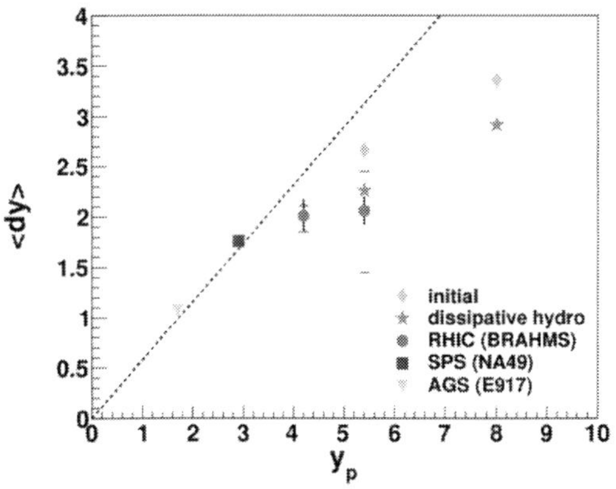

Figure 2. Mean rapidity losses at AGS [21], SPS [22] and RHIC [20, 23] experiments and those in dissipative hydrodynamic calculations for RHIC and LHC. The dotted line is a linear extrapolation of the low energy results [20].

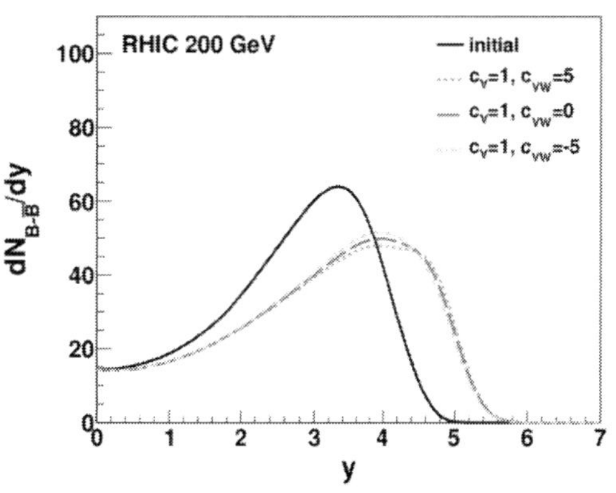

Figure 3. The net baryon distribution of the initial condition and dissipative hydrodynamic results at RHIC with the cross coefficient $c_{VW} = 5, 0$ and -5 are shown in solid, dotted, dashed and dash-dotted curves, respectively.

the net baryon components at the later stage as the hydrodynamic flow accelerates them. Also the Soret effect would be small in high-energy nuclear collisions. One would need to include transverse dynamics and realistic transport coefficients for more quantitative discussions.

Acknowledgments

The author is grateful for the valuable comments by Hatsuda T. The work of A.M. is supported by the Grant-in-Aid for JSPS Research Fellows (No. 10J07647).

References

[1] Yagi K, Hatsuda T and Miake Y 2005 *Camb. Monogr. Part. Phys. Nucl. Phys. Cosmol.* **23** 1
[2] Schenke B, Jeon S and Gale C 2011 *Phys. Rev. Lett.* **106** 042301
[3] Busza W and Goldhaber A S 1984 *Phys. Lett.* B **139** 235
[4] Videbaek F and Hansen O 1995 *Phys. Rev.* C **52** 2684
[5] Monnai A 2012 *Phys. Rev.* C **86** 014908
[6] Israel W and Stewart J M 1979 *Annals Phys.* **118** 341
[7] Monnai A and Hirano T 2010 *Nucl. Phys.* A **847** 283
[8] Borsanyi S, Endrodi G, Fodor Z, Jakovac A, Katz S D, Krieg S, Ratti C and Szabo K K 2010 *JHEP* **1011** 077
[9] Borsanyi S, Fodor Z, Katz S D, Krieg S, Ratti C and Szabo K 2012 *JHEP* **1201** 138
[10] Kovtun P, Son D T and Starinets A O 2005 *Phys. Rev. Lett.* **94** 111601
[11] Natsuume M and Okamura T 2008 *Phys. Rev.* D **77** 066014 [Erratum-ibid. D **78** 089902]
[12] Hosoya A, Sakagami M A and Takao M 1984 *Annals Phys.* **154** 229
[13] Drescher H J and Nara Y 2007 *Phys. Rev.* C **75** 034905
[14] Drescher H J and Nara Y 2007 *Phys. Rev.* C **76** 041903
[15] Kharzeev D, Levin E and Nardi M 2004 *Nucl. Phys.* A **730** 448 [Erratum-ibid. A **743** 329]
[16] Kharzeev D, Levin E and Nardi M 2005 *Nucl. Phys.* A **747** 609
[17] Mehtar-Tani Y and Wolschin G 2009 *Phys. Rev. Lett.* **102** 182301
[18] Mehtar-Tani Y and Wolschin G 2009 *Phys. Rev.* C **80** 054905
[19] Adler S S *et al.* [PHENIX Collaboration] 2004 *Phys. Rev.* C **69** 034909
[20] Bearden I G *et al.* [BRAHMS Collaboration] 2004 *Phys. Rev. Lett.* **93** 102301
[21] Back B B *et al.* [E917 Collaboration] 2001 *Phys. Rev. Lett.* **86** 1970
[22] Appelshauser H *et al.* [NA49 Collaboration] 1999 *Phys. Rev. Lett.* **82** 2471
[23] Arsene I C *et al.* [BRAHMS Collaboration] 2009 *Phys. Lett.* B **677** 267

Journal of Physics: Conference Series **432** (2013) 012012 doi:10.1088/1742-6596/432/1/012012

Fluctuations of conserved charges at finite temperature from lattice QCD

S. Krieg[1,2], for the Wuppertal-Budapest collaboration
S. Borsanyi[1], Z. Fodor[1,2,3], S. D. Katz[1,3], C. Ratti[4], K. K. Szabo[1]

[1] Fachbereich C – Physik, Bergische Universität Wuppertal, D-42119 Wuppertal, Germany
[2] IAS, JSC, Forschungszentrum Jülich GmbH, D-52425 Jülich, Germany
[3] Institute for Theoretical Physics, Eötvös University, H-1117 Budapest, Hungary
[4] Università degli Studi di Torino and INFN, Sezione di Torino, I-10125 Torino, Italy

E-mail: s.krieg@fz-juelich.de

Abstract. Recent results on fluctuations of conserved charges are reviewed and compared between different collaborations, where good agreement is found. Uncertainties of the lattice QCD simulations involved are discussed. Dedicated simulations addressing some of these uncertainties are shown, which indicate that, in the parameter regions studied so far, they are under control.

1. Introduction

The Quantum Chromodynamics (QCD) phase diagram is presently studied at various ultra-relativistic heavy-ion experiments, such as CERN SPS, RHIC at Brookhaven National Laboratory, ALICE at the LHC, and, in the future, FAIR at GSI, Darmstadt. One particular subject of these studies is the QCD transition between the hot quark-gluon-plasma to the hadronic 'phase', which is predicted to be an analytic crossover by Lattice QCD simulations [1]. A lot of effort is invested to find observables that would signal the QCD phase transition in an unambiguous way. Correlations and fluctuations of conserved charges have been proposed for this purpose [2, 3], since these quantities have significantly different values in a confined or a deconfined system.

Fluctuations of conserved charges can be obtained as linear combinations of diagonal and non-diagonal quark number susceptibilities, which can be calculated on the lattice at zero chemical potential [4, 5]. These observables can give us an insight into the nature of the matter under study [4, 6]. Diagonal susceptibilities measure the response of the quark number density to changes in the chemical potential, and show a rapid rise in the vicinity of the phase transition. Non-diagonal susceptibilities give us information about the correlation between different flavors. They are supposed to vanish in a non-interacting quark-gluon plasma (QGP). It was shown in Ref. [7] that correlations between different flavors are nonzero in perturbative QCD at large temperatures due to the presence of flavor-mixing diagrams: a quantitative analysis of this observable allows one to draw conclusions about the presence of bound states in the QGP [8]. Another observable which was proposed to this purpose, and which can be obtained from a combination of diagonal and non-diagonal quark number susceptibilities, is the baryon-strangeness correlator [9].

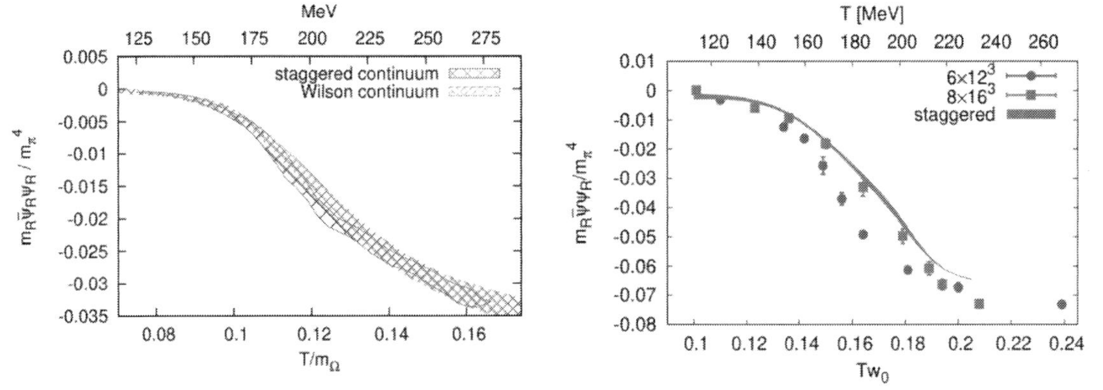

Figure 1. Crosschecks between different lattice actions. *Left panel:* The continuum extrapolated chiral condensate obtained from simulations with $N_f = 2 + 1$ flavors of quarks at a pion mass of 545 MeV, using Wilson and staggered fermions [11]. *Right panel:* The continuum extrapolated chiral condensate obtained from simulations with $N_f = 2$ flavors of quarks at a pion mass of 350 MeV, using overlap and staggered quarks [12].

In the present contribution we will first discuss uncertainties of lattice QCD simulations and show results from dedicated simulations providing cross-checks for our calculations. We will then discuss continuum extrapolated results of our collaboration on some of the aforementioned observables. We compare our results to the predictions of the HRG model with resonances up to 2.5 GeV mass at small temperatures, and of the Hard Thermal Loop (HTL) resummation scheme at large temperatures, where available. Furthermore, we will compare our results to those of the hotqcd collaboration.

In our simulations we use $N_f = 2 + 1$ dynamical staggered quark flavors, in a temperature regime between 125 and 400 MeV [10] and set the light and strange quark masses to their physical values. Lattices with $N_t = 6,\ 8,\ 10,\ 12$ are used. Continuum extrapolations are performed for all observables under study. Further details can be found in [10].

2. Discussion of uncertainties

Simulations in finite temperature Lattice QCD typically use the staggered formalism. In this formalism one is faced with the problem that the Dirac operator does not implement a single quark flavor, but rather four so-called "tastes", which correspond to four flavors in the continuum limit. The tastes' degeneracy is lifted by non-physical (continuum irrelevant) so-called taste-breaking interactions. As a consequence, the pion sector is distorted and contains, in addition to the one would-be Goldstone-boson, 15 additional "pions". This results in an rms pion mass which is larger than that of the "real" pion. The impact of this effect on the thermodynamic observables has been recently discussed in the HRG framework [13, 14].

Improved action reduce the taste-breaking effects considerably, mostly by using "smearing", i.e. by averaging over gauge links aiming to reduce ("UV" related) lattice artifacts. While this procedure is very successful at taming the distorted pion sector, one still has to deal with the fact that one is simulating more flavors than one actually wants. This is addressed by the so-called "rooting trick" in which the fermionic determinant is taken to be an appropriate root (e.g. square root or fourth root) of the staggered determinant. Accumulating practical evidence suggests that this is a viable procedure, but, as of today, a proof of its correctness is still missing.

One way to check for problems related to these above issues is to repeat simulations with other lattice actions that are free of such uncertainties and to compare the results. Ideally,

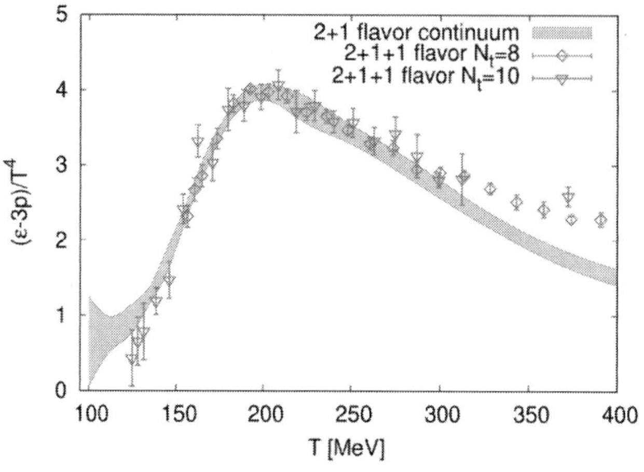

Figure 2. $N_F = 2+1$ Equation of State of QCD compared with individual data points obtained with an additional dynamical charm ($N_f = 2+1+1$). At temperatures above ≈ 300 MeV the charm quark becomes relevant.

one would want to do that for all results, if it were not for the computational costs this does require. Therefore, a reasonable first step is to perform cross-checks at some parameter values, such as with larger pion masses (and thus at lower costs), but still in the continuum limit. Such a calculation using Wilson fermions is shown on the left panel of Fig. 1. Here, simulations were performed at a pion mass of about 545 MeV and a direct comparison for the continuum extrapolated chiral condensate is shown (Details on the simulation and on the renormalization procedure used can be found in [11]). The results agree within errors. Another, albeit not continuum extrapolated, comparison for the same observable is made on the right panel of Fig. 1, where the staggered results are compared to ones obtained with overlap fermions. Here, the pion mass is even lower. Since the continuum limit is still missing, one cannot make any final statements, however, the nice agreement of the fine data set (8×16^3) with the staggered continuum estimate suggest that also here, agreement between the different formulations can be expected (the coarse data set is likely not in the scaling region). Note, that both the staggered and overlap simulations in this latter comparison did not include a dynamical strange quark. Since here we only want to perform a cross-check of the staggered formalism, this is also not required. Related to this latter comment, for some quantities at temperatures above ca. 300 MeV, certainly for Equation of State of QCD, a dynamical charm quark is required, as Fig. 2 shows. This is, however, most likely not the case for fluctuations, as shown in Fig. 3 [15].

3. Fluctuations

3.1. Details of the lattice simulations

The lattice action used in our calculations is the same one we used in [16, 17], namely a tree-level Symanzik improved gauge, and a stout-improved staggered fermion action (see Ref. [18] for details). The stout-smearing procedure [19], as mentioned earlier, yields an improved discretization of the fermion-gauge vertex and reduces taste breaking artifacts. Analogously to our action, using an alternative link-smearing scheme, the HISQ action [20] suppresses the taste breaking in a similar way. The latter is used by the hotQCD collaboration in its latest studies [21, 22, 23]. For details about the simulation algorithm we refer the reader to [17].

In analogy with what we did in [16, 17], we set the scale at the physical point by simulating

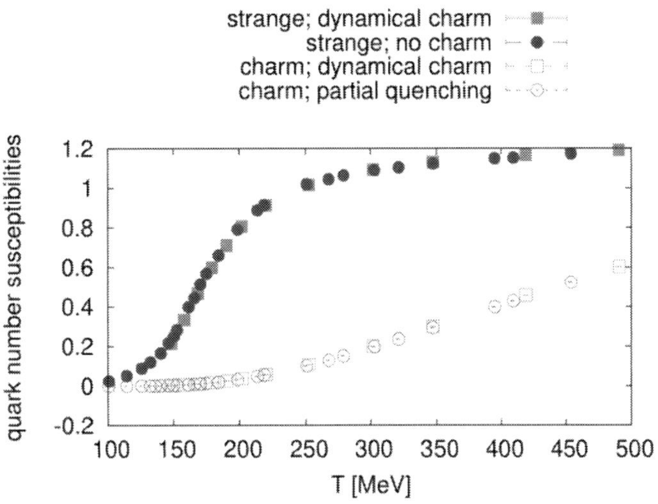

Figure 3. Strange and charm susceptibilities with and without dynamical charm.

at $T = 0$ with physical quark masses [17] and reproducing the kaon and pion masses and the kaon decay constant. This gives an uncertainty of about 2% in the scale setting.

3.2. *Observables under study*

The baryon number B, strangeness S and electric charge Q fluctuations can be obtained, at vanishing chemical potentials, from the QCD partition function. The relationships between the quark chemical potentials and those of the conserved charges are as follows:

$$\begin{aligned} \mu_u &= \frac{1}{3}\mu_B + \frac{2}{3}\mu_Q; \\ \mu_d &= \frac{1}{3}\mu_B - \frac{1}{3}\mu_Q; \\ \mu_s &= \frac{1}{3}\mu_B - \frac{1}{3}\mu_Q - \mu_S. \end{aligned} \tag{1}$$

Starting from the QCD pressure,

$$\frac{p}{T^4} = \frac{1}{VT^3} \ln Z(V, T, \mu_B, \mu_S, \mu_Q) \tag{2}$$

we can define the moments of charge fluctuations as follows:

$$\chi_{lmn}^{BSQ} = \frac{\partial^{l+m+n} p/T^4}{\partial(\mu_B/T)^l \partial(\mu_S/T)^m \partial(\mu_Q/T)^n}. \tag{3}$$

We will concentrate on the quadratic fluctuations

$$\chi_2^X = \frac{1}{VT^3}\langle N_X^2 \rangle \tag{4}$$

and on the correlators among different charges or quark flavors:

$$\chi_{11}^{XY} = \frac{1}{VT^3}\langle N_X N_Y \rangle. \tag{5}$$

Given the relationships between chemical potentials (1) the diagonal susceptibilities of the conserved charges can be obtained from quark number susceptibilities in the following way:

$$\begin{aligned}
\chi_2^B &= \frac{1}{9}\left[\chi_2^u + \chi_2^d + \chi_2^s + 2\chi_{11}^{us} + 2\chi_{11}^{ds} + 2\chi_{11}^{ud}\right] , \\
\chi_2^Q &= \frac{1}{9}\left[4\chi_2^u + \chi_2^d + \chi_2^s - 4\chi_{11}^{us} + 2\chi_{11}^{ds} - 4\chi_{11}^{ud}\right] , \\
\chi_2^I &= \frac{1}{4}\left[\chi_2^u + \chi_2^d - 2\chi_{11}^{ud}\right] , \\
\chi_2^S &= \chi_2^s .
\end{aligned} \tag{6}$$

If we do not wish to take further derivatives, we can take all three chemical potentials (u, d, s) to zero. In this case, nothing distinguishes between the u and d derivative: this gives slightly simplified formulae:

$$\begin{aligned}
\chi_2^B &= \frac{1}{9}\left[2\chi_2^u + \chi_2^s + 4\chi_{11}^{us} + 2\chi_{11}^{ud}\right] , \\
\chi_2^Q &= \frac{1}{9}\left[5\chi_2^u + \chi_2^s - 2\chi_{11}^{us} - 4\chi_{11}^{ud}\right] , \\
\chi_2^I &= \frac{1}{2}\left[\chi_2^u - \chi_{11}^{ud}\right] .
\end{aligned} \tag{7}$$

The baryon-strangeness correlator, which was proposed in Ref. [9] as a diagnostic to understand the nature of the degrees of freedom in the QGP, has the following expression in terms of quark number susceptibilities:

$$C_{BS} = -3\frac{\langle N_B N_S \rangle}{\langle N_S^2 \rangle} = 1 + \frac{\chi_{11}^{us} + \chi_{11}^{ds}}{\chi_2^s} . \tag{8}$$

3.3. Results

The first observables we discuss are the diagonal light and strange quark number susceptibilities: their behavior as functions of the temperature is shown in the two panels of Fig. 4. The different symbols correspond to different values of N_t, from 8 to 16. The red band is the continuum extrapolation, obtained from the unimproved data, not from the improved ones shown in the figure. The continuum extrapolation is performed through a parabolic fit in the variable $(1/N_t)^2$, over five N_t values from 6 to 16. The band shows the spread of the results of other possible fits. The comparison between the improved data and the continuum bands in the figure shows the success of the improvement program throughout the entire temperature range. Both observables show a rapid rise in a certain temperature range, and reach approximately 90% of the ideal gas value at large temperatures. However, the temperature around which the susceptibilities rise is approximately 15-20 MeV larger for strange quarks than for light quarks. Besides, the light quark susceptibility shows a steeper rise with temperature, compared to the strange quark one. They approach each other at high temperatures. The pattern of temperature dependence is strongly related to the actual quark mass. The difference between the light and strange susceptibilities computed at the physical pion mass is more pronounced than in earlier works with heavier pion masses [24, 25]. This is more evident in Fig. 5: in the left panel we show the continuum extrapolation of both susceptibilities on the same plot. In the right panel we show the ratio χ_s/χ_u: it reaches 1 only around 300 MeV, while for smaller temperatures it is < 1. It is worth noticing that all these observables agree with the corresponding HRG model predictions for temperatures below the transition.

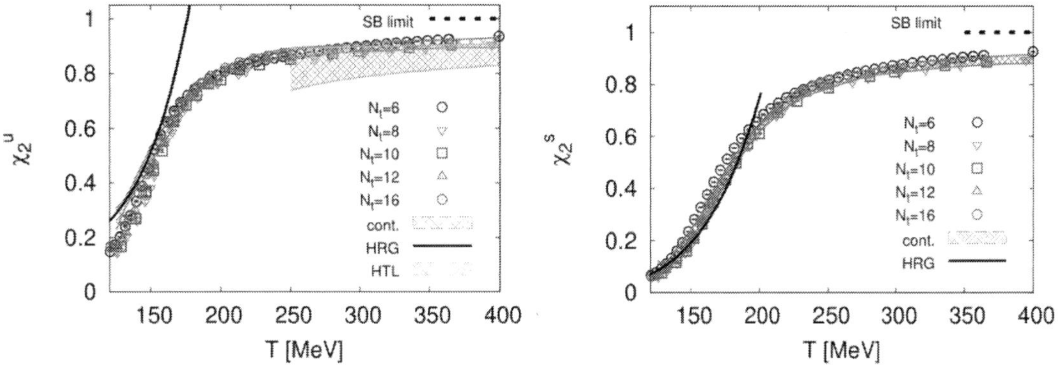

Figure 4. *Left panel:* diagonal light quark susceptibility as a function of the temperature. *Right panel:* diagonal strange quark susceptibility as a function of the temperature. In both panels, the different symbols correspond to different N_t values. The red band is the continuum extrapolation. The black curve is the HRG model prediction for these observables. The dashed line shows the ideal gas limit. The light blue band in the left panel is the HTL prediction taken from Ref. [7].

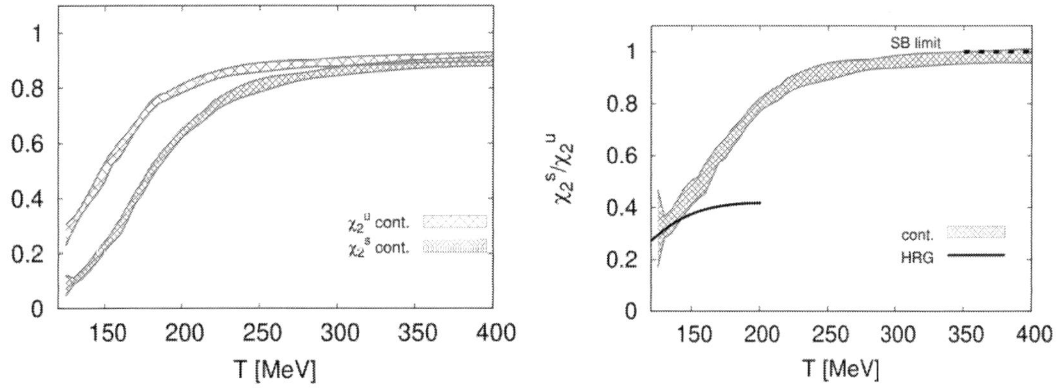

Figure 5. *Left panel:* direct comparison between the continuum limit of light and strange quark susceptibilities. *Right panel:* ratio χ_2^s/χ_2^u as a function of the temperature. The red band is the lattice continuum result. The black, solid curve is the HRG model prediction. The dashed line indicates the ideal gas limit.

The non-diagonal us susceptibility measures the degree of correlation between different flavors. This observable vanishes in the limit of an ideal, non-interacting QGP. However, flavor-mixing diagrams in perturbative QCD yield a finite value for this correlation also at large temperatures [7]. We show our result in Fig. 6. χ_{11}^{us} is non-zero in the entire temperature range under study. It has a dip in the vicinity of the transition, where the correlation between u and s quarks turns out to be maximal. It agrees with the HRG model prediction in the hadronic phase. This correlation stays finite and large for a certain temperature range above the transition temperature T_c [16, 17, 14]. A quantitative comparison between lattice results and predictions for a purely partonic QGP state can give us information about the probability of bound states survival above T_c [8].

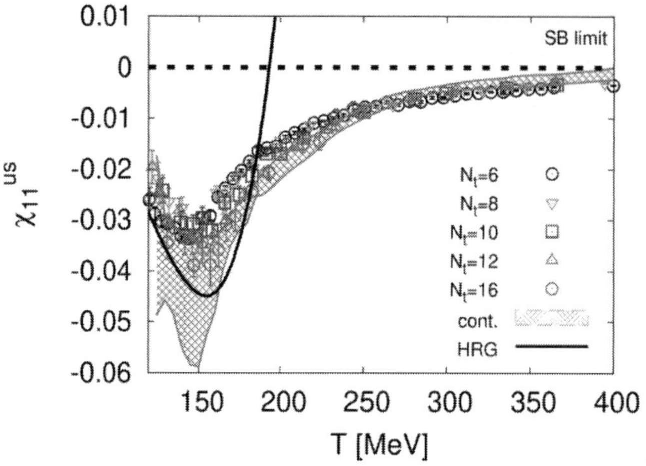

Figure 6. Non-diagonal u-s correlator as a function of the temperature. The different symbols correspond to different N_t values. The red curve is the continuum extrapolated result. The black curve is the HRG model prediction. The dashed line indicates the ideal gas limit for this observable.

Quadratic baryon number, electric charge and isospin fluctuations can be obtained from the above partonic susceptibilities through Eqs. (7). We show our results for these observables in Fig. 7 and in the left panel of Fig. 8. In the low-temperature, hadronic phase we have a very good agreement with the HRG model predictions. In the vicinity of the phase transition, these quantities all show a rapid rise with temperature, in analogy with what already observed for the light and strange quark susceptibilities. At large temperature they reach approximately 90% of their respective ideal gas values. A comparison between all diagonal susceptibilities, rescaled by their corresponding Stefan-Boltzmann limits, is shown in the right panel of Fig. 8, from which it is evident that they all show similar features in their temperature dependence, even if the temperature at which they rise is larger for the strangeness and baryon number susceptibilities.

The baryon-strangeness correlator C_{BS} defined in Eq. (8) was proposed long ago [9] as a diagnostic for strongly interacting matter. It is supposed to be equal to one for a non-interacting QGP, while it is temperature-dependent and generally smaller than one in a hadronic system. We show our result for this observable in Fig. 9. At the smallest temperatures it agrees with the HRG model result, and it shows a rapid rise across the phase transition. It reaches the ideal gas limit much faster than the other observables under study, yet there is a window of about 100 MeV above T_c, where its value is still smaller than one. In analogy with χ_{11}^{us}, this observable seems to leave some room for bound states survival above T_c.

In Fig. 10, finally, we compare our results to those of the hotQCD collaboration. In the temperature region, were both collaborations have data, the results agree within errors.

4. Conclusions

We have discussed uncertainties of Lattice QCD simulations and have presented continuum results of our collaboration on diagonal and non-diagonal quark number susceptibilities. Our results were obtained simulating $N_F = 2+1$ dynamical staggered quark flavors with physical masses, in a temperature range between 125 and 400 MeV. The continuum extrapolations were based on $N_t = 6, 8, 10, 12$ and 16 lattices. We calculated the systematic errors by varying over the ambiguities of the possible extrapolations.

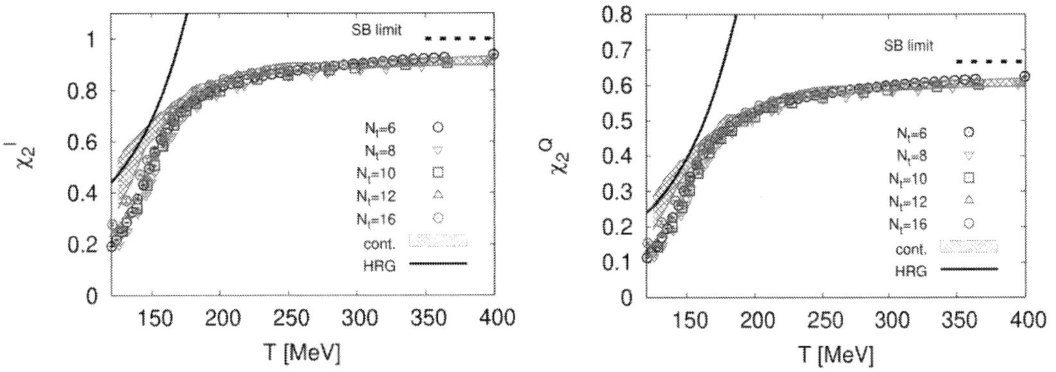

Figure 7. *Left panel:* isospin susceptibility as a function of the temperature. *Right panel:* electric charge susceptibility as a function of the temperature. In both panels, the different dots correspond to different N_t values. The red band is the continuum extrapolation. The black curve is the HRG model prediction for these observables. The dashed line shows the ideal gas limit.

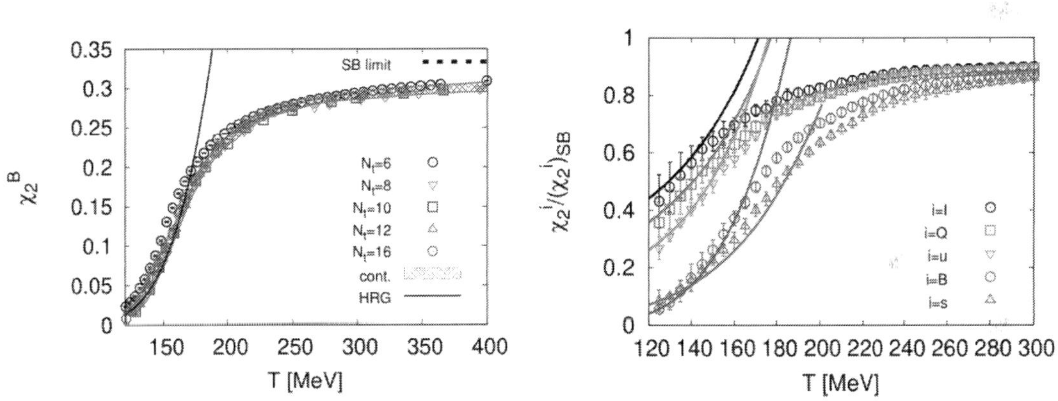

Figure 8. *Left panel:* quadratic fluctuation of baryon number as a function of the temperature. The different symbols correspond to different N_t values, the red band is the continuum extrapolation and the black, solid curve is the HRG model result. The ideal gas limit is shown by the black, dashed line. *Right panel:* comparison between all diagonal susceptibilities, rescaled by the corresponding ideal gas limit, as functions of the temperature.

All observables consistently show a very good agreement with the HRG model predictions for temperatures below the phase transition.

The diagonal fluctuations have some common features: they all show a rapid rise in the vicinity of the phase transition, and reach approximately 90% of the corresponding ideal gas value at large temperatures. The rise of both strange quark and baryon number susceptibilities is shifted to temperatures about 20 MeV higher than those for light quark, charge and isospin susceptibilities. Non-diagonal flavor and charge correlators remain different from their ideal gas values for a certain window of temperatures above the transition, thus not excluding the possibility of bound state survival above T_c. We have compared our results to those by the hotQCD collaboration and found them to be in agreement within errors.

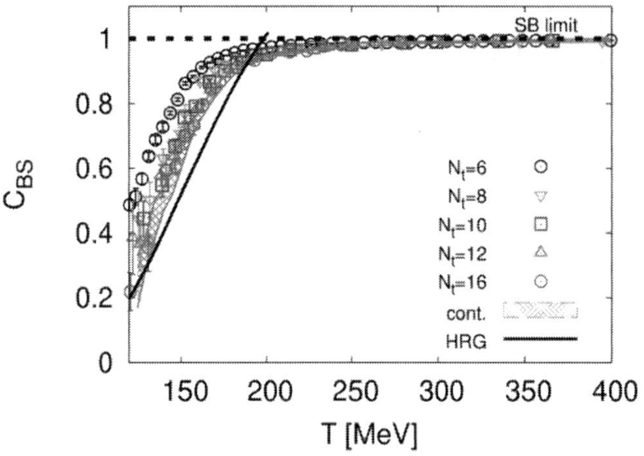

Figure 9. Baryon-strangeness correlator as a function of the temperature. The different symbols correspond to different N_t values, the red band is the continuum extrapolation and the black, solid curve is the HRG model result. The ideal gas limit is shown by the black, dashed line.

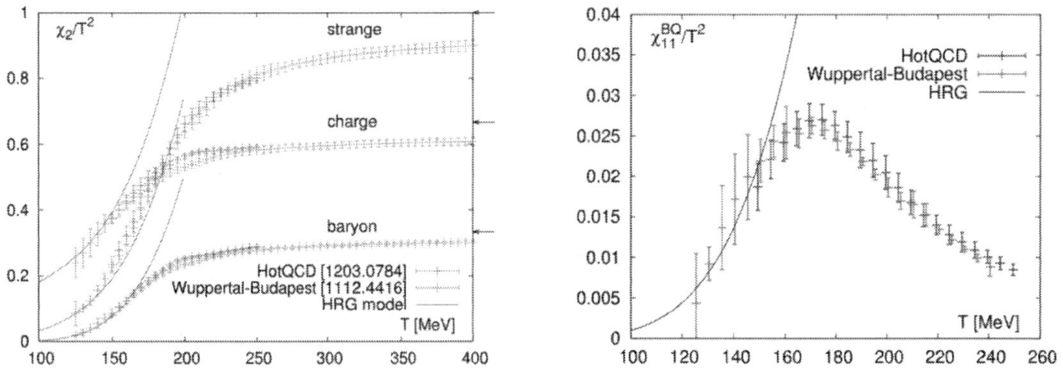

Figure 10. Diagonal and off-diagonal correlators of conserved charges as a function of temperature. Only continuum extrapolated lattice data are shown. There is a good agreement between hotQCD [26] and Wuppertal-Budapest [10] data, obtained using the HISQ and stout actions, respectively. The broader temperature region by the Wuppertal-Budapest collaboration allows a comparison with the HRG model's prediciton, and was used to benchmark resummation schemes at high temperature [27].

Acknowledgments

Computations were carried out at the Universities Wuppertal and Budapest on GPU clusters [28], on the Wuppertal QPACE machine and at Forschungszentrum Juelich. This work is supported in part by the Deutsche Forschungsgemeinschaft grants FO 502/2 and SFB- TR 55, by the EU (FP7/2007-2013)/ERC no. 208740 and by the Italian Ministry of Education, Universities and Research under the Firb Research Grant RBFR0814TT.

References

[1] Aoki Y, Endrodi G, Fodor Z, Katz S and Szabo K 2006 *Nature* **443** 675–678 (*Preprint* hep-lat/0611014)

[2] Asakawa M, Heinz U W and Muller B 2000 *Phys.Rev.Lett.* **85** 2072–2075 (*Preprint* hep-ph/0003169)
[3] Jeon S and Koch V 2000 *Phys.Rev.Lett.* **85** 2076–2079 (*Preprint* hep-ph/0003168)
[4] Gottlieb S A, Liu W, Toussaint D, Renken R and Sugar R 1988 *Phys.Rev.* **D38** 2888–2896
[5] Gavai R, Potvin J and Sanielevici S 1989 *Phys.Rev.* **D40** 2743
[6] McLerran L D 1987 *Phys.Rev.* **D36** 3291
[7] Blaizot J, Iancu E and Rebhan A 2001 *Phys.Lett.* **B523** 143–150 (*Preprint* hep-ph/0110369)
[8] Ratti C, Bellwied R, Cristoforetti M and Barbaro M 2012 *Phys.Rev.* **D85** 014004 (*Preprint* 1109.6243)
[9] Koch V, Majumder A and Randrup J 2005 *Phys.Rev.Lett.* **95** 182301 (*Preprint* nucl-th/0505052)
[10] Borsanyi S, Fodor Z, Katz S D, Krieg S, Ratti C *et al.* 2012 *JHEP* **1201** 138 (*Preprint* 1112.4416)
[11] Borsanyi S, Durr S, Fodor Z, Hoelbling C, Katz S D *et al.* 2012 *JHEP* **1208** 126 (*Preprint* 1205.0440)
[12] Borsanyi S, Delgado Y, Durr S, Fodor Z, Katz S D *et al.* 2012 *Phys.Lett.* **B713** 342–346 (*Preprint* 1204.4089)
[13] Huovinen P and Petreczky P 2010 *Nucl.Phys.* **A837** 26–53 (*Preprint* 0912.2541)
[14] Borsanyi S *et al.* (Wuppertal-Budapest Collaboration) 2010 *JHEP* **1009** 073 (*Preprint* 1005.3508)
[15] Borsanyi S, Endrodi G, Fodor Z, Katz S D, Krieg S *et al.* 2011 *PoS* **LATTICE2011** 201 (*Preprint* 1204.0995)
[16] Aoki Y, Fodor Z, Katz S and Szabo K 2006 *Phys.Lett.* **B643** 46–54 (*Preprint* hep-lat/0609068)
[17] Aoki Y, Borsanyi S, Durr S, Fodor Z, Katz S D *et al.* 2009 *JHEP* **0906** 088 (*Preprint* 0903.4155)
[18] Aoki Y, Fodor Z, Katz S and Szabo K 2006 *JHEP* **0601** 089 (*Preprint* hep-lat/0510084)
[19] Morningstar C and Peardon M J 2004 *Phys.Rev.* **D69** 054501 (*Preprint* hep-lat/0311018)
[20] Follana E *et al.* (HPQCD Collaboration, UKQCD Collaboration) 2007 *Phys.Rev.* **D75** 054502 (*Preprint* hep-lat/0610092)
[21] Bazavov A and Petreczky P (HotQCD Collaboration) 2009 *PoS* **LAT2009** 163 (*Preprint* 0912.5421)
[22] Bazavov A and Petreczky P (HotQCD collaboration) 2010 *J.Phys.Conf.Ser.* **230** 012014 (*Preprint* 1005.1131)
[23] Bazavov A, Bhattacharya T, Cheng M, DeTar C, Ding H *et al.* 2012 *Phys.Rev.* **D85** 054503 (*Preprint* 1111.1710)
[24] Gavai R and Gupta S 2006 *Phys.Rev.* **D73** 014004 (*Preprint* hep-lat/0510044)
[25] Cheng M, Hendge P, Jung C, Karsch F, Kaczmarek O *et al.* 2009 *Phys.Rev.* **D79** 074505 (*Preprint* 0811.1006)
[26] Bazavov A *et al.* (HotQCD Collaboration) 2012 *Phys.Rev.* **D86** 034509 (*Preprint* 1203.0784)
[27] Andersen J O, Mogliacci S, Su N and Vuorinen A 2012 (*Preprint* 1210.0912)
[28] Egri G I, Fodor Z, Hoelbling C, Katz S D, Nogradi D *et al.* 2007 *Comput.Phys.Commun.* **177** 631–639 (*Preprint* hep-lat/0611022)

Journal of Physics: Conference Series **432** (2013) 012013 doi:10.1088/1742-6596/432/1/012013

QCD bulk thermodynamics and conserved charge fluctuations with HISQ fermions

Christian Schmidt (for HotQCD[1] and BNL-Bielefeld[2] Collaborations)
Fakultät für Physik, Universität Bielefeld, Postfach 100131, D-33501 Bielefeld, Germany
E-mail: schmidt@physik.uni-bielefeld.de

Abstract. After briefly reviewing recent progress by the HotQCD collaboration in studying the 2+1 flavor QCD equation of state, we will focus on results on fluctuations of conserved charges by the BNL-Bielefeld and HotQCD collaborations. Higher order cumulants of the net-charge distributions are increasingly dominated by a universal scaling behavior, which arises due to a critical point of QCD in the chiral limit. Considering cumulants up to the 6th order, we observe that they generically behave as expected from universal scaling laws, which is quite different from cumulants calculated within the hadron resonance gas model. Taking ratios of these cumulants, we obtain volume independent results that can be compared to the experimental measurements. We will argue that the freeze-out chemical potentials and the freeze-out temperature, usually obtained by a HRG model fit to the measured hadronic yields, can also be obtained in a model independent way from ab-initio lattice QCD calculations by utilizing observables related to conserved charge fluctuations. Further, we will show that the freeze-out strangeness and electric charge chemical potentials can be fixed by imposing strangeness neutrality and isospin asymmetry constraints in the lattice QCD calculations, in order to accommodate conditions met in heavy ion collisions. All results have been obtained with the highly improved staggered quark action (HISQ) and almost physical quark masses on lattices with temporal extent of $N_\tau = 6, 8, 10, 12$.

1. Introduction

Calculating the phase diagram of strongly interacting matter is one of the most important and outstanding problems of non-perturbative QCD. A generic plot of the QCD phase diagram based on model calculations and model independent symmetry arguments is shown in Fig.1. To clarify whether or not there is a critical point in the phase diagram is relevant for cosmology, astro physics and heavy ion collisions. Currently various experimental programs aim at probing the phase diagram at nonzero baryon number density, such as the beam energy scan at RHIC, NA61 at the SPS as well as future experiments at FAIR and NICA. In Fig.1 we have indicated the approximate initial conditions of the fireballs as they are created in some of those experiments. Also shown is a sketch of the trajectories on which the fireball evolves in the temperature (T) and

[1] Collaborations members are: A. Bazavov, T. Bhattacharya, M. Buchoff, M. Cheng, N. Christ, C. DeTar, H.-T. Ding, S.Gottlieb, R. Gupta, P. Hegde, U. Heller, C. Jung, F. Karsch, E. Laermann, L. Levkova, Z. Lin, R. Mawhinney, S. Mukherjee, P. Petreczky, D. Renfrew, C. Schmidt, C. Schroeder, W. Söldner, R. Soltz, R. Sugar, D. Toussaint, P. Vranas.

[2] Collaboration members are: A. Bazavov, H.-T. Ding, P. Hegde, O. Kaczmarek, F. Karsch, E. Laermann, S. Mukherjee, P. Petreczky, C. Schmidt, D. Smith, W. Soeldner, M. Wagner.

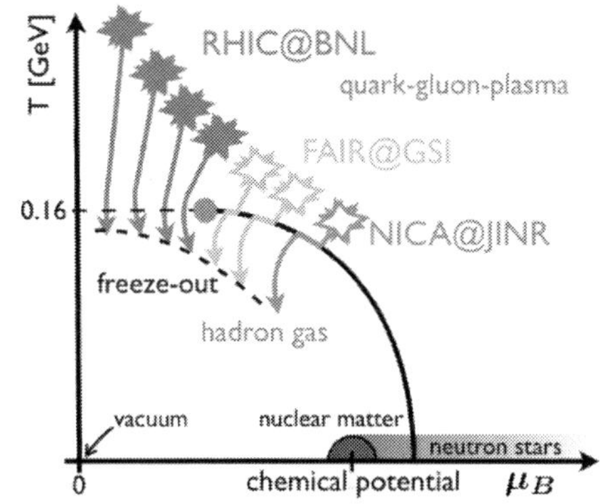

Figure 1. Generic phase diagram of QCD, based on model calculations and model independent symmetry arguments. Also indicated are the approximate initial conditions of the fireballs that are created in heavy ion experiments as well as their hydrodynamic trajectories on which they evolve, until they reach chemical freeze-out.

baryon chemical potential (μ_B) plane, during its expansion and cooling. It has been shown that this evolution can be very well described by hydrodynamics, with a surprisingly low viscosity [1]. While the initial conditions have to be modeled [2], the hydrodynamic evolution relies, at least up to first order, on the QCD equation of state (EoS) and two additional transport coefficients. The EoS can be obtained by first principle lattice calculations and we report on recent strategies of the HotQCD collaboration to calculate the EoS with the highly improved staggered quark (HISQ) action. Note that due to the notorious sign problem lattice QCD calculations are currently restricted to zero baryon chemical potential. However, the effect of a nonzero chemical potential can be taken into account by a controlled Taylor expansion in μ_B/T [3, 4].

We continue with a discussion of fluctuations of conserved charges such as net baryon number, electric charge and strangeness. As one expects a critical point in the phase diagram at nonzero density, fluctuation observables are the most natural choice to detect critical behavior in QCD. Their divergent behavior can give insight in the structure of the phase diagram and has been proposed as a signal for the critical point [5]. Moreover, as we argue here, they can be used to extract the chemical freeze-out conditions as observed in heavy ion collisions [6, 7]. After hadronization of the fireball, we expect a strongly interacting gas of hadrons with a fluctuating chemical decomposition. However, from a given point on the trajectory, the particle abundances stay fixed. This point is called chemical freeze-out. The coordinates of this point in the (T, μ_B)-diagram are usually obtained by fitting the experimentally measured particle abundances to the hadron resonance gas (HRG) model. Those fits have been quite successful in the past [8]. It has been shown that the freeze-out parameters that are obtained by HRG model fits from various experiments at different collision energies fall on a unique curve that can be parametrized as a function of the center of mass energy ($\sqrt{s_{NN}}$) [9].

2. The highly improved staggered quark (HISQ) action

Lattice QCD simulations utilize a discretized version of the QCD action, which at non-zero values of the lattice spacing is not uniquely defined. In fact, many lattice actions exist, which differ in the way they handle fermion doublers, to what extent the chiral symmetry is preserved and also in the amount of discretization effects they possess. A widely used class of lattice actions is given by various improved versions of so called staggered actions. They are numerically cheap and preserve a $U(1)$ subgroup of the $SU(N_f)_L \times SU(N_f)_R$ chiral symmetry that is manifest in the QCD Lagrangian at zero quark masses. Results that are presented here have been obtained with the HISQ action [10], that strongly reduces lattice discretization effects over the naive staggered action in two perspectives. Firstly, the discretization effects stemming from the covariant derivative term are improved by incorporating in addition to the usual nearest neighbor discretization of the derivative a higher order three-link term (Naik-term). This term dramatically reduces discretization effects at high temperature. Secondly, the flavor (taste) symmetry breaking, which is inherent to staggered quark actions, is reduced by a two step smearing procedure, where the smeared links are projected back to the gauge group $SU(3)$ after the first step. The taste breaking results in an unphysical splitting of the pion spectrum at given lattice spacing, which then leads to an averaged pion mass (root-mean-square mass $m_\pi^{\rm RMS}$). Although the lightest pion mass (Goldstone pion) is kept fixed through out our simulations at different lattice spacings the $m_\pi^{\rm RMS}$ varies. It can be seen as a measure of the taste breaking effects and is shown in Fig. 2 (left). As one can see, among the three compared staggered actions

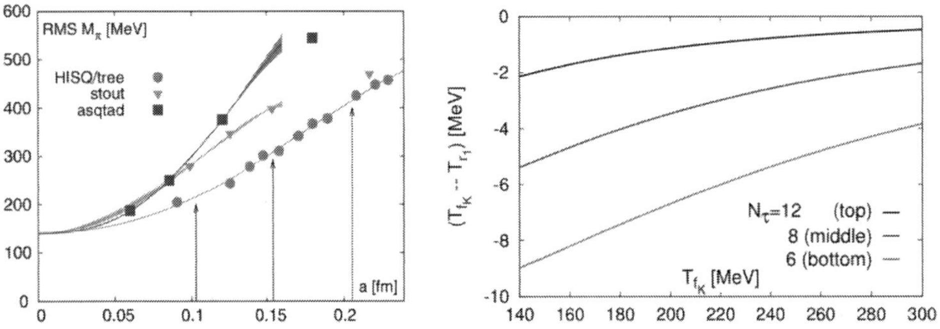

Figure 2. Root-mean-square mass of the pion spectrum as function of the lattice spacing (a) for the HISQ/tree, stout and asqtad action (left). Arrows indicate the lattice spacings corresponding to our $N_\tau = 12, 8$ and 6 lattices (from left to right) at $T = 200$ MeV. The right panel shows the temperature ambiguity as obtained from two different scales for $N_\tau = 12, 8$ and 6 lattices.

the HISQ action performs best. The improvement of the taste breaking is most important at low and intermediate temperatures and, as discused in Sec. 4.3, drastically modifies the electric charge fluctuations in the hadron gas phase.

The results that are presented here have been obtained with (2+1)-flavor of HISQ fermions, with a strange quark mass (m_s) that was tuned to its physical value and light quark masses that are fixed by the quark mass ratio $m_l/m_s = 1/20$. This is slightly heavier than the physical value $m_l/m_s \approx 1/27$. We calculate on lattices with temporal extent of $N_\tau = 6, 8, 10, 12$ and spacial extent $N_\sigma = 4N_\tau$. For the $T = 0$ calculations needed to renormalize the EoS calculations, we use $N_\tau \geq N_\sigma$. The temperature scale was set by two independent experimentally accessible quantities, the Sommer scale [11] (r_1) and the Kaon decay constant (f_K). The ambiguity in the temperature obtained by those two quantities at finite N_τ is shown in Fig. 2 (right).

3. Status of the EoS calculation

The quantity most convenient to calculate on the lattice is the trace anomaly in units of the fourth power of the temperature $\Theta^{\nu\nu}/T^4$. This is given by the derivative of p/T^4 with respect to the temperature, *i.e.*

$$\frac{\Theta^{\nu\nu}(T)}{T^4} \equiv \frac{\epsilon - 3p}{T^4} = T\frac{\partial}{\partial T}(p/T^4). \tag{1}$$

Since the pressure is given by the logarithm of the partition function, $p/T = V^{-1}\ln Z$, the calculation of the trace anomaly requires only the evaluation of straightforward expectation values. Note, however, that we have to subtract the corresponding $T = 0$ result in order to remove the UV divergencies, *i.e.*, for each parameter set of couplings and masses we have to perform an additional $T = 0$ calculation, which makes the EoS determination numerically very demanding. All further thermodynamic quantities can be deduced from the trace anomaly, *e.g.* the pressure is obtained as

$$\frac{p(T)}{T^4} - \frac{p(T_0)}{T_0^4} = \int_{T_0}^{T} dT' \, \frac{\Theta^{\nu\nu}(T)}{T'^5}. \tag{2}$$

In Fig. 3 we show our current status of the $\Theta^{\nu\nu}(T)/T^4$ calculations from $N_\tau = 6, 8, 10$ and 12 lattices [12]. Also shown is a continuum estimate, obtained with a different staggered type action,

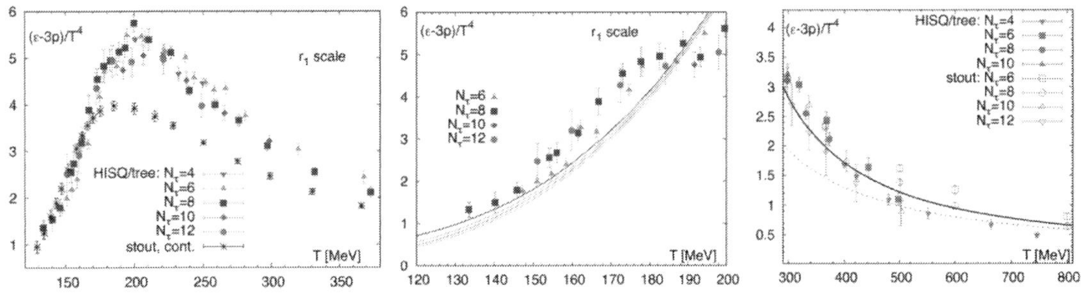

Figure 3. Comparison of the HISQ interaction measure on $N_\tau = 6, 8, 10$ and 12 lattices with the stout continuum estimate [13] (left). The HISQ interaction measure at low (middle) and high (right) temperatures. Solid lines in the middle panel indicate HRG results, dashed and solid lines in the right panel show 1-loop and 2-loop perturbative results, respectively.

the stout action, taken from Ref. [13]. As one can see, results obtained with the two actions currently are not in agreement in the temperature region $180 \lesssim T \lesssim 300$ MeV. The discrepancies are most pronounced at $T \approx 200$ MeV, where the trace anomaly has its maximum. Note, that the HISQ data is not yet extrapolated to the continuum, which may reduce or eliminate the discrepancy. It is, however, obvious that for the pressure that is obtained by the integral in Eq. (2), we encounter differences of (10-15)% for $T \gtrsim 300$ MeV between the present HISQ data at $N_\tau = 12$ and the stout continuum estimate [13].

The middle panel in Fig. 3 shows the low temperature region of the interaction measure. The solid lines indicate corresponding HRG results, where from top to bottom the pion mass was set to its physical value and to the root-mean-square mass $m_\pi^{\rm RMS}$ corresponding to our $N_\tau = 12, 10, 8$ and 6 lattices. As one can see, the trace anomaly calculated in the HRG model is not very sensitive to the pion mass already at temperatures above 130 MeV. Furthermore, one sees that the lattice data approach the HRG model at $T \sim 150 - 160$ MeV.

Fig. 3 (right) shows the trace anomaly obtained with the HISQ and stout actions in the high temperature regime. Also shown here are the perturbative results at 1-loop and 2-loop order, indicated by the dashed and solid lines, respectively. We find good agreement with the lattice data already above $T \gtrsim 300$ MeV.

4. Conserved charge fluctuations

We will now continue with the discussion of fluctuations of conserved charges, which can be measured in experiments and calculated on the lattice. They are well suited to studying critical behavior but can also be used to determine freeze-out conditions.

4.1. Cumulants of conserved charge fluctuations

On the lattice we calculate derivatives of the partition function with respect to baryon (B), electric charge (Q) and strangeness (S) chemical potentials, which are also known as generalized susceptibilities and are defined as

$$\left(VT^3\right) \cdot \chi_{ijk}^{BQS}(T) = \left(\partial^{i+j+k} \ln Z(T, \mu_B, \mu_Q, \mu_S)\right) \Big/ \left(\partial \hat{\mu}_B^i \partial \hat{\mu}_Q^j \partial \hat{\mu}_S^k\right) , \quad (3)$$

with $\hat{\mu}_X = \mu_X/T$ and $X = B, Q, S$. The lattice studies are performed at $\mu_X = 0$ and sufficiently close to the thermodynamic limit. The generalized susceptibilities are intensive quantities. They can also be interpreted as Taylor expansion coefficients of $\ln Z$ and, furthermore, be related to the cumulants of the fluctuations of the net charges (N_X), which are measured in heavy ion collisions. *E.g.*, for the diagonal fluctuations one obtains

$$\left(VT^3\right) \cdot \chi_2^X = \left\langle (\delta N_X)^2 \right\rangle , \quad (4)$$

$$\left(VT^3\right) \cdot \chi_4^X = \left\langle (\delta N_X)^4 \right\rangle - 3 \left\langle (\delta N_X)^2 \right\rangle^2 , \quad (5)$$

$$\left(VT^3\right) \cdot \chi_6^X = \left\langle (\delta N_X)^6 \right\rangle - 15 \left\langle (\delta N_X)^4 \right\rangle \left\langle (\delta N_X)^2 \right\rangle + 30 \left\langle (\delta N_X)^2 \right\rangle^3 , \quad (6)$$

with $\delta N_X = N_X - \langle N_X \rangle$. In Fig. 4 (left and middle) we show our current results for the diagonal fluctuations of net baryon number and electric charge, obtained with the HISQ action on lattices with temporal extent of $N_\tau = 6$ and 8.

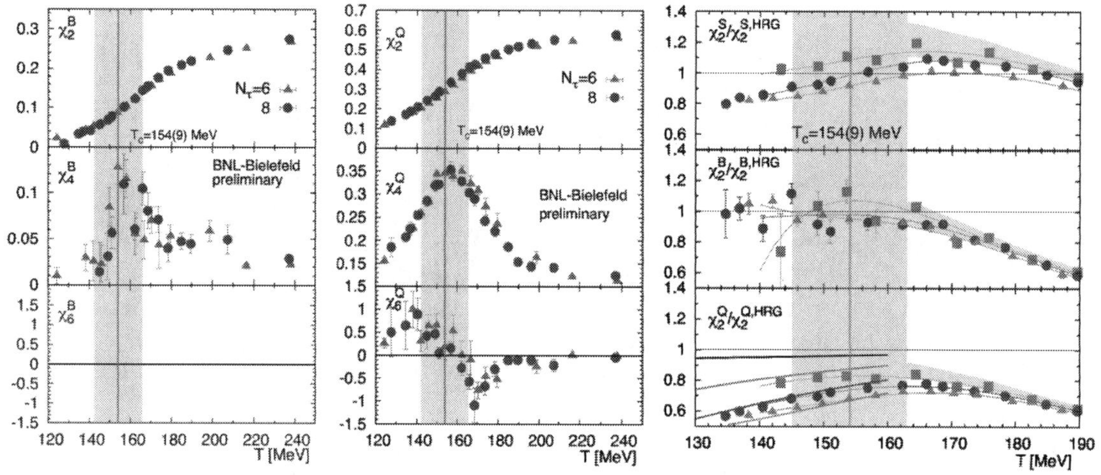

Figure 4. Cumulants of net baryon number fluctuations (left) and net electric charge fluctuations (middle) up to the sixth order; quadratic cumulant of strangeness, baryon number and electric charge normalized by the HRG results (right). Different symbols denote different lattice spacings: $N_\tau = 6$ (triangles), 8 (circles) and 12 (squares). Solid lines on the right panel indicate spline fits to the data. For the electric charge fluctuations additional solid lines indicate HRG results with a modified pion mass (see text for explanation). The vertical bar indicates the QCD transition temperature as obtained in [17].

4.2. O(4) critical behavior

In the chiral limit of QCD with two massless quarks the transition from the hadronic world to the quark gluon plasma is presumably of second order (see Fig. 1). At vanishing chemical potential the critical point is expected to be in the universality class of the three dimensional $O(4)$ symmetric spin model and is expected to persist even in the presence of a physical strange quark mass. It is illuminating to perform a universal scaling analysis that is connected with the critical point in the chiral limit. As a finite chemical potential does not alter chiral symmetry breaking, it enters to leading order only in the temperature like scaling field t, defined as

$$t = t_0^{-1}\left((T-T_0)/T_0 + \kappa_B \hat{\mu}_B^2 + \kappa_S \hat{\mu}_S^2 + \kappa_{BS}\hat{\mu}_B\hat{\mu}_S\right). \tag{7}$$

Here $t_0, T_0, \kappa_B, \kappa_S, \kappa_{BS}$ are non-universal normalization constants. It is easy to see that at $\mu_X = 0$ the contribution from the singular part of the partition functions to the generalized susceptibilities follows the pattern [14],

$$\chi_{2n}^X \sim |t|^{2-n-\alpha}, \tag{8}$$

where α is the critical exponent of the specific heat, which is small and negative. The fourth order cumulants will thus develop a cusp in the chiral limit, whereas the sixth order cumulants are divergent with amplitudes that have different signs below and above T_c. The lattice data is qualitatively consistent with that picture as can be seen in Fig. 4 (left and middle). A more detailed scaling analysis using a recent parametrization of the $O(4)$-scaling function of the specific heat [15] is work in progress. It will allow to obtain various non-universal normalization constants that map QCD to the universal $O(4)$-symmetric theory such as $T_0, t_0, h_0, \kappa_X, \kappa_{XY}$. Some of them are of immediate interest, such as the transition temperature in the chiral limit T_0 or the curvature κ_X that characterizes the change of the transition temperature in the direction of the chemical potential μ_X. Recently the HotQCD Collaboration and the BNL-Bielefeld Collaboration performed a scaling analysis of the chiral condensate, the chiral susceptibility [16, 17] as well as a mixed susceptibility [18]. That also determined the normalization constant $z_0 = h_0^{1/\beta\delta}/t_0$, where β and δ are critical exponents, which fixes the quark mass dependence of the QCD transition (at $\mu_X = 0$). This was used by HotQCD in order to determine the crossover temperature $T_c = 154(9)$ MeV [17] at physical quark masses.

4.3. Comparison with the Hadron Resonance Gas

Below the QCD transition temperature we can compare our lattice results with the statistical Hadron Resonance Gas (HRG) model, which describes the hadronization process in heavy ion collisions quite successfully [8]. For the second order cumulants of baryon number, electric charge and strangeness fluctuations this comparison is performed in Fig. 4 (right) [19]. Here we plot the lattice data normalized by the corresponding HRG results. Light blue bands indicate the continuum extrapolations based on lattices with temporal extent $N_\tau = 6, 8, 12$. We find that the continuum extrapolations of χ_2^B and χ_2^Q approach the HRG from below and are in agreement with the HRG at temperatures up to $T \sim (150-160)$ MeV. The strangeness fluctuations, however, seem to overshoot the HRG and eventually approach the HRG at lower temperatures from above. Note that for the electric charge fluctuations additional solid lines at low temperature indicate HRG results with a modified pion mass. We have chosen the pion masses such that they agree with the averaged pion mass (root-mean-square mass m_π^{RMS}) of the pion spectrum on the lattice at given N_τ. In general, we find that the modified HRG provides a good approximation to the electric charge fluctuations below T_c.

A comparison of the HRG with higher order cumulants is at present not very meaningful as we do not have a continuum result yet. Moreover, for the electric charge fluctuations, which can in principle be immediately compared to the experimental results, the distorted pion spectrum on the lattice becomes increasingly problematic, as the higher order cumulants are increasingly sensitive to an increased m_π^{RMS}. One thus has to use even finer lattice spacings such as $N_\tau = 16$ in order to control this systematic effect. In fact, we find that the HRG results for χ_4^Q with a

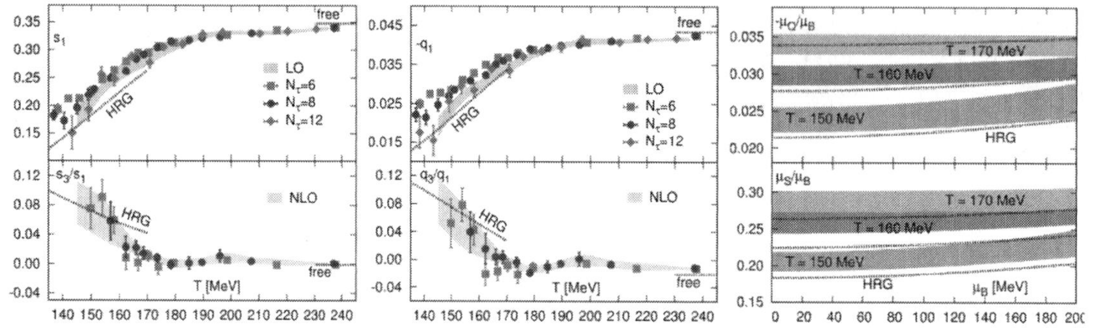

Figure 5. The leading and next-to-leading order expansion coefficients of the strangeness (left) and the negative of the electric charge chemical potentials (middle) versus temperature for $r = 0.4$. For s_1 and q_1 the LO-bands show results for the continuum extrapolation. For s_3 and q_3 we give an estimate for continuum results (NLO bands) based on spline interpolations of the $N_\tau = 8$ data. Dashed lines at low temperature are from the HRG model and at high temperature from a massless, 3-flavor quark gas. The right hand panel shows NLO results for μ_S/μ_B and μ_Q/μ_B as function of μ_B for three values of the temperature.

pion mass corresponding to the $m_\pi^{\rm RMS}$ of our $N_\tau = 6$ and 8 lattices are indistinguishable from that of a HRG with an infinitely heavy pion mass.

4.4. Determination of freeze-out conditions

In oder to compare the lattice QCD results with experimentally measured fluctuations one has to eliminate the unknown fireball volume by taking ratios of cumulants. We have recently proposed a detailed method to extract the remaining freeze-out parameters, *i.e.* the freeze-out temperature (T^f) and the freeze-out chemical potential (μ_B^f), from a comparison of lattice and experimental results of ratios of cumulants of the electric charge fluctuations [6, 7]. In order to resemble the conditions met in heavy ion conditions we fix the strangeness and electric charge chemical potentials by demanding net strangeness neutrality and the correct iso-spin asymmetry:

$$\langle N_S \rangle = 0, \qquad \langle N_Q \rangle = r \langle N_B \rangle . \tag{9}$$

Here r is a constant. We choose $r = 0.4$, which approximates well the situation met in Au-Au and Pb-Pb collisions. We fulfill the conditions given in Eq. 9 by expanding the net densities as well as the chemical potentials $\hat\mu_S$ and $\hat\mu_Q$ in terms of $\hat\mu_B$. We then solve order by order for the expansion coefficients of the chemical potentials. In next-to-leading order (NLO) we obtain

$$\hat\mu_S = s_1(T)\hat\mu_B + s_3(T)\hat\mu_B^3 + \mathcal{O}(\hat\mu_B^5), \qquad \hat\mu_Q = q_1(T)\hat\mu_B + q_3(T)\hat\mu_B^3 + \mathcal{O}(\hat\mu_B^5), \tag{10}$$

where the results for $s_1(T)$, $s_3(T)$, $q_1(T)$, $q_3(T)$ can easily be expressed in terms of the generalized susceptibilities χ_{ijk}^{BQS}. In Fig. 5 (left and middle) we show those expansion coefficients. The upper panel shows the leading order (LO), whereas the lower panels show the ratio NLO to LO coefficients. The band in the upper panels indicates the continuum extrapolation based on the $N_\tau = 6, 8$ and 12 data, the band in the lower panels is a continuum estimate based on $N_\tau = 6$ and 8 data. We find that the NLO contributions are negligible in the high temperature region and below 10% in the temperature interval relevant for the analysis of freeze-out conditions, *i.e.*, $T \approx (160 \pm 10)$ MeV. In fact, in this temperature range the leading order lattice QCD results deviate from HRG model calculations expanded to the same order by less than 15%. Note that one can also investigate the convergence properties of the HRG model itself. In the HRG model

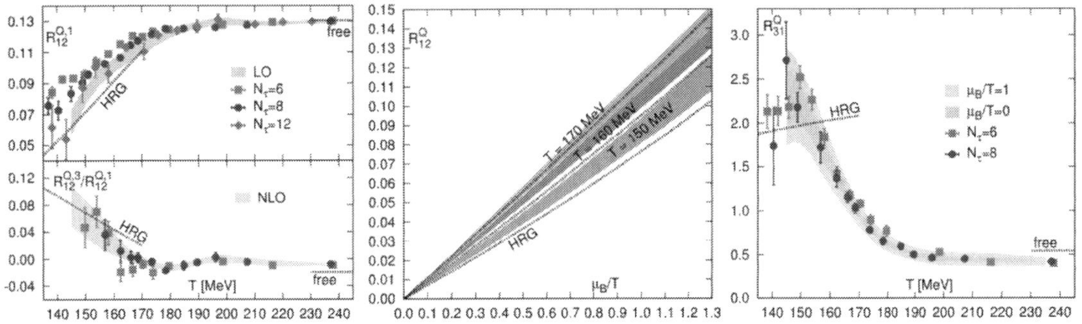

Figure 6. The left panel shows LO (top) and NLO (bottom) expansion coefficients of R_{12}^Q for for $r = 0.4$. The bands and lines are as in Fig. 5(left).On the middle panel we plot R_{12}^Q versus μ_B/T, including the NLO contribution, for three values of the temperature. R_{31}^Q versus temperature is shown on the right panel for $\mu_B = 0$. The wider band on the data set for $N_\tau = 8$ shows an estimate of the magnitude of NLO corrections.

the NLO expansion reproduces the full HRG result for $\hat{\mu}_Q$ and $\hat{\mu}_S$ to better than 1.0% for all values of $\hat{\mu}_B \lesssim 1.3$. Altogether, we thus expect that the NLO truncated QCD expansion is a good approximation to the complete QCD results for $\hat{\mu}_Q$ and $\hat{\mu}_S$ for $\mu_B \lesssim 200$ MeV.

Our results for the strangeness and electric charge chemical potentials at NLO as function of μ_B and T are shown in Fig. 5 (right). While μ_S/μ_B varies between 0.2 and 0.3 in the interval 150 MeV $\lesssim T \lesssim$ 170 MeV, the absolute value of μ_Q/μ_B is an order of magnitude smaller. Both ratios are almost constant for $\mu_B \lesssim 200$ MeV, which is consistent with HRG model calculations.

We do now consider ratios of cumulants $R_{nm}^X = \chi_n^X/\chi_m^X$, which to a large extent eliminate the dependence of cumulants on the freeze-out volume. After fixing μ_Q and μ_S by the constraints given in Eq. 9, they only depend on μ_B and T. Ratios with $n+m$ even are non-zero for $\mu_B = 0$, while the odd-even ratios are in leading order proportional μ_B and thus vanish for $\mu_B = 0$. Ratios with $n+m$ even or odd thus provide complementary information on T^f and μ_B^f. The simplest of those ratios are given by

$$R_{12}^X \equiv \frac{M_X}{\sigma_X^2} = \frac{\chi_1^X}{\chi_2^X} = \hat{\mu}_B \left(R_{12}^{X,1} + R_{12}^{X,3}\, \hat{\mu}_B^2 + \mathcal{O}(\hat{\mu}_B^4) \right) , \tag{11}$$

$$R_{31}^X \equiv \frac{S_X \sigma_X^3}{M_X} = \frac{\chi_3^X}{\chi_1^X} = R_{31}^{X,0} + R_{31}^{X,2}\, \hat{\mu}_B^2 + \mathcal{O}(\hat{\mu}_B^4) . \tag{12}$$

Here we expressed the cumulant ratios also in terms of the mean value (M_X), the variance (σ_X) and the skewness (S_X), which characterize the shape of the distribution of X. These ratios can be calculated in QCD as well as in the HRG model [14], and eventually can be compared to experimental data in order to determine T^f and μ_B^f . Here we have evaluated them up to the NLO contribution $\mathcal{O}(\hat{\mu}_B^3)$ in the Taylor expansion of R_{12}^X and to LO for R_{31}^X. For the discussion of results presented below, we concentrate on net electric charge fluctuations, *i.e.* on R_{12}^Q and R_{31}^Q. While net baryon number fluctuations are experimentally accessible only through measurements of net proton number fluctuations, which may cause some difficulties [21, 22], electric charge fluctuations may be easier to analyze. They are thus better suited for a comparison with the experimental data.

In Fig. 6 (left) we show our results on the LO and NLO expansion coefficients of R_{12}^Q. The bands on the lower and upper panel have the same meaning as in Fig. 5 (left). We find that the NLO corrections to R_{12}^Q are below 10%, which makes the LO result a good approximation for a large range of $\hat{\mu}_B$. Systematic errors arising from the truncation of the Taylor series for R_{12}^Q at NLO may again be estimated by comparing the full result in the HRG model calculation with the corresponding truncated results. Here we find for $T = (160 \pm 10)$ MeV and $\hat{\mu}_B \lesssim 1.3$ that the difference is less than 1.0%. Moreover, we estimated that taste violation effects in the NLO calculation lead to systematic errors that are at most 5% and thus will be negligible in R_{12}^Q . Taylor series truncated at NLO are thus expected to give a good approximation to the full result for a wide range of baryon chemical potentials.

In Fig. 6 (middle) we show the full μ_B and T dependence of R_{12}^Q, including the NLO contribution. Obviously the ratio R_{12}^Q shows a strong sensitivity on μ_B but varies little with T in the temperature range $T = (160 \pm 10)$ MeV. For the determination of (T^f, μ_B^f) a second, complimentary information is needed. To this end we use the ratio R_{31}^Q, which is strongly dependent on T but receives corrections only at $\mathcal{O}(\hat{\mu}_B^2)$. The leading order result for this ratio is shown in Fig. 6 (right). Apparently this ratio shows a characteristic temperature dependence for $T \gtrsim 155$ MeV that is quite different from that of HRG model calculations. The NLO correction to this ratio vanishes in the high temperature limit and at low T the HRG model also suggests small corrections. In fact, in the HRG model the LO contributions to R_{31}^Q differ by less than 2% from the exact results on the freeze-out curve for $\mu_B \lesssim 200$ MeV. The broader band in Fig. 6 (right) indicates an estimate of the NLO contribution at $\hat{\mu}_B = 1$ from our $N_\tau = 8$ calculations.

We now are in the position to extract μ_B^f and T_f from R_{12}^Q and R_{31}^Q which eventually will be measured in the beam energy scan at RHIC [23]. A large value for R_{31}^Q, *i.e.* $R_{31}^Q \simeq 2$ would suggest a low freeze-out temperature $T \lesssim 155$ MeV, while a value $R_{31}^Q \simeq 1$ would suggest a large freeze-out temperature, $T \sim 170$ MeV. A value of $R_{31}^Q \simeq 1.5$ would correspond to $T \sim 160$ MeV. A measurement of R_{31}^Q thus suffices to determine the freeze-out temperature. In the HRG model parametrization of the freeze-out curve [9] the favorite value for T^f in the beam energy range 200 GeV $\geq \sqrt{s_{NN}} \geq 39$ GeV indeed varies by less than 2 MeV and is about 165 MeV. At this temperature the values for R_{31}^Q calculated in the HRG model and in QCD differ quite a bit, as is obvious from Fig. 6 (right). While $R_{31}^Q \simeq 2$ in the HRG model, one finds $R_{31}^Q \simeq 1.2$ in QCD at $T = 165$ MeV. Values close to the HRG value are compatible with QCD calculations only for $T \lesssim 157$ MeV. We thus expect to either find freeze-out temperatures that are about 5% below HRG model results or values for R_{31}^Q that are significantly smaller than the HRG value. A measurement of this cumulant ratio at RHIC thus will allow to determine T^f and probe the consistency with HRG model predictions.

For any of these temperature values a comparison of an experimental value for R_{12}^Q with Fig.6 (middle) will allow to determine μ_B^f. To be specific, at $T = 160$ MeV we expect to find $\mu_B^f = (20 - 30)$ MeV, if R_{12}^Q lies in the range $0.012 - 0.020$, $\mu_B^f = (50 - 70)$ MeV for $0.032 \leq R_{12}^Q \leq 0.045$ and $\mu_B^f = (80 - 120)$ MeV for $0.05 \leq R_{12}^Q \leq 0.08$. These parameter ranges are expected [9] to cover the regions relevant for RHIC beam energies $\sqrt{s_{NN}} = 200$ GeV, 62.4 GeV and 39 GeV, respectively. As is evident from Fig. 6 (middle) the values for μ_B^f will shift to smaller (larger) values when T^f turns out to be larger (smaller) than 160 MeV. A more refined analysis of (T^f, μ_B^f) will become possible, once the ratios R_{12}^Q and R_{31}^Q have been measured experimentally.

5. Summary

We have reviewed the current status of our equation of state calculations with the HISQ action, based on lattices of temporal extension $N_\tau = 6, 8, 10, 12$. Recent results for the trace anomaly indicate a discrepancy of roughly 20% in the peak region ($T \approx 200$ MeV) between HotQCD [12] and the Budapest-Wuppertal results [13]. However, final conclusions can only be drawn after the continuum extrapolation of our data, which we will attempt as soon as the $N_\tau = 10$ and

12 data have reached higher statistical significance. We have further discussed fluctuations of conserved charges. We have shown that higher order cumulants of conserved charge fluctuations become more and more sensitive to critical behavior. For the second order fluctuations we have explicitly discussed their approach to the HRG in the low temperature region. Finally, we proposed a method to extract freeze-out parameters, *i.e.* the freeze-out temperature (T^f) and the freeze-out chemical potential (μ_B^f), from a comparison of lattice and experimental results of ratios of cumulants of the electric charge fluctuations [6, 7]. At nonzero baryon density the method is based on a next to leading order (NLO) Taylor expansion of two different ratios. In general we find that the NLO contributions are below 10% in the μ_B/T range relevant for BES at RHIC down to collision energies of $\sqrt{s_{NN}} \gtrsim 20$ GeV.

Acknowledgments

The numerical generation of gauge field configurations has been performed on BlueGene/L computers at the New York Center for Computational Sciences (NY-CCS) at Brookhaven National Laboratory and on clusters of the USQCD collaboration in JLab and FNAL. The evaluations of fluctuation observables have been performed on the USQCD owned GPU-cluster at JLab and the GPU-cluster at the University of Bielefeld. We further acknowledge support by contract DE-AC02-98CH10886 with the U.S. Department of Energy, the Bundesministerium für Bildung und Forschung under grant 06BI9001, the Gesellschaft für Schwerionenforschung under grant BILAER, the Deutsche Forschungsgemeinschaft under grant GRK881 and the EU Integrated Infrastructure Initiative Hadron-Physics 2.

References

[1] P. Romatschke and U. Romatschke, Phys. Rev. Lett. **99**, 172301 (2007) [arXiv:0706.1522 [nucl-th]].
[2] C. Gale, S. Jeon, B. Schenke, P. Tribedy and R. Venugopalan, arXiv:1210.5144 [hep-ph].
[3] C. R. Allton, *et al.*, Phys. Rev. D **66** (2002) 074507 [hep-lat/0204010].
[4] C. Schmidt, Eur. Phys. J. C **61**, 537 (2009).
[5] M. A. Stephanov, K. Rajagopal and E. V. Shuryak, Phys. Rev. D **60**, 114028 (1999) [hep-ph/9903292].
[6] A. Bazavov *et al.*, arXiv:1208.1220 [hep-lat], to appear in Phys. Rev. Lett.
[7] S. Mukherjee, Quark Matter 2012.
[8] P. Braun-Munzinger, K. Redlich, J. Stachel, In *Hwa, R.C. (ed.) et al.: Quark gluon plasma*, 491-599.
[9] J. Cleymans, H. Oeschler, K. Redlich and S. Wheaton, Phys. Rev. C **73**, 034905 (2006).
[10] E. Follana *et al.* [HPQCD and UKQCD Collaborations], Phys. Rev. D **75**, 054502 (2007).
[11] R. Sommer, Nucl. Phys. B **411**, 839 (1994) [hep-lat/9310022].
[12] A. Bazavov [HotQCD Collaboration], arXiv:1210.6312 [hep-lat].
[13] S. Borsanyi *et al.*, [BW Collaboration], JHEP **1011**, 077 (2010), [arXiv:1007.2580[hep-lat]].
[14] S. Ejiri, F. Karsch, K. Redlich, Phys. Lett. **B633**, 275 (2006).
[15] J. Engels and F. Karsch, Phys. Rev. D **85**, 094506 (2012) [arXiv:1105.0584 [hep-lat]].
[16] S. Ejiri *et al.*, Phys. Rev. D **80**, 094505 (2009) [arXiv:0909.5122 [hep-lat]].
[17] A. Bazavov *et al.*, Phys. Rev. D **85**, 054503 (2012) [arXiv:1111.1710 [hep-lat]].
[18] O. Kaczmarek *et al.*, Phys. Rev. D **83**, 014504 (2011) [arXiv:1011.3130 [hep-lat]].
[19] A. Bazavov, *et al.*, Phys. Rev. D **86**, 034509 (2012) [arXiv:1203.0784 [hep-lat]].
[20] M. M. Aggarwal *et al.* [STAR Collaboration], Phys. Rev. Lett. 105, 022302 (2010).
[21] M. Kitazawa, M. Asakawa, Phys. Rev. C **85**, 021901 (2012).
[22] A. Bzdak, V. Koch, V. Skokov, arXiv:1203.4529 [hep-ph].
[23] L. Kumar [STAR Collaboration], Quark Matter 2012.

Extreme QCD 2012 (xQCD) IOP Publishing
Journal of Physics: Conference Series **432** (2013) 012014 doi:10.1088/1742-6596/432/1/012014

Bottomonium from lattice QCD as a probe of the Quark-Gluon Plasma

G. Aarts, C. Allton
Department of Physics, Swansea University, Swansea, United Kingdom

A. Kelly, J.-I. Skullerud
Department of Mathematical Physics, National University of Ireland Maynooth, Maynooth, County Kildare, Ireland

S. Kim
Department of Physics, Sejong University, Seoul 143-747, Korea

T. Harris, S. M. Ryan
School of Mathematics, Trinity College, Dublin 2, Ireland

M. P. Lombardo
INFN-Laboratori Nazionali di Frascati, I-00044, Frascati (RM) Italy

M. B. Oktay
Physics Department, University of Utah, Salt Lake City, Utah, USA

D. K. Sinclair
HEP Division, Argonne National Laboratory, 9700 South Cass Avenue,Argonne, Illinois 60439, USA

Abstract. We study the temperature dependence of bottomonium for temperatures in the range $0.4T_c < T < 2.1T_c$, using non-relativistic dynamics for the bottom quark and full relativistic lattice QCD simulations for $N_f = 2$ light flavors. We consider the behaviour of the correlators in Euclidean space, we analyze the associated spectral functions and we study the dependence on the momentum. Our results are amenable to a successful comparison with effective field theories. They help build a coherent picture of the behaviour of bottomonium in the plasma, consistent which the current LHC results.

Journal of Physics: Conference Series **432** (2013) 012014 doi:10.1088/1742-6596/432/1/012014

1. Introduction

In this note we review the results on the spectrum of bottomonium in the quark-gluon plasma obtained by our FASTSUM collaboration [1–3] using a hybrid approach which combines exact relativistic dynamics for the light quarks with a non-relativistic approach for the bottom quark.

Recently, the first detailed results on the sequential Upsilon suppression in PbPb collision at LHC energies have appeared [4] (see also several talks at this meeting [5]). These results call for a quantitative understanding of the behaviour of bottomonium at high temperature. Due to their considerable experimental and theoretical interest these studies have been carried out in the past with a variety of techniques. We refer to recent excellent reviews [6, 7], as well as to our published papers, for background material and a complete set of references to early studies; here we limit ourselves to an exposition of our approach and results.

In the next Section we outline our approach to the problem. Next, we present the information we have obtained from the correlators alone. We continue by showing the results on the spectral functions. We finally discuss momentum dependence and we close with a short summary.

2. Five steps to Bottomonium

Our strategy is to treat as accurately as possible the light quarks, and use non-relativistic QCD (NRQCD) for the b quarks. In this way we hope to keep the systematic effects under control. Moreover, the accuracy of the results can be improved following the standard lattice techniques. In short outline, this is our 'five steps approach to bottomonium' strategy:

i) Use full relativistic dynamics for up and down quarks (the inclusion of the strange quark is in progress). Gauge configurations with two degenerate dynamical light Wilson-type quark flavors are produced on highly anisotropic lattices, in a range of temperatures comprised between $0.4T_c$ and $2.1T_c$. Details of the lattice action and parameters can be found in Refs. [8, 9].

ii) We use NRQCD for the bottom quarks. Since NRQCD relies on the scale separation $M \gg T$ and we study temperatures up to $2T_c \simeq 400$ MeV, its application in our temperature range is fully justified. The light quarks, instead, feel the full relativistic dynamics. We computed NRQCD propagators on our configurations using a mean-field improved action with tree-level coefficients, which includes terms up to and including $\mathcal{O}(v^4)$, where v is the typical velocity of a bottom quark in bottomonium.

iii) Analyze bottomonium correlators in Euclidean space. In standard conditions we expect an exponential decay. An analytic model for the Euclidean thermal behaviour is not known — we contented ourselves to study what to expect when quarks are no longer bound. Consider free quarks in continuum NRQCD with energy $E_{\mathbf{p}} = \mathbf{p}^2/2M$. The correlators for the S and P waves are

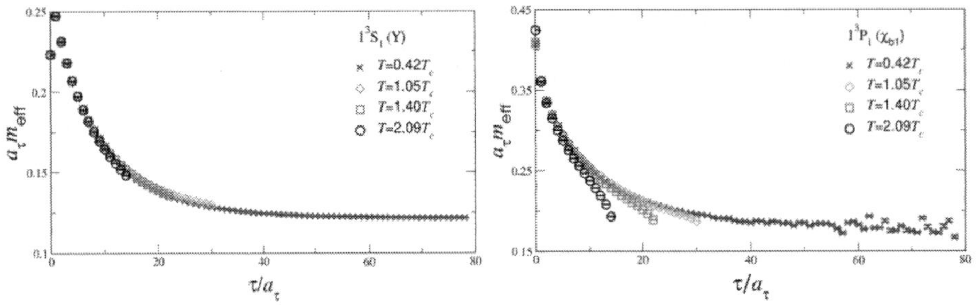

Figure 1. Effective mass plots for the Υ (left) and χ_{b1} (right) using point sources for various temperatures: note the different temperature dependence (Ref. [1]).

then of the form [12]

$$G_S(\tau) \sim \int \frac{d^3p}{(2\pi)^3} \exp(-2E_{\mathbf{p}}\tau) \sim \tau^{-3/2}, \tag{1}$$

$$G_P(\tau) \sim \int \frac{d^3p}{(2\pi)^3} \mathbf{p}^2 \exp(-2E_{\mathbf{p}}\tau) \sim \tau^{-5/2}, \tag{2}$$

i.e., they decay as a power for large euclidean time. Of course, interactions and finite lattice spacing and volume effects are expected to modify this in the realistic case. Moreover, again, this is a limiting high-temperature behaviour which should not be expected to hold even if the quarks unbind in the plasma.
iv) Obtain spectral functions. In NRQCD, the spectral relation reduces to

$$G(\tau) = \int_{-2M}^{\infty} \frac{d\omega'}{\pi} \exp(-\omega'\tau)\rho(\omega') \qquad \text{(NRQCD)}, \tag{3}$$

even at nonzero temperature (note $\omega = 2M + \omega'$). As a result, all problems associated with thermal boundary conditions are absent. We will extract NRQCD spectral functions from the euclidean correlators using the Maximum Entropy Method (MEM) [17].
v) Extract physics from the results, discuss phenomenology, validate effective models. We underscore that any meaningful comparison with experimental results and continuum physics requires control over systematic and lattice artifacts.

3. Υ and χ_b correlators in the plasma

We discuss here the results one can obtain from the analysis of correlators in Euclidean time, focusing on the Υ and χ_b states [1].

In Fig. 1, standard effective masses, defined by $m_{\text{eff}}(\tau) = -\log[G(\tau)/G(\tau - a_\tau)]$, are shown for both the Υ and the χ_{b1} propagators at various temperatures. Single exponential decay should yield a τ-independent plateau. In both cases we find that at the lowest temperature, $T = 0.42T_c$, exponential behaviour is

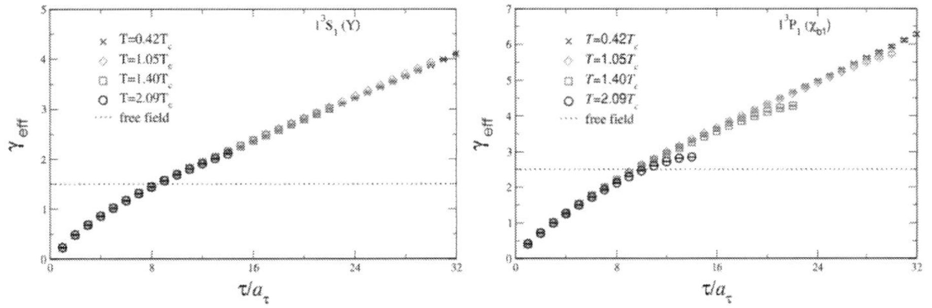

Figure 2. Effective exponents $\gamma_{\rm eff}(\tau)$ for the Υ (left) and χ_{b1} (right), as a function of Euclidean time for various temperatures. The dotted line indicates the non-interacting result in the continuum (Ref. [1]).

visible provided one goes to late euclidean times. At higher temperature we can see first indications that the Υ is not sensitive to the quark-gluon plasma up to $T \simeq 2T_c$, while the χ_b may melt at much lower temperatures, as exponential behaviour is no longer visible.

To visualize the approach to quasi-free behaviour we construct effective power plots [1], using the definition $\gamma_{\rm eff}(\tau) = -\tau\frac{G'(\tau)}{G(\tau)} = -\tau\frac{G(\tau+a_\tau)-G(\tau-a_\tau)}{2a_\tau G(\tau)}$, where the prime denotes the (discretized) derivative. For a power decay, $G(\tau) \sim \tau^{-\gamma}$, this yields a constant result, $\gamma_{\rm eff}(\tau) = \gamma$. The results are shown in Fig. 2. We confirm again that the Υ displays a very mild temperature dependence, while for the χ_{b1} the effective power tends to flatten out. As discussed, in the continuum limit and in the free case one should observe a flat behaviour. The observed flattening might thus suggest an approach to a free field behaviour, however several caveats apply. Also shown are the effective exponents in the continuum non-interacting limit. In the case of the χ_{b1}, we observe that the effective exponent tends towards the non-interacting result at the highest temperature we consider.

To extend and validate these results, we proceed to the calculation of the spectral functions.

4. Spectral functions

Our main results for the spectral functions [2] can be seen in Fig. 3, which shows that as the temperature is increased the ground state peaks of the Υ and η_b remain visible. One caveat applies also here: the apparent width at zero temperature is most likely due to a lattice/MEM artifact, and this calls for an analysis of the discretization effects on the spectral function, which is one of our ongoing projects. The peaks associated to the excited states become suppressed at higher temperature and are no longer discernible at $T/T_c \sim 1.68$. The temperature dependence of the position and width of the ground state peaks is compared to analytical predictions obtained within the EFT formalism [18]. We note that the survival of the $\Upsilon(1S)$ state and suppression of excited states is consistent with the recent experimental results [4].

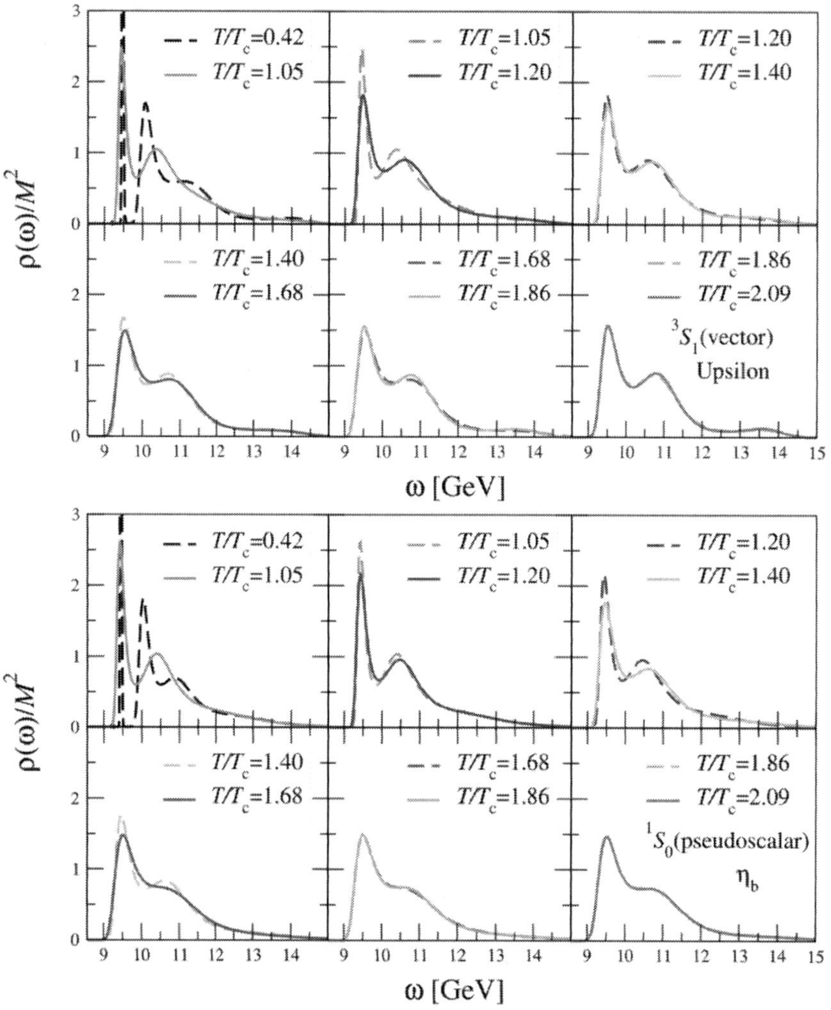

Figure 3. Spectral functions $\rho(\omega)$, normalised with the heavy quark mass, in the vector (Υ) channel (upper panel) and in the pseudoscalar (η_b) channel (lower panel) for all temperature available. The subpanels are ordered from cold (top left) to hot (bottom right). Every subpanel contains two adjacent temperatures to facilitate the comparison (Ref. [2]).

From these spectral functions we can determine the mass (from the peak position) and (an upper bound on) the width of the ground state at each temperature. In Fig. 4 we show the temperature dependence of the mass shift ΔE, normalised by the heavy quark mass and the temperature dependence of the width, normalised by the temperature.

We now contrast our results with analytic predictions derived assuming a weakly coupled plasma. According to Ref. [18], the thermal contribution to the

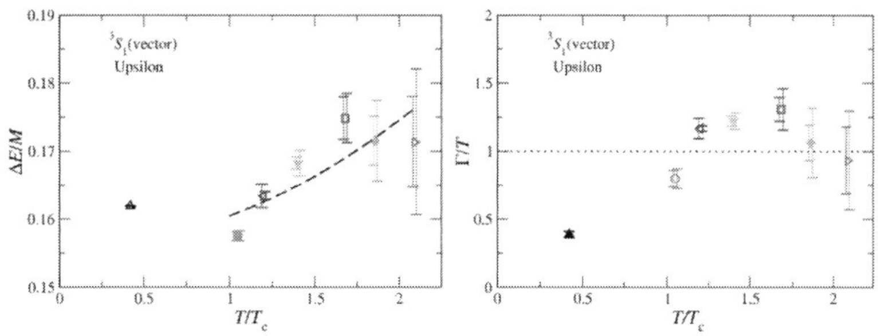

Figure 4. Position of the ground state peak ΔE, normalised by the heavy quark mass (left), and the upper limit on the width of the ground state peak, normalised by the temperature (right), as a function of T/T_c in the vector (Υ) channel. Similar results have been obtained for the pseudoscalar (η_b) channel (Ref. [2]).

width is given, at leading order in the weak coupling and large mass expansion, by

$$\frac{\Gamma}{T} = \frac{1156}{81}\alpha_s^3 \simeq 14.27\alpha_s^3, \tag{4}$$

i.e., the width increases linearly with the temperature. If we take as an estimate from our results that $\Gamma/T \sim 1$, we find that this corresponds to $\alpha_s \sim 0.4$, which is a reasonable result. (It would be of interest to compute α_s directly on our configurations.) In the same spirit the thermal mass shift is given in Ref. [18] by

$$\delta E_{\text{thermal}} = \frac{17\pi}{9}\alpha_s \frac{T^2}{M} \simeq 5.93\alpha_s \frac{T^2}{M}. \tag{5}$$

In these simulations we have $T_c \sim 220$ MeV, $M \sim 5$ GeV. Taking these values together with $\alpha_s \sim 0.4$ as determined above, Eq. (5) becomes

$$\frac{\delta E_{\text{thermal}}}{M} = 5.93\alpha_s \left(\frac{T_c}{M}\right)^2 \left(\frac{T}{T_c}\right)^2 \sim 0.0046 \left(\frac{T}{T_c}\right)^2. \tag{6}$$

In order to contrast our results with this analytical prediction, we have compared the temperature dependence of the peak positions to the simple expression

$$\frac{\Delta E}{M} = c + 0.0046 \left(\frac{T}{T_c}\right)^2, \tag{7}$$

where c is a free parameter. This is shown by the dashed line in Fig. 4 (left panel). The numerical results and the analytic ones are not inconsistent, within the large errors, which encourages us to pursue further these studies.

5. Momentum Dependence

We extend now our study of bottomonium spectral functions in the quark-gluon plasma to nonzero momentum [3]. Note that effective theories analysis predict

Journal of Physics: Conference Series **432** (2013) 012014 doi:10.1088/1742-6596/432/1/012014

$\mathbf{n}$	(1,0,0)	(1,1,0)	(1,1,1)	(2,0,0)	(2,1,0)	(2,1,1)	(2,2,0)
$\lvert\mathbf{p}\rvert$ (GeV)	0.634	0.900	1.10	1.23	1.38	1.52	1.73
$v\ [\Upsilon(^3S_1)]$	0.0670	0.0951	0.116	0.130	0.146	0.161	0.183
$v\ [\eta_b(^1S_0)]$	0.0672	0.0954	0.117	0.130	0.146	0.161	0.183

Table 1. Nonzero momenta used in this study. Also indicated are the corresponding velocities $v = |\mathbf{p}|/M_S$ of the ground states in the vector (Υ) and pseudoscalar (η_b) channels, using the ground state masses determined previously [1], $M_\Upsilon = 9.460$ GeV and $M_{\eta_b} = 9.438$ GeV.

significant momentum effects at large momenta, and that current CMS results have been obtained at large momenta. So there is both phenomenological and experimental motivation for these studies.

The momenta and velocities that are accessible on the lattice are constrained by the discretization and the spatial lattice spacing. The lattice dispersion relation reads

$$a_s^2\mathbf{p}^2 = 4\sum_{i=1}^{3}\sin^2\frac{p_i}{2}, \qquad p_i = \frac{2\pi n_i}{N_s}, \qquad -\frac{N_s}{2} < n_i \leq \frac{N_s}{2}. \tag{8}$$

To avoid lattice artifacts, only momenta with $n_i < N_s/4$ are used: we consider the combinations (and permutations thereof) given in table 1. The largest momentum, using $\mathbf{n} = (2,2,0)$, is $|\mathbf{p}| \simeq 1.73$ GeV, corresponding to $v = |\mathbf{p}|/M_S \simeq 0.2$. Therefore, the range of velocities we consider is non-relativistic.

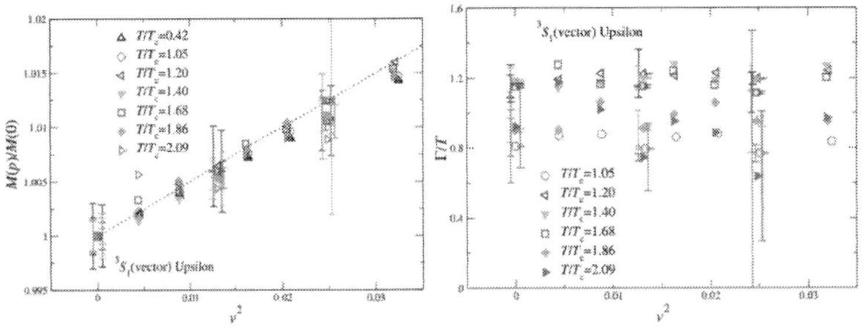

Figure 5. Position of the ground state peak $M(\mathbf{p})/M(0)$ (left) and the upper limit on the width of the ground state peak, normalized by the temperature, Γ/T (right), as a function of the velocity squared (v^2) in the vector (Υ) channel Analogous results have been obtained for the the pseudoscalar (η_b) channel. The dotted line in the left figure represents $M(\mathbf{p})/M(0) = 1 + \frac{1}{2}v^2$ (Ref. [3]).

As in the zero momentum case, we extract masses and widths from our correlators using MEM, and we display them in Fig. 5. We observe that the peak position increases linearly with v^2, as expected. Assuming the lowest-order, non-relativistic expression $M(\mathbf{p}) = M(0) + \mathbf{p}^2/2M(0)$, one finds

$$\frac{M(\mathbf{p})}{M(0)} = 1 + \frac{\mathbf{p}^2}{2M^2(0)} = 1 + \frac{1}{2}v^2, \tag{9}$$

which is indicated with the dotted lines in the left figures.

The dependence on the velocity can be compared with EFT predictions. In Ref. [19] a study of the velocity dependence was carried out in the context of QED, working in the rest frame of the bound state (i.e. the heat bath is moving). In order to compare with our setup, we consider the case in which the temperature is low enough for bound states to be present and that the velocities are non-relativistic. In that case, one finds [19], in the rest frame of the bound state and at leading order in the EFT expansion,

$$\frac{\Gamma_v}{\Gamma_0} = \frac{\sqrt{1-v^2}}{2v} \log\left(\frac{1+v}{1-v}\right), \tag{10}$$

where Γ_0 is the width at rest. Interpreting the width as an inverse lifetime, one can express this result in the rest frame of the heat bath by dividing with the Lorentz factor $\gamma = 1/\sqrt{1-v^2}$. An expansion for non-relativistic velocities then yields

$$\frac{\Gamma_v}{\Gamma_0} = 1 - \frac{2v^2}{3} + \mathcal{O}\left(v^4\right). \tag{11}$$

If we apply this result to our study of bottomonium, we find that the effect of the nonzero velocity shows up as a correction at the percent level (recall that $v^2 \lesssim 0.04$), which is beyond our level of precision but consistent with the observed v independence within errors. Similarly, additional thermal effects in the dispersion relation are currently beyond our level of precision. In summary, the observation in our low-momentum range are consistent with Ref. [19], and in order to observe the predicted non-trivial momentum dependence we need to explore larger momenta.

6. Summary

We have presented our results on bottomonium in the quark-gluon plasma, for temperatures up to $2.1T_c$, at the threshold of the region currently explored by LHC heavy-ions experiments. Our analysis uses full relativistic gauge field configurations, and a non-relativistic approach for the bottom quarks. Our goal is to reach a quantitative, systematic control on the results, which is called for by the new level of experimental accuracy. Related to this, we aim at understanding in detail the range of validity of the available effective models. This is a task of foremost importance, since an accurate analytic modeling is needed as an input to real time evolution.

Our results so far were based on a set of gauge field configurations obtained with two flavors of dynamical quarks. We have used anisotropic lattices in order to maximize the number of points in Euclidean time — a much needed feature for a successful MEM analysis. In this way even on our hottest lattices

($T = 2.1T_c$) we have been able to analyze correlators and proceed to the calculation of the spectral functions.

On these lattice ensembles we have observed that the temperature affects only very mildly the Υ and η_b correlators, suggesting the survival of the fundamental states in these channels up to $T \simeq 2T_c$. A subsequent calculation of the spectral functions carried out with a MEM analysis has confirmed the survival of the fundamental states and revealed the melting of the excited states in the same channels above $1.4T_c$. The correlators for the χ_b and h_b indicate a quick melting of these states above T_c.

On the methodological side, we have argued that the peculiar nature of non-relativistic dynamics greatly simplifies the MEM analysis and enhances its reliability. We have invested a considerable effort in monitoring the systematic effects associated with the MEM analysis, which is described in our papers, and we feel that we have them well under control. The special form of the NRQCD spectral relations shows that the NRQCD spectral function is an inverse Laplace transform of the correlator. In a nutshell, this makes the MEM analysis easier than in the full relativistic case, and perhaps opens the way to a direct, more simple evaluation of the spectral functions themselves.

Our activities in the near future are evolving along different lines. Firstly, new results which will be available in the near future include a MEM analysis for the χ_b and h_b channels.

Secondly, we wish to approach more closely the continuum limit, with a physical matter content. To this end, simulations with a reduced lattice spacing, and the inclusion of the strange quarks, are already in progress. On the NRQCD side, on the new $N_f = 2 + 1$ lattices the quark mass is tuned precisely using the kinetic mass and the lattice NRQCD dispersion relation. Moreover we are considering using the 1-loop determinations of the $\sigma \cdot B$ term in the NRQCD coefficients which has recently produced an impressive improvement in the determination of the (zero temperature) hyperfine splitting [20] . All this will enhance the control over various lattice systematic errors. In a longer term perspective, relativistic beauty might also become feasible on these finer lattices, mirroring the approach that is already being followed for charmonium [8, 21]. This will provide a definitive check of the NRQCD approach.

Next, as for the momentum dependence, we have mentioned that both experimental and phenomenological activities call for calculations at larger momenta: a sizable dependence is expected at large momenta and large momenta are those studied by CMS. Large momentum studies will be possible on our finer lattices and this is being planned.

As a final remark, leading to further developments, an important aspect of our study is the comparison with effective theories. Within our limited precision, and limited explored range, the comparison has been satisfactory. This motivates us to enhance this analysis in the future. To carry out this program, besides the improvement already described, we will have to explore a wider set of parameters. One economic way of doing this is to vary the heavy quark mass, studying artificial mesons: this has been very helpful at zero temperature in cross checking lattice results and effective model predictions — for instance the results for the light meson spectrum are being compared with chiral perturbation theory and the heavy meson spectrum with heavy quark effective theory. In the same spirit we would like to know the dependence of the bottomonium spectrum on the heavy quark mass, and temperature, and compare it with various effective models in their respective ranges of applicability. Remember in fact that different effective models hold true in correspondence with different relative values of the relevant scales such as $M, M\alpha_s, T, M\alpha_s^2, m_D$, so upon varying mass and temperatures one should observe a crossover between different behaviours, thus building confidence in using the corresponding models. This program has started and its first, preliminary results have been reported at Lattice 2012 [22].

Acknowledgments
We wish to thank the Organisers for their suggestion that MPL replace the intended invited speaker — S.M. Ryan — who had to cancel her participation, so that our work could still be presented at the meeting.

References

[1] G. Aarts, S. Kim, M. P. Lombardo, M. B. Oktay, S. M. Ryan, D. K. Sinclair, J. -I. Skullerud, Phys. Rev. Lett. **106** (2011) 061602
[2] G. Aarts, C. Allton, S. Kim, M. P. Lombardo, M. B. Oktay, S. M. Ryan, D. K. Sinclair and J. I. Skullerud, JHEP **1111** (2011) 103
[3] G. Aarts, C. Allton, S. Kim, M. P. Lombardo, M. B. Oktay, S. M. Ryan, D. K. Sinclair and J. -I. Skullerud, arXiv:1210.2903 [hep-lat].
[4] S. Chatrchyan *et al.* [CMS Collaboration], arXiv:1208.2826 [nucl-ex].
[5] G. Breto-Rangel, P. Braun-Munzinger, Nu Xu, M. Strickland, P. Petreczki, this Volume
[6] T. Hatsuda, Review talk at *Quark Matter 2012*, to appear in the Proceedings
[7] O. Kaczmarek, Review talk at *Hard Probes 2012*, arXiv:1208.4075 [hep-lat].
[8] G. Aarts, C. Allton, M. B. Oktay, M. Peardon and J. I. Skullerud, Phys. Rev. D **76** (2007) 094513
[9] R. Morrin, A. Ó Cais, M. Peardon, S. M. Ryan and J. I. Skullerud, Phys. Rev. D **74** (2006) 014505
[10] M. Laine, O. Philipsen, P. Romatschke and M. Tassler, JHEP **0703** (2007) 054
[11] M. Laine, JHEP **0705** (2007) 028
[12] Y. Burnier, M. Laine and M. Vepsäläinen, JHEP **0801** (2008) 043
[13] N. Brambilla, J. Ghiglieri, A. Vairo and P. Petreczky, Phys. Rev. D **78** (2008) 014017
[14] N. Brambilla, M. A. Escobedo, J. Ghiglieri and A. Vairo, JHEP **1112** (2011) 116
[15] A. Beraudo, J. P. Blaizot and C. Ratti, Nucl. Phys. A **806** (2008) 312
[16] A. Beraudo, J. P. Blaizot, P. Faccioli and G. Garberoglio, Nucl. Phys. A **846** (2010) 104
[17] M. Asakawa, T. Hatsuda, Y. Nakahara, Prog. Part. Nucl. Phys. **46** (2001) 459-508
[18] N. Brambilla, M. A. Escobedo, J. Ghiglieri, J. Soto, A. Vairo, JHEP **1009** (2010) 038
[19] M. A. Escobedo, J. Soto and M. Mannarelli, Phys. Rev. D **84** (2011) 016008
[20] T. C. Hammant, A. G. Hart, G. M. von Hippel, R. R. Horgan and C. J. Monahan, Phys. Rev. Lett. **107**, 112002 (2011)
[21] A. Kelly, D. B. Mehta, M. B. Oktay, J.-I. Skullerud, *Charm diffusion from 2-flavor lattice QCD*, in preparation; A. Kelly, *Spectral functions of charmonium in 2+1 flavour lattice QCD*, poster at Strong and Electroweak Matter 2012.
[22] S. Kim, G. Aarts, C. Allton, M. P. Lombardo, M. B. Oktay, S. M. Ryan, D. K. Sinclair and J. -I. Skullerud, arXiv:1210.7586 [hep-lat].

Bottomonia in the Quark Gluon Plasma

Michael Strickland
Physics Department, Gettysburg College, Gettysburg, PA 17325 United States

Abstract. I review recent calculations of the suppression of bottomonium states in heavy ion collisions. A non-relativistic potential is used which is complex valued. This allows one to extract the binding energies and decay widths of the ground and excited states of bottomonium as a function of the typical plasma particle momentum and momentum-space anisotropy. The decay widths determined are used as input and integrated over space-time taking into account the dynamical evolution of the typical particle momentum and momentum-space anisotropy. The suppression of $\Upsilon(1s)$, $\Upsilon(2s)$, $\Upsilon(3s)$, χ_{b1}, and χ_{b2} is obtained as a function of centrality, rapidity, and transverse momentum. The obtained results are compared with data from the STAR and CMS collaborations.

1. Introduction

In November of 2010 the Large Hadron Collider (LHC) at the European Organization for Nuclear Research (CERN) achieved its first relativistic heavy ion collisions of lead nuclei (Pb-Pb). The goal of such experiments is to recreate conditions which only existed in the early universe before the formation of hadrons. The LHC is searching for evidence of the creation of a primordial state of matter called the quark gluon plasma (QGP) which comprised the entire universe until approximately 10^{-5} seconds after the big bang when the temperature of the universe was on the order of 10^{12} Kelvin. The LHC experiment is attempting to recreate such temperatures by colliding lead nuclei at $\sqrt{s_{NN}} = 2.76$ TeV/nucleon whereas the lower energy Relativistic Heavy Ion Collider (RHIC) at Brookhaven National Laboratory has been colliding gold nuclei at $\sqrt{s_{NN}} = 200$ GeV/nucleon. In addition to simply crossing the threshold for the creation of a quark gluon plasma, the experiments are also making detailed measurements of the collision products in order to determine fundamental properties of the QGP. The resulting data are being compared to theoretical calculations based on quantum chromodynamics (QCD) which describes the interactions of quarks and gluons.

At high temperatures one expects the emergence of Debye screening of the interaction between quarks and gluons. This leads to the dissolution of hadronic bound states [1]. A particularly interesting subset of hadronic states consists of those comprised of heavy quarks, since the spectrum of low lying states can be found using potential-based non-relativistic treatments. Based on such potential models there were early predictions [2, 3] that J/ψ production would be suppressed in heavy ion collisions relative to the corresponding production in proton-proton collisions scaled by the number of nucleons participating in the collision. In recent years there have been important theoretical advances in the understanding of heavy quark states at finite temperature using analytic techniques [4–12] and lattice QCD [13–21]. Most recently, interest has shifted to bound states of bottom and anti-bottom quarks (bottomonium) for the following reasons

(i) Bottom quarks ($m_b \simeq 4.2$ GeV) are more massive than charm quarks ($m_c \simeq 1.3$ GeV) and as a result the heavy quark effective theories underpinning phenomenological applications are on much surer footing.

(ii) Due to their higher mass, the effects of initial state nuclear suppression are expected to be smaller than for the charmonium states [22].

(iii) The masses of bottomonium states ($m_\Upsilon \approx 10$ GeV) are much higher than the temperatures ($T \lesssim 1$ GeV) generated in relativistic heavy ion collisions. As a result, bottomonium production will be dominated by initial hard scatterings.

(iv) Since bottom quarks and anti-quarks are relatively rare within the plasma, the probability for regeneration of bottomonium states through recombination is much smaller than for charm quarks.

As a result one expects the bottomonium system to be a cleaner probe of the quark gluon plasma than the charmonium system for which the modeling has necessarily become quite involved. For this reason we will focus on the bottomonium states and only consider the thermal suppression of these states, ignoring initial state effects and any possible thermal generation or recombination. In this conference proceedings we will review recent theoretical calculations of bottomonium suppression at energies probed in relativistic heavy ion collisions at RHIC and LHC. We will present an overview of the important aspects of the calculation and refer the reader to Refs. [23, 24] for details. We will compare our predictions for inclusive $\Upsilon(1s)$ and $\Upsilon(2s)$ suppression with recent data to from the CMS and STAR collaborations.

2. Theoretical methods

In the last few years there have been important theoretical developments in the theoretical treatment of heavy quarkonium in the quark gluon plasma. First among these are the first-principles calculations of imaginary-valued contributions to the heavy quark potential. The first calculation of the leading-order perturbative imaginary part of the potential due to gluonic Landau damping was performed by Laine et al. [5]. Subsequently, an additional imaginary-valued contribution to the potential coming from singlet to octet transitions has also been computed using the effective field theory approach [9]. These imaginary-valued contributions to the potential are related to quarkonium decay processes in the plasma. The consequences of such imaginary parts on heavy quarkonium spectral functions [25, 26], perturbative thermal widths [5, 27], quarkonia at finite velocity [28], in a T-matrix approach [4, 8, 29–31], and in stochastic real-time dynamics [32] have recently been studied.

Additionally, there have been significant advances in the dynamical models used to simulate plasma evolution. In particular, there has been a concerted effort to understand the effects of plasma momentum-space anisotropies generated by the rapid longitudinal expansion of the matter along the beamline direction. The resulting dynamical models are now able to describe the anisotropic hydrodynamical evolution using full (3+1)-dimensional simulations [33–38]. This is important because momentum-space anisotropies can have a significant impact on quarkonium suppression since in regions of high momentum-space anisotropy one expects reduced quarkonium binding [7, 10–12, 39]. In Refs. [23, 24] the dynamical evolution of the anisotropic plasma was combined with the real and imaginary parts of the binding energy obtained using modern complex-valued potentials.

2.1. Anisotropic potential model and binding energies

Early on it was shown that ideal relativistic hydrodynamics is able to reproduce the soft collective flow of the matter and single particle spectra produced at RHIC [40–43]. Based on this, there was a concerted effort to develop a systematic framework for describing the soft collective motion.

This effort resulted in a number of works dedicated to the development and application of relativistic viscous hydrodynamics to relativistic heavy ion collisions [44–62].

One of the weakness of the traditional viscous hydrodynamics approach is that it relies on an implicit assumption that the system is close to thermal equilibrium which implies that the system is also very close to being isotropic in momentum space. However, one finds during the application of these methods that this assumption breaks down at the earliest times after the initial impact of the two nuclei due to large momentum-space anisotropies in the p_T-p_L plane which can persist for many fm/c [63]. In addition, one finds that near the transverse and longitudinal edges of the system these momentum-space anisotropies are large at all times [35, 36, 63]. Similar conclusions have been obtained in the context of strongly coupled systems where it has been shown using the conjectured AdS/CFT correspondence one achieves viscous hydrodynamical behavior at times when the system still possesses large momentum-space anisotropies and that these anisotropies remain large throughout the evolution [64–70]. Based on these results one is motivated to apply a dynamical framework that can accommodate potentially large momentum-space anisotropies.

In order to take into account plasma momentum-space anisotropy, the phase-space distribution of gluons in the local rest frame is assumed to be given by [7, 71–74]

$$f(t,\mathbf{x},\mathbf{p}) = f_{\rm iso}\left(\sqrt{\mathbf{p}^2+\xi(\mathbf{p}\cdot\mathbf{n})^2}/p_{\rm hard}\right), \tag{1}$$

where $f_{\rm iso}$ is an isotropic distribution which in thermal equilibrium is given by a Bose-Einstein distribution, ξ is the momentum-space anisotropy parameter, and $p_{\rm hard}$ is a momentum scale which specifies the typical momentum of the particles in the plasma and can be identified with the temperature in the limit of thermal isotropic ($\xi=0$) equilibrium. The two parameters $p_{\rm hard}$ and ξ can, in general, depend on proper time and position; however, we do not indicate this explicitly for compactness of notation.

In general one finds that the heavy quark potential has real and imaginary parts, $V = \Re[V]+i\Im[V]$. One can determine the real part of the heavy-quark potential in the non-relativistic limit from the Fourier transform of the 00-component of the static gluon propagator. In an anisotropic plasma with a distribution function given by Eq. (1) one finds [10], at leading order in the strong coupling constant,

$$\begin{aligned} V(\mathbf{r},\xi) &= -g^2 C_F \int \frac{d^3\mathbf{p}}{(2\pi)^3}\, e^{i\mathbf{p}\cdot\mathbf{r}} \Delta^{00}(\omega=0,\mathbf{p},\xi)\,, &(2)\\ &= -g^2 C_F \int \frac{d^3\mathbf{p}}{(2\pi)^3}\, e^{i\mathbf{p}\cdot\mathbf{r}} \frac{\mathbf{p}^2+m_\alpha^2+m_\gamma^2}{(\mathbf{p}^2+m_\alpha^2+m_\gamma^2)(\mathbf{p}^2+m_\beta^2)-m_\delta^4}\,, &(3)\end{aligned}$$

where g is the strong coupling constant and $C_F = (N_c^2-1)/(2N_c)$ is the quadratic Casimir of the fundamental representation of $SU(N_c)$. The mass scales m_α, m_β, m_γ, and m_δ are listed in Ref. [10]. One can factorize the denominator of (3) by introducing

$$2m_\pm^2 \equiv M^2 \pm \sqrt{M^4 - 4(m_\beta^2(m_\alpha^2+m_\gamma^2)-m_\delta^4)}\,, \tag{4}$$

with $M^2 \equiv m_\alpha^2+m_\beta^2+m_\gamma^2$ [71]. This allows us to write

$$V(\mathbf{r},\xi) = -g^2 C_F \int \frac{d^3\mathbf{p}}{(2\pi)^3}\, e^{i\mathbf{p}\cdot\mathbf{r}} \frac{\mathbf{p}^2+m_\alpha^2+m_\gamma^2}{(\mathbf{p}^2+m_+^2)(\mathbf{p}^2+m_-^2)}\,. \tag{5}$$

In general one must evaluate (5) numerically. The integration can be reduced to a two-dimensional integral over a polar angle, θ, and the length of the three-momentum, p. However, there can be poles in the integration domain due to the fact that m_-^2 can be negative for certain polar angles and momenta [71]. These poles are first order and can be dealt with using a principle-part prescription. After evaluating the integral numerically, one finds that the resulting potential is very well described by a Debye-screened Coulomb potential with an anisotropic Debye mass μ

$$\Re[V(r)] = -\frac{g^2 C_F}{4\pi r} e^{-\mu r} , \tag{6}$$

where

$$\left(\frac{\mu}{m_D}\right)^{-4} = 1 + \xi \left(a - \frac{2^b(a-1) + (1+\xi)^{1/8}}{(3+\xi)^b}\right)\left(1 + \frac{c(\theta)(1+\xi)^d}{(1+e\xi^2)}\right) , \tag{7}$$

The coefficients a-e are fixed by (a) requiring that the small and large anisotropy limits of the analytic potential are reproduced and (b) fitting to the numerical results obtained by direct integration of Eq. (5) [24].

Here we will focus on a model in which the real part of the potential is obtained from internal energy of the system since models based on the free energy seem to be incapable of reproducing either the LHC or RHIC $R_{AA}[\Upsilon]$. The real part of the potential is given by [24]

$$\Re[V] = -\frac{a}{r}\,(1+\mu\, r)\, e^{-\mu\, r} + \frac{2\sigma}{\mu}\left[1 - e^{-\mu\, r}\right] - \sigma\, r\, e^{-\mu\, r} - \frac{0.8\,\sigma}{m_Q^2 r} , \tag{8}$$

where $a = 0.385$ and $\sigma = 0.223$ GeV^2 [75] and the last term is a temperature- and spin-independent finite quark mass correction taken from Ref. [76]. In this expression $\mu = \mathcal{G}(\xi,\theta) m_D$ is an anisotropic Debye mass where $\mathcal{G}$ is a rather complicated function (see Eq. (7) above) which depends on the degree of plasma momentum-space anisotropy, ξ, and the angle of the line connecting the quark-antiquark pair with respect to the beamline direction, θ, and m_D is the isotropic Debye mass [24]. In the limit $\xi \to 0$ one has $\mathcal{G} = 1$. In the figures in the results section, results obtained with Eq. (8) are often labeled as "Potential Model B".

The imaginary part of the potential $\Im[V]$ is obtained from a leading order perturbative calculation which was performed in the small anisotropy limit

$$\Im[V] = -\alpha_s C_F T \left\{\phi(r/m_D) - \xi\left[\psi_1(r/m_D,\theta) + \psi_2(r/m_D,\theta)\right]\right\} , \tag{9}$$

where ϕ, ψ_1, and ψ_2 can be expressed in terms of hypergeometric functions [39]. After combining the real and imaginary parts of the potential, the 3d Schrödinger equation is solved numerically to obtain the real and imaginary parts of the binding energy as a function of ξ and $p_{\rm hard}$ [12, 77]. The imaginary part of the binding energy is related to the width of the state

$$\Gamma(\tau,\mathbf{x}_\perp,\varsigma) = \begin{cases} 2\Im[E_{\rm bind}(\tau,\mathbf{x}_\perp,\varsigma)] & \Re[E_{\rm bind}(\tau,\mathbf{x}_\perp,\varsigma)] > 0 , \\ 10 \text{ GeV} & \Re[E_{\rm bind}(\tau,\mathbf{x}_\perp,\varsigma)] \le 0 , \end{cases} \tag{10}$$

where $\varsigma = \text{arctanh}(z/t)$ is the spatial rapidity. The value of 10 GeV in the second case is chosen to be large in order to quickly suppress states which are fully unbound (which is the case when the real part of the binding energy is negative).

2.2. Dynamical Model

The dynamical model used gives the spatio-temporal evolution of the typical transverse momentum of the plasma partons, $p_{\rm hard}(\tau,\mathbf{x})$, and the plasma momentum-space anisotropy, $\xi(\tau,\mathbf{x})$, both of which are specified in the local rest frame of the plasma. The widths obtained from solution of the 3d Schrödinger equation are then integrated and exponentiated to compute the relative number of states remaining at a given proper time. This quantity is then averaged

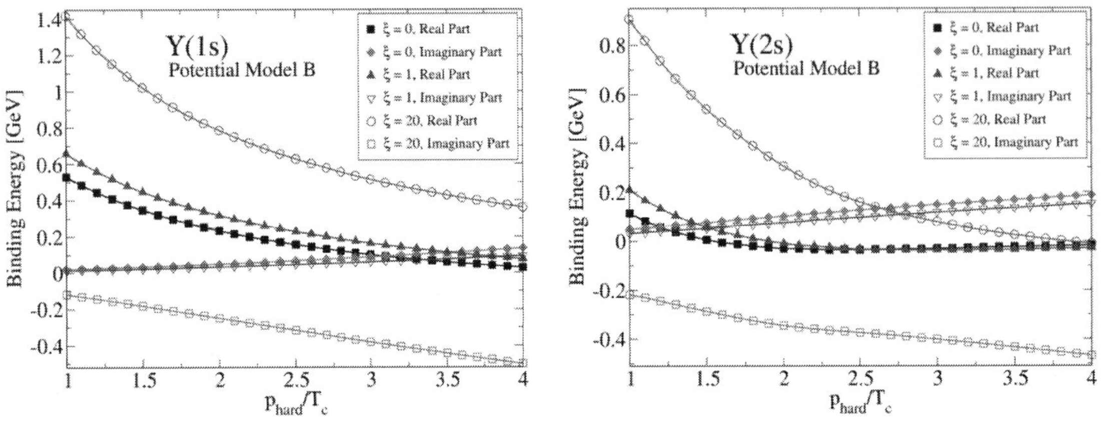

Figure 1. Real and imaginary parts of the $\Upsilon(1s)$ (left) and $\Upsilon(2s)$ (right) binding energies as a function of the hard momentum scale, $p_{\rm hard}$ for different levels of momentum-space anisotropy ξ.

over the transverse plane taking into account the local conditions in the plasma and weighting by the spatial probability distribution for bottomonium production which is given by the number of binary collisions computed in the Glauber model with a Woods-Saxon distribution for each nucleus. For the temporal integration the initial time is set by the formation time of the state in question. The resulting R_{AA} is a function of the transverse momentum, p_T, the rapidity ς, and the nuclear impact parameter b. To compare to experimental results transverse momentum cuts are applied assuming a $n(p_T) = n_0 E_T^{-4}$ spectrum. In addition, any cuts on the rapidity due to detector acceptance and centrality are applied as needed. For details of the dynamical model and R_{AA} computation we refer the reader to Ref. [24].

2.3. Initial conditions

For the initial conditions we use a Woods-Saxon distribution for each nucleus and determine the transverse dependence of the initial temperature via the third root of the number of participants (wounded nucleons). In the spatial rapidity direction we have investigated two possible temperature profiles: (a) a broad plateau containing a boost-invariant central region with Gaussian limited-fragmentation at large rapidity [78]

$$n(\varsigma) = n_0 \exp(-(|\varsigma| - \varsigma_{\rm flat}/2)^2/2\sigma_\varsigma^2)\, \Theta(|\varsigma| - \varsigma_{\rm flat}/2)\,, \tag{11}$$

where $\varsigma_{\rm flat} = 10$ is the width of the central rapidity plateau, $\sigma_\varsigma = 0.5$ is the width of the limited fragmentation tails, and n_0 is the number density at central rapidity [57]; and (b) a Gaussian motivated by low-energy fits to pion spectra

$$n(\varsigma) = n_0 \exp(-\varsigma^2/2\sigma_\varsigma^2) \quad \text{with} \quad \sigma_\varsigma^2 = 0.64 \cdot 8\, c_s^2 \ln\left(\sqrt{s_{NN}}/2m_p\right)/3(1-c_s^4)\,, \tag{12}$$

where $c_s = 1/\sqrt{3}$ is the sound velocity, $m_p = 0.938$ GeV is the proton mass, and $\sqrt{s_{NN}}$ is the nucleon-nucleon center-of-mass energy [24, 79]. The temperature distribution is given by $T \sim n^{1/3}$. We note that Eq. (11) has the advantage that it has been tuned to successfully describe the rapidity dependence of the elliptic flow in LHC heavy ion collisions.

3. Results and Conclusions

In Fig. 1 we show the real and imaginary parts of the $\Upsilon(1s)$ and $\Upsilon(2s)$ binding energies as a function of the hard momentum scale, $p_{\rm hard}$, for $\xi \in \{0, 1, 20\}$. Defining the disassociation scale

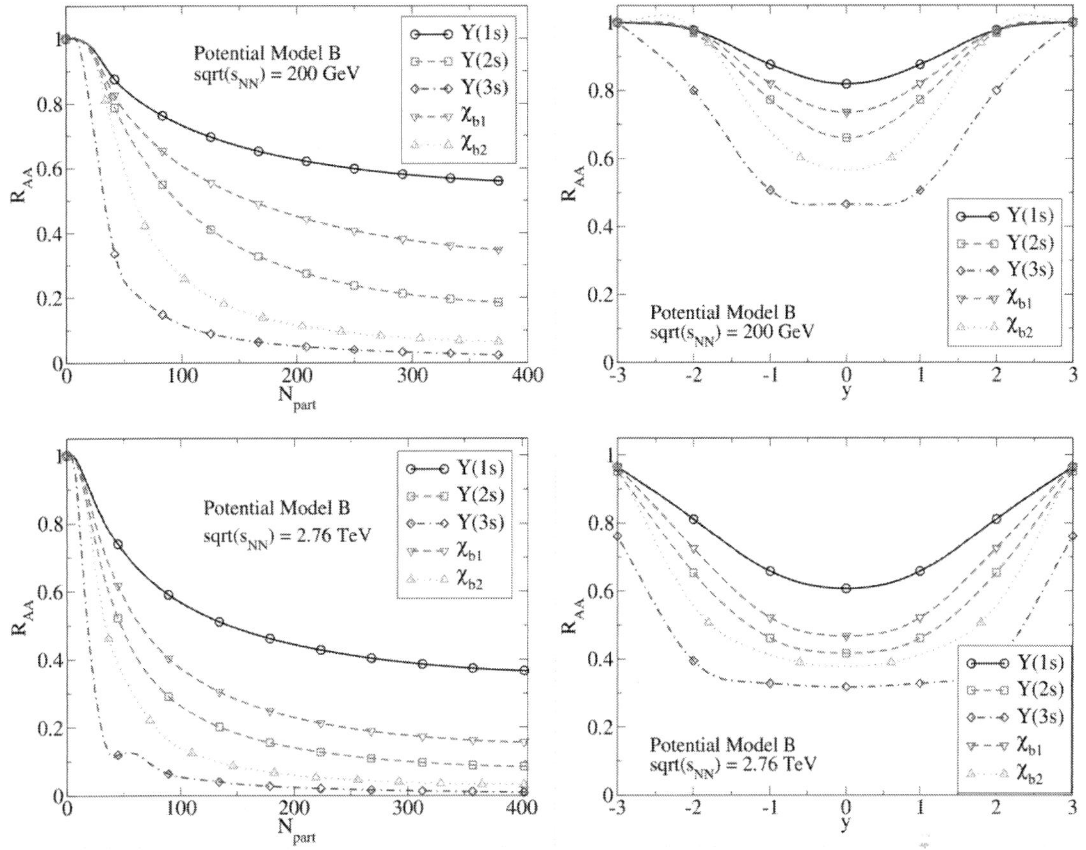

Figure 2. RHIC and LHC suppression factor R_{AA} for the $\Upsilon(1s)$, $\Upsilon(2s)$, $\Upsilon(3s)$, χ_{b1}, and χ_{b2} states as a function of the number of participants (left) and rapidity (right). In the top plots we used $\sqrt{s_{NN}} = 200$ GeV and in the bottom plots we used $\sqrt{s_{NN}} = 2.76$ TeV. In both cases, we assumed a shear viscosity to entropy density ratio of $4\pi\eta/\mathcal{S} = 1$, and implemented cuts of $0 < p_T < 20$ GeV and (left) rapidity $|y| < 0.5$ for RHIC and $|y| < 2.4$ for LHC (right) centrality 0-100%.

as the value of $p_{\rm hard}$ at which the real and imaginary parts of the binding energy become equal, one finds isotropic dissociation scales of 593 MeV for the $\Upsilon(1s)$ and 228 MeV for the $\Upsilon(2s)$. In addition, as one can see from this figure, as the level of momentum space anisotropy increases, one finds that the disassociation scale increases. This means that in regions of the plasma where there is a high degree of momentum-space anisotropy, the states will be less suppressed.

In Fig. 2 we show the predicted suppression factor R_{AA} for the $\Upsilon(1s)$, $\Upsilon(2s)$, $\Upsilon(3s)$, χ_{b1}, and χ_{b2} states as a function of the number of participants (left) and rapidity (right). In this figure we see clear signs of sequential suppression, with the higher excited states having stronger suppression than the ground state. However, we note that even for states that are "melted" at relatively low central temperatures, we still obtain a non-vanishing suppression factor for these states. This is due to the fact that near the edges, where the temperature is lower, one does not see suppression of the states. Upon performing the geometrical average, we see that a large fraction of the states produced can survive even when the central temperature of the plasma is above their naive dissociation temperature.

In Fig. 3 (left) we plot $R_{AA}[\Upsilon(1s + 2s + 3s)]$ and compare with experimental data from the

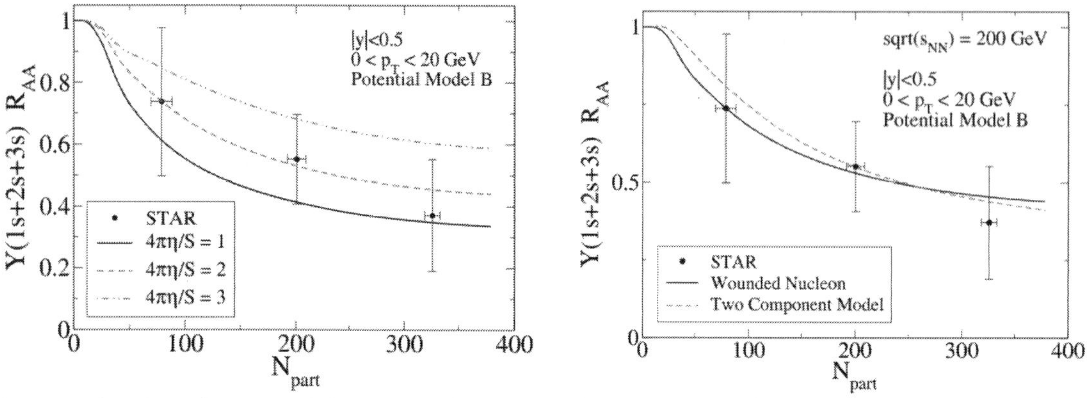

Figure 3. RHIC $\Upsilon(1s + 2s + 3s)$ suppression factor (left) compared with experimental data from the STAR Collaboration [80]. The three different lines correspond to different assumptions for the shear viscosity to entropy ratio $4\pi\eta/\mathcal{S} \in \{1, 2, 3\}$. In both plots we used $\sqrt{s_{NN}} = 200$ GeV and implemented cuts of $0 < p_T < 20$ GeV and $|y| < 0.5$. On the right, the solid black line is the result obtained assuming wounded nucleon initial conditions and the dashed red line is the result obtained used a two component model with $\alpha = 0.145$.

STAR Collaboration [80]. As can be seen from this figure, the model does a reasonably good job of reproducing the existing STAR data for $R_{AA}[\Upsilon(1s+2s+3s)]$. From the left panel we can obtain an estimate of $\eta/\mathcal{S}$: $0.08 < \eta/\mathcal{S} < 0.24$. In Fig. 3 (right) we show the results obtained for $R_{AA}[\Upsilon(1s+2s+3s)]$ at RHIC energies for two different types of initial conditions (Glauber and mixed Glauber plus binary collision scaling). For both lines shown in Fig. 3 (right) we have assumed $4\pi\eta/\mathcal{S} = 2$. Because changing the initial condition type affects particle multiplicities we have adjusted the initial temperature at RHIC energies from 433 MeV to 461 MeV in order to keep the charged particle multiplicity fixed at $dN_{ch}/dy = 620$. As can be seen from Fig. 3 (right), for peripheral collisions there is a larger dependence on the choice of initial condition type, while for central collisions the result obtained is not much affected by the choice of initial condition. This is to be contrasted with the dependence on the assumed value of $\eta/\mathcal{S}$ which affects the suppression at all centralities.

When considering the suppression of the $\Upsilon(1s)$ and $\Upsilon(2s)$ states it is important to include the effect of feed-down from higher excited states. In pp collisions only approximately 51% of $\Upsilon(1s)$ states come from direct production and similarly for the $\Upsilon(2s)$. One can compute the inclusive suppression of a state using $R_{AA}^{\rm full}[\Upsilon(ns)] = \sum_{i\,\in\,{\rm states}} f_i\, R_{i,AA}$ where f_i are the feed-down fractions and $R_{i,AA}$ is the direct suppression of each state which decays into the $\Upsilon(ns)$ state being considered. Here we will use $f_i = \{0.510, 0.107, 0.008, 0.27, 0.105\}$ for the $\Upsilon(1s)$, $\Upsilon(2s)$, $\Upsilon(3s)$, χ_{b1}, and χ_{b2} feed-down to $\Upsilon(1s)$, respectively [82]. For the inclusive $\Upsilon(2s)$ production we use $f_i = \{0.500, 0.500\}$ for the $\Upsilon(2s)$ and $\Upsilon(3s)$ states, respectively. For details of the computation of the direct R_{AA} for each state see Ref. [24].

In Fig. 4 we compare to recent data on the inclusive $\Upsilon(1s)$ and $\Upsilon(2s)$ suppression available from the CMS collaboration [83]. For this figure we used a broad rapidity plateau as the initial density profile as specified in Eq. (11). The central temperatures were taken to be $T_0 = \{520, 504, 494\}$ MeV at $\tau_0 = 0.3$ fm/c for $4\pi\eta/\mathcal{S} = \{1, 2, 3\}$, respectively, in order to fix the final charged multiplicity to $dN_{\rm ch}/dy = 1400$ in each case. As can been seen from this figure, the predictions agree reasonably well with the available data. The data seem to prefer the

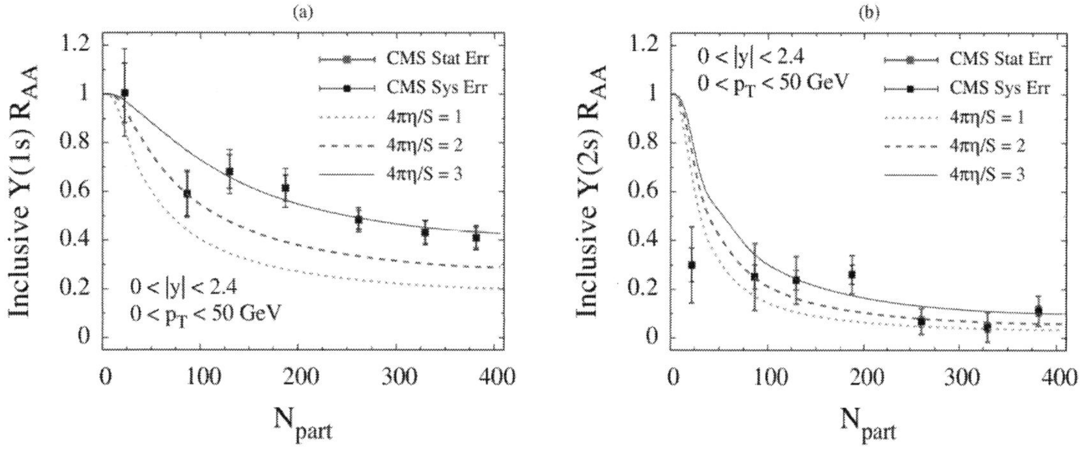

Figure 4. Predictions for the central rapidity inclusive (a) $\Upsilon(1s)$ and (b) $\Upsilon(2s)$ suppression including feed-down as a function of N_{part} along with recent data from the CMS collaboration. The three different lines correspond to different assumptions for the shear viscosity to entropy ratio $4\pi\eta/\mathcal{S} \in \{1, 2, 3\}$.

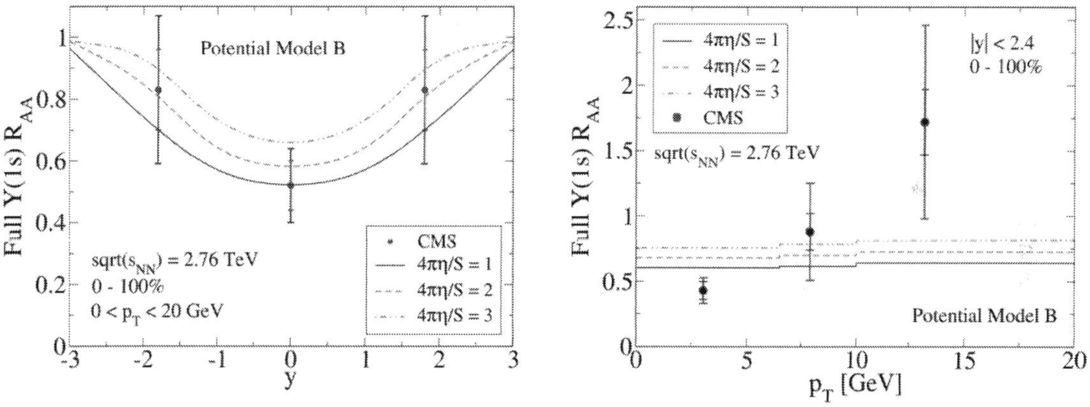

Figure 5. LHC inclusive suppression factor R_{AA} for the $\Upsilon(1s)$ including feed down effects as a function of rapidity and transverse momentum compared to experimental data are from the CMS Collaboration [81].

largest value of η/S shown; however, there is a $\pm 14\%$, $\pm 21\%$ 1s, 2s global uncertainty reported by CMS, making it hard to draw firm conclusions.

In Fig. 5 we show the inclusive suppression factor $R_{AA}^{\text{full}}[\Upsilon(1s)]$. Comparing to the available CMS data [81] we can obtain an estimate for $\eta/\mathcal{S}$ at LHC energies: $0.08 < \eta/\mathcal{S} < 0.24$ which is the same range obtained from the STAR data obtained with gold-gold collisions at lower energies. As before, more precisely determining $\eta/\mathcal{S}$ will require more data from the LHC which should be forthcoming in the near future.

4. Conclusions

In this proceedings contribution we have reviewed recent calculations of bottomonium suppression in RHIC and LHC heavy ion collisions. We presented comparisons of prior predictions with data from the CMS Collaboration for inclusive $\Upsilon(1s)$ and $\Upsilon(2s)$ suppression and data from the STAR Collaboration for inclusive $\Upsilon(1s+2s+3s)$ suppression. The underlying calculations employed a complex-valued potential which incorporates both screening and in-medium dissociation of the states under consideration. We solved for the resulting real and imaginary parts of the binding energy of the $\Upsilon(1s)$, $\Upsilon(2s)$, $\Upsilon(3s)$, χ_{b1}, and χ_{b2} states. We then folded this information together with the real-time evolution of both the typical momentum of the plasma particles ($p_{\rm hard}$) and their momentum-space anisotropy (ξ). We demonstrated that the resulting inclusive suppression of Υ states is in good agreement with available data. We note in closing that there are now very interesting developments concerning quarkonium in an anisotropic medium obtained employing the conjectured AdS/CFT correspondence [84–86]. These methods offer some hope to determine the temperature and momentum-space anisotropy dependence of the long range part of the potential from first principles, albeit in a theory that is not QCD.

Acknowledgments

This work was supported by NSF grant No. PHY-1068765 and the Helmholtz International Center for FAIR LOEWE program.

References

[1] Shuryak E V 1980 *Phys. Rept.* **61** 71–158
[2] Matsui T and Satz H 1986 *Phys. Lett.* **B178** 416
[3] Karsch F, Mehr M T and Satz H 1988 *Z. Phys.* **C37** 617
[4] Grandchamp L, Lumpkins S, Sun D, van Hees H and Rapp R 2006 *Phys.Rev.* **C73** 064906 (*Preprint* hep-ph/0507314)
[5] Laine M, Philipsen O, Romatschke P and Tassler M 2007 *JHEP* **03** 054 (*Preprint* hep-ph/0611300)
[6] Laine M 2007 *JHEP* **05** 028 (*Preprint* 0704.1720)
[7] Dumitru A, Guo Y and Strickland M 2008 *Phys. Lett.* **B662** 37–42 (*Preprint* 0711.4722)
[8] Rapp R, Blaschke D and Crochet P 2010 *Prog.Part.Nucl.Phys.* **65** 209–266 (*Preprint* 0807.2470)
[9] Brambilla N, Ghiglieri J, Vairo A and Petreczky P 2008 *Phys. Rev.* **D78** 014017 (*Preprint* 0804.0993)
[10] Dumitru A, Guo Y, Mocsy A and Strickland M 2009 *Phys.Rev.* **D79** 054019 (*Preprint* 0901.1998)
[11] Burnier Y, Laine M and Vepsalainen M 2009 *Phys.Lett.* **B678** 86–89 (*Preprint* 0903.3467)
[12] Margotta M, McCarty K, McGahan C, Strickland M and Yager-Elorriaga D 2011 *Phys.Rev.* **D83** 105019 (*Preprint* 1101.4651)
[13] Umeda T, Nomura K and Matsufuru H 2005 *Eur. Phys. J.* **C39S1** 9–26 (*Preprint* hep-lat/0211003)
[14] Asakawa M and Hatsuda T 2004 *Phys. Rev. Lett.* **92** 012001 (*Preprint* hep-lat/0308034)
[15] Datta S, Karsch F, Petreczky P and Wetzorke I 2004 *Phys. Rev.* **D69** 094507 (*Preprint* hep-lat/0312037)
[16] Aarts G, Allton C, Oktay M B, Peardon M and Skullerud J I 2007 *Phys. Rev.* **D76** 094513 (*Preprint* 0705.2198)
[17] Hatsuda T 2006 *PoS* **LAT2006** 010
[18] Jakovac A, Petreczky P, Petrov K and Velytsky A 2007 *Phys. Rev.* **D75** 014506 (*Preprint* hep-lat/0611017)
[19] Aarts G, Kim S, Lombardo M, Oktay M, Ryan S *et al.* 2011 *Phys.Rev.Lett.* **106** 061602 (*Preprint* 1010.3725)
[20] Aarts G, Allton C, Kim S, Lombardo M P, Oktay M B *et al.* 2012 (*Preprint* 1210.2903)
[21] Ding H T 2012 (*Preprint* 1210.5442)
[22] Rakotozafindrabe A, Ferreiro E G, Fleuret F, Lansberg J P and Matagne N 2012 *PoS* **QNP2012** 159 (*Preprint* 1207.3193)
[23] Strickland M 2011 *Phys.Rev.Lett.* **107** 132301 (*Preprint* 1106.2571)
[24] Strickland M and Bazow D 2012 *Nucl.Phys.* **A879** 25–58 (*Preprint* 1112.2761)
[25] Burnier Y, Laine M and Vepsalainen M 2008 *JHEP* **0801** 043 (*Preprint* 0711.1743)
[26] Miao C, Mocsy A and Petreczky P 2011 *Nucl. Phys.* **A855** 125–132 (*Preprint* 1012.4433)
[27] Brambilla N, Escobedo M A, Ghiglieri J, Soto J and Vairo A 2010 *JHEP* **09** 038 (*Preprint* 1007.4156)
[28] Escobedo M A, Soto J and Mannarelli M 2011 *Phys.Rev.* **D84** 016008 (*Preprint* 1105.1249)
[29] Riek F and Rapp R 2011 *New J. Phys.* **13** 045007 (*Preprint* 1012.0019)
[30] Emerick A, Zhao X and Rapp R 2012 *Eur.Phys.J.* **A48** 72 (*Preprint* 1111.6537)
[31] Zhao X and Rapp R 2011 *Nucl.Phys.* **A859** 114–125 (*Preprint* 1102.2194)
[32] Akamatsu Y and Rothkopf A 2012 *Phys.Rev.* **D85** 105011 (*Preprint* 1110.1203)
[33] Florkowski W and Ryblewski R 2011 *Phys.Rev.* **C83** 034907 (*Preprint* 1007.0130)

[34] Martinez M and Strickland M 2010 *Nucl. Phys.* **A848** 183–197 (*Preprint* 1007.0889)
[35] Ryblewski R and Florkowski W 2011 *J.Phys.G* **G38** 015104 (*Preprint* 1007.4662)
[36] Martinez M and Strickland M 2011 *Nucl.Phys.* **A856** 68–87 (*Preprint* 1011.3056)
[37] Martinez M, Ryblewski R and Strickland M 2012 *Phys.Rev.* **C85** 064913 (*Preprint* 1204.1473)
[38] Ryblewski R and Florkowski W 2012 *Phys.Rev.* **C85** 064901 (*Preprint* 1204.2624)
[39] Dumitru A, Guo Y and Strickland M 2009 *Phys.Rev.* **D79** 114003 (*Preprint* 0903.4703)
[40] Huovinen P, Kolb P F, Heinz U W, Ruuskanen P V and Voloshin S A 2001 *Phys. Lett.* **B503** 58–64 (*Preprint* hep-ph/0101136)
[41] Hirano T and Tsuda K 2002 *Phys. Rev.* **C66** 054905 (*Preprint* nucl-th/0205043)
[42] Tannenbaum M J 2006 *Rept. Prog. Phys.* **69** 2005–2060 (*Preprint* nucl-ex/0603003)
[43] Kolb P F and Heinz U W 2003 *In* Hwa, R.C. (ed.) et al.: Quark gluon plasma 634–714 (*Preprint* nucl-th/0305084)
[44] Muronga A 2002 *Phys. Rev. Lett.* **88** 062302 (*Preprint* nucl-th/0104064)
[45] Muronga A 2004 *Phys. Rev.* **C69** 034903 (*Preprint* nucl-th/0309055)
[46] Muronga A and Rischke D H 2004 (*Preprint* nucl-th/0407114)
[47] Baier R, Romatschke P and Wiedemann U A 2006 *Phys.Rev.* **C73** 064903 (*Preprint* hep-ph/0602249)
[48] Romatschke P and Romatschke U 2007 *Phys. Rev. Lett.* **99** 172301 (*Preprint* 0706.1522)
[49] Baier R, Romatschke P, Son D T, Starinets A O and Stephanov M A 2008 *JHEP* **04** 100 (*Preprint* 0712.2451)
[50] Dusling K and Teaney D 2008 *Phys. Rev.* **C77** 034905 (*Preprint* 0710.5932)
[51] Luzum M and Romatschke P 2008 *Phys. Rev.* **C78** 034915 (*Preprint* 0804.4015)
[52] Song H and Heinz U W 2009 *J.Phys.G* **G36** 064033 (*Preprint* 0812.4274)
[53] El A, Xu Z and Greiner C 2010 *Phys. Rev.* **C81** 041901 (*Preprint* 0907.4500)
[54] Denicol G, Kodama T and Koide T 2010 *J.Phys.G* **G37** 094040 (*Preprint* 1002.2394)
[55] Denicol G, Koide T and Rischke D 2010 *Phys.Rev.Lett.* **105** 162501 (*Preprint* 1004.5013)
[56] Schenke B, Jeon S and Gale C 2011 *Phys.Rev.Lett.* **106** 042301 (*Preprint* 1009.3244)
[57] Schenke B, Jeon S and Gale C 2011 *Phys.Lett.* **B702** 59–63 (*Preprint* 1102.0575)
[58] Bozek P 2011 *Phys.Lett.* **B699** 283–286 (*Preprint* 1101.1791)
[59] Niemi H, Denicol G S, Huovinen P, Molnar E and Rischke D H 2011 *Phys.Rev.Lett.* **106** 212302 (*Preprint* 1101.2442)
[60] Niemi H, Denicol G, Huovinen P, Molnar E and Rischke D 2012 *Phys.Rev.* **C86** 014909 (*Preprint* 1203.2452)
[61] Bozek P and Wyskiel-Piekarska I 2012 *Phys.Rev.* **C85** 064915 (*Preprint* 1203.6513)
[62] Denicol G, Niemi H, Molnar E and Rischke D 2012 *Phys.Rev.* **D85** 114047 (*Preprint* 1202.4551)
[63] Martinez M and Strickland M 2009 *Phys. Rev.* **C79** 044903 (*Preprint* 0902.3834)
[64] Chesler P M and Yaffe L G 2009 *Phys.Rev.Lett.* **102** 211601 (*Preprint* 0812.2053)
[65] Chesler P M and Yaffe L G 2010 *Phys.Rev.* **D82** 026006 (*Preprint* 0906.4426)
[66] Heller M P, Janik R A and Witaszczyk P 2012 *Phys.Rev.Lett.* **108** 201602 (*Preprint* 1103.3452)
[67] Heller M P, Janik R A and Witaszczyk P 2012 *Phys.Rev.* **D85** 126002 (*Preprint* 1203.0755)
[68] Heller M P, Mateos D, van der Schee W and Trancanelli D 2012 *Phys.Rev.Lett.* **108** 191601 (*Preprint* 1202.0981)
[69] Wu B and Romatschke P 2011 *Int.J.Mod.Phys.* **C22** 1317–1342 (*Preprint* 1108.3715)
[70] Chesler P M and Teaney D 2011 4 pages, one figure (*Preprint* 1112.6196)
[71] Romatschke P and Strickland M 2003 *Phys. Rev.* **D68** 036004 (*Preprint* hep-ph/0304092)
[72] Mrowczynski S, Rebhan A and Strickland M 2004 *Phys. Rev.* **D70** 025004 (*Preprint* hep-ph/0403256)
[73] Romatschke P and Strickland M 2004 *Phys.Rev.* **D70** 116006 (*Preprint* hep-ph/0406188)
[74] Schenke B and Strickland M 2006 *Phys. Rev.* **D74** 065004 (*Preprint* hep-ph/0606160)
[75] Petreczky P 2010 *J.Phys.G* **G37** 094009 (*Preprint* 1001.5284)
[76] Bali G S, Schilling K and Wachter A 1997 *Phys. Rev.* **D56** 2566–2589 (*Preprint* hep-lat/9703019)
[77] Strickland M and Yager-Elorriaga D 2010 *J. Comput. Phys.* **229** 6015–6026 (*Preprint* 0904.0939)
[78] Strickland M 2012 (*Preprint* 1207.5327)
[79] Bleicher M 2005 (*Preprint* hep-ph/0509314)
[80] Reed R 2011 *J.Phys.G* **G38** 124185 (*Preprint* 1109.3891)
[81] CMS Collaboration (CMS) 2011 Quarkonium production in PbPb collisions at $\sqrt{s_{\rm NN}}$=2.76 TeV CMS-PAS-HIN-10-006
[82] Affolder A A *et al.* (CDF) 2000 *Phys.Rev.Lett.* **84** 2094–2099 (*Preprint* hep-ex/9910025)
[83] Chatrchyan S *et al.* (CMS Collaboration) 2012 (*Preprint* 1208.2826)
[84] Giataganas D 2012 *JHEP* **1207** 031 (*Preprint* 1202.4436)
[85] Rebhan A and Steineder D 2012 *JHEP* **1208** 020 (*Preprint* 1205.4684)
[86] Chernicoff M, Fernandez D, Mateos D and Trancanelli D 2012 (*Preprint* 1208.2672)

Extreme QCD 2012 (xQCD) IOP Publishing
Journal of Physics: Conference Series **432** (2013) 012016 doi:10.1088/1742-6596/432/1/012016

Making the most of Taylor expansion and imaginary μ

E. Laermann, F. Meyer
Fakultät für Physik, Universität Bielefeld,D–33615 Bielefeld, Germany

M. P. Lombardo
INFN-Laboratori Nazionali di Frascati, I–00044, Frascati (RM) Italy

Abstract. We present preliminary results for the curvature of the pseudocritical line and susceptibilities in $N_f = 2 + 1$ flavor QCD. The computations are carried out on lattice sizes of $16^3 \times 4$, at matching parameters of early work of the Bielefeld group. Emphasis is placed on the control of systematic errors, by cross-validating results obtained by use of the Taylor expansion and measurements at imaginary chemical potential. To this end, we generalize the magnetic equation of state to the analysis of the number density, and we extend it to imaginary values of the chemical potential.

1. Introduction

Most likely, a large fraction of the participants in this meeting are familiar with the challenges faced by lattice QCD at non-zero baryon density [1]: the region of quark chemical potential μ_q exceeding $m_\pi/2$ remains prohibitively difficult, due to a severe sign problem [2].

For smaller chemical potential a variety of methods have been proposed to circumvent these difficulties, which have been quite successful in many respects. However, even in the easy region $\mu_q < m_\pi/2$ there are challenges to face: for instance, it has been noted that most current estimates of the QCD endpoint lie dangerously close to the 'forbidden region' $\mu_q > m_\pi/2$; the qualitative agreement among different results on the slope of the critical line still needs to be corroborated by some convincing continuum estimates obtained with different methods. And, there is a growing interest in the values of higher order susceptibilities, particularly at larger values of the chemical potential, while the current results are limited to the lowest order ones [3]. So there is room for improvement, which can perhaps be achieved by mixing and matching different approaches.

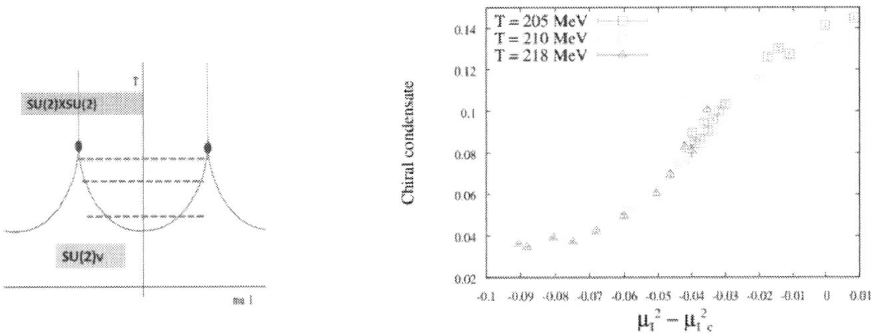

Figure 1. A sketchy view of the phase diagram of QCD for two massless flavors in the temperature, imaginary chemical potential plane, with the superimposed trajectories of our simulations. The blue points indicate the Roberge-Weiss endpoint (left). The chiral condensate measured along such trajectories, plotted against the rescaled chemical potential $\mu^2 - \mu_c^2(T)$ (right). Note the (approximate) universal behaviour of the condensate.

2. Two complemetary approaches: Taylor expansion and imaginary chemical potential

One way to handle the physics of small chemical potential relies on Taylor expansion [4]. This approach is well understood mathematically, however it only works inside its convergency circle, whose radius is unknown. Moreover even inside the circle it is hard to judge when convergence has been achieved. Alternatively, analytic continuation from imaginary chemical potential – where conventional simulations are possible – to real chemical potential can in principle be extended to the entire analytic domain, but the danger here is a possible lack of control over the continuation itself: for instance, terms which are subleading for imaginary chemical potential might well become leading for real chemical potential. The strengths of the two methods are somewhat complementary, and their weaknesses different: this suggests to develop a hybrid method which combines the two, and this note reports on our ongoing effort in this direction [5].

3. Results

Our study for staggered 2+1 flavors on lattices of size $16^3 \times 4$ clearly is exploratory. However, the quark mass values are close to the ones realized in nature. The Goldstone pion mass is tuned to about 220 MeV and the kaon acquires its physical mass. This corresponds to a (degenerate) light to strange quark mass ratio of $m_l/m_s = 1/10$. The action utilized is the p4fat3 action for which Taylor coefficients computed at $\mu = 0$ are available in our range of temperatures and at our quark masses [4]. The computations were carried out at three temperatures above the critical temperature for two massless flavors for the considered lattice spacings: $T = 205, 210, 218$ MeV. In Fig.1, left we sketch the simulations' trajectories, alongside the critical line in the chiral limit, in the μ_I, T plane. The sketch evidences the μ_I periodicity as well as the symmetry under reflection $\mu_I \to -\mu_I$; the actual simulations were performed only at positive μ_I, and within the first half–period. The blue points mark the endpoint of the Roberge Weiss phase transition. The light quark chiral condensate, plotted in Fig.1, right, is showing a fairly rapid rise when μ_I is

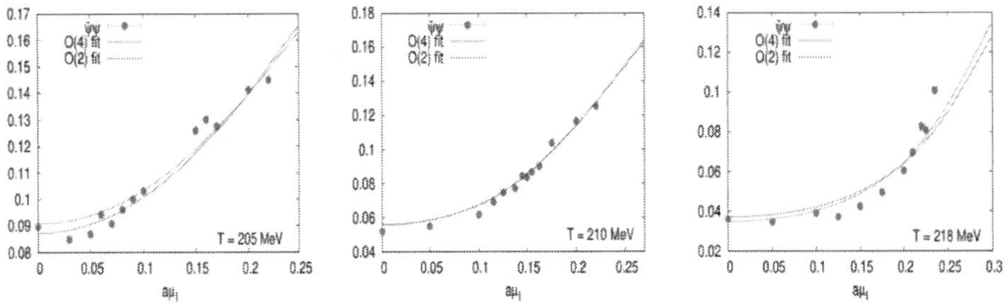

Figure 2. The fits motivated by the magnetic Equation of State for the chiral condensate,superimposed to the data point.

increased, signaling the pseudocritical point for the transition to the broken phase (marked $SU(2)_V$ in the diagram). We plot the results against a rescaled chemical potential to confirm the feasibility of a scaling study based on the magnetic Equation of State.

3.1. *The magnetic Equation of State*

Consider now the magnetic equation of state [6]

$$M = h^{1/\delta} f_G(z) \qquad z = th^{\frac{-1}{\beta\delta}} \tag{1}$$

where M and h are identified with the light chiral condensate and the light bare mass: $M = m_s\langle\bar{\psi}\psi\rangle_l$, $h = \frac{1}{h_0}\frac{m_l}{m_s}$ respectively, f_G is a scaling function which only depends on the universality class of the transition, and $t = 0$ defines the critical line. Our light mass does not change: if scaling (i.e. Eq. 1) holds true the chiral condensate results plotted versus $\mu^2 - \mu_c^2(T)$, where $\mu_c^2(T)$ is the critical chemical potential for each temperature T, should fall on a single curve. In Fig.1 we see that this is indeed happening, demonstrating the universal behaviour of the chiral condensate at least in the range of parameters we are exploring. $\mu_c^2(T)$ is found by searching for the optimal match between results for different temperatures, and compares favorably with alternative estimates.

3.2. *Curvature of the critical line from the magnetic EoS*

For small chemical potential t can be explicitely written as

$$t = \frac{1}{t_0}\frac{T - T_c}{T_c} + k(\mu/T)^2 \tag{2}$$

The standard condition $f_G(0) = f'_G(0) = 1$ fixes the normalization t_0 and h_0 which have been computed in Ref. [6]. As a working hypothesis we assume that t_0 and h_0 do not depend on the temperature, and we check a posteriori that this indeed is the case. k – the curvature of the critical line – can then be computed from the Equation of State, using the explicit forms available for the two plausible universality classes $O(2)$ and $O(4)$, and the results contrasted with those obtained in Ref. [7].

From the EoS for the order parameter one can derive the EoS for the magnetic susceptibility, as well as for the number density n and its susceptibility [8], leading to alternative estimates of k. To fix the notation, let us write the

Table 1. The results for the curvature of the critical line k (Eq.(2)) from the fits to the magnetic EoS for the chiral condensate ($k_{\bar{\psi}\psi}$), and from the fits to the magnetic EoS for the number density (k_n).

	$k_{\bar{\psi}\psi}$			k_n		
T(MeV)	205	210	218	205	210	218
O(2)	0.026(1)	0.031(1)	0.025(3)	0.021(1)	0.021(1)	0.016(1)
O(4)	0.022(1)	0.028(1)	0.025(3)	0.039(1)	0.036(1)	0.023(2)

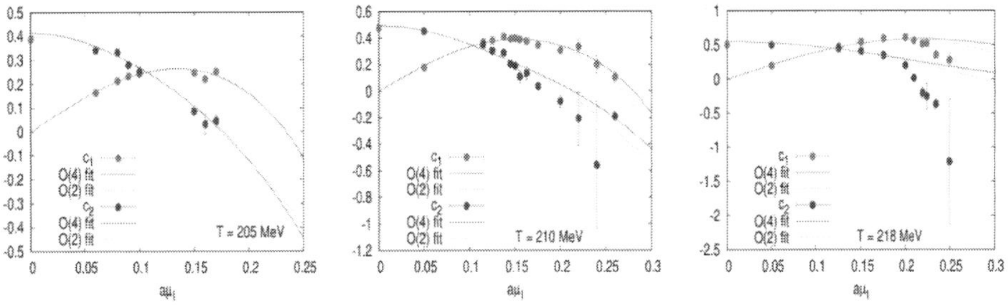

Figure 3. The fits motivated by the magnetic Equation of State for c_1 and c_2 (related to quark number and its susceptibility respectively, see text) superimposed to the data points.

Taylor expansion of the pressure in terms of the quark chemical potentials around a generic $\mu_0 = (\mu_{u,0}, \mu_{d,0}, \mu_{s,0})$ for our $N_f = 2+1$ system:

$$\frac{p}{T^4}(\hat{\mu}) = \sum_{k,l,n} c_{kln}(\hat{\mu}_u - \hat{\mu}_{u,0})^k(\hat{\mu}_d - \hat{\mu}_{d,0})^l(\hat{\mu}_s - \hat{\mu}_{s,0})^n$$

with the abbreviation $\hat{\mu}_q = \mu_q/T$ and the coefficients

$$c_{kln} = \frac{1}{k!l!n!}\frac{\partial^k}{\partial\hat{\mu_u}^k}\frac{\partial^l}{\partial\hat{\mu_d}^l}\frac{\partial^n}{\partial\hat{\mu_s}^n}\left(\frac{p}{T^4}\right) \tag{3}$$

evaluated at μ_0. We will in particular discuss the EoS for $c_1 = c_{100}$ and $c_2 = c_{200}$, which are related to the particle number and its susceptibility. In Fig.2 and Fig.3 we show the results of the EoS fits and in Table 1 we record the k values returned by the fits for the chiral condensate and the quark number (the coefficient c_1). While the fit results, at least for O(2), do not seem to be inconsistent, it is clear that at values for μ_I close to the Roberge-Weiss limit other scaling studies need to be performed, work for the future.

3.3. The Taylor expansion

How far in μ we can trust the Taylor expansion? A direct comparison of the number density of e.g. the u quark given by the partial sums of its Taylor series:

$$\frac{n_u}{T^3}(\hat{\mu}) = \sum_{k,l,n} k c_{kln}(\hat{\mu}_u - \hat{\mu}_{u,0})^{k-1}(\hat{\mu}_d - \hat{\mu}_{d,0})^l(\hat{\mu}_s - \hat{\mu}_{s,0})^n. \tag{4}$$

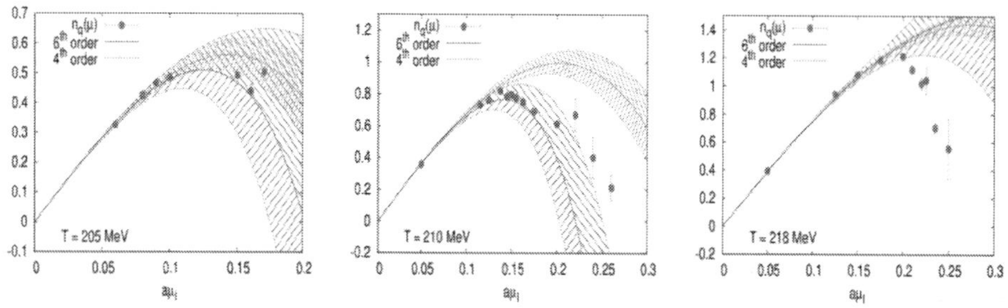

Figure 4. The results for the number density at imaginary chemical potential, contrasted with the partial sums of the Taylor series of the indicated order.

with results obtained directly at imaginary μ might give the answer. We show the results for increasing temperatures in Fig.4: for $T = 205$ MeV a good convergence is achieved indicating a small contribution from the 8th order terms and superior ones. At the highest temperature instead the results suggest that a 6th expansion is not capable to describe the data at $\mu_I \gtrsim 0.2$. The simultaneous consideration of Taylor expansion and imaginary μ results naturally suggest a strategy to extract higher order terms of the expansions – or, equivalently, higher order susceptibilities – withouth having to implement the cumbersome higher order derivatives. At the same time one can place lower bounds on the radius of convergence simply by assessing by eye the extent of the region where actual convergence is achieved.

4. Parting comments

We have presented an overview of results obtained by performing simulations at zero and non–zero imaginary chemical potential. We have noted that an analysis based on the magnetic equation of state for the chiral condensate, the number density and their susceptibilities at imaginary chemical potential gives information on the universality class of the transition for two massless flavors, and on the slope of the chiral line. We have argued that the simultaneous consideration of results from the Taylor expansion at $\mu = 0$ and at imaginary μ can place a lower bound on the radius of convergence at $\mu = 0$, and facilitate the computation of higher order coefficients.

References

[1] G. Aarts, PoS LATTICE **2012** 017 (2012).
[2] K. Splittorff, PoS LAT **2006**, 023 (2006).
[3] M. P. Lombardo, PoS LATTICE **2012**, 016 (2012).
[4] M. Cheng et al., Phys. Rev. D79 (2009) 074505.
[5] F. Meyer, Master thesis, Bielefeld 2012.
[6] S. Ejiri et al., Phys. Rev. D80 (2009) 094505.
[7] O. Kaczmarek et al., Phys. Rev. D83 (2011) 014504.
[8] J. Engels and F. Karsch, Phys. Rev. D85 (2012) 094506.

Some exact results on the QCD critical point

Yoshimasa Hidaka[1] **and Naoki Yamamoto**[2,3,4]

[1]Theoretical Research Division, Nishina Center, RIKEN, Wako 351-0198, Japan
[2]Yukawa Institute for Theoretical Physics, Kyoto University, Kyoto 606-8502, Japan
[3]Institute for Nuclear Theory, University of Washington, Seattle, Washington 98195-1550, USA
[4]Maryland Center for Fundamental Physics, Department of Physics, University of Maryland, College Park, MD 20742-4111, USA

E-mail: hidaka@riken.jp, nyama@umd.edu

Abstract. We show, in a model-independent manner, that the QCD critical point can appear only inside the pion condensation phase of the phase-quenched QCD as long as the contribution of flavor-disconnected diagrams is negligible. The sign problem is known to be maximally severe in this region, implying that the QCD critical point is reachable by the present lattice QCD techniques only if there is an enhancement of the flavor-disconnected contribution at finite baryon chemical potential.

1. Introduction and conclusion

The phase structure of quantum chromodynamics (QCD) at finite temperature T and finite baryon chemical potential μ_B has been a subject of considerable interest. Among others, of particular interest is the possible QCD critical point, which is the terminal point of the first-order chiral phase transition (for a review, see [1]). Such critical points may appear not only in hot matter between the hadron phase and quark-gluon plasma [2] but also in dense matter between the hadron phase and color superconductor [3, 4], and are potentially observable in heavy-ion collision experiments at RHIC and FAIR. Due to the sign problem, however, it has not yet been established in first principles lattice QCD simulations whether the QCD critical points really exist or not.

The first attempt to locate the QCD critical point was made by Fodor and Katz based on the reweighting method [5]. As pointed out by Splittorff [6], however, their critical points happen to be located on the critical line of the phase-quenched QCD with the fermion determinant $\det \mathcal{D}$ replaced by $|\det \mathcal{D}|$ (see Figure 1 for the phase diagram), where the reweighting method breaks down. See also [9, 10] for other but related problems. Indeed in every model calculation (at least in the mean-field approximation), such as the random matrix model [12], Nambu-Jona-Lasinio (NJL) model [13], and Polyakov-loop extended NJL model [14], critical points are somehow observed only inside the critical line of the phase-quenched theory where the sign problem is maximally severe [11, 6]. The question is whether this is the case in real QCD; if so, the QCD critical point is quite difficult to access by the present lattice QCD techniques.

In this article we show, in a model-independent manner, that the QCD critical point cannot exist under the following conditions [15]:

(i) The contribution of flavor-disconnected diagrams is negligible compared with that of connected diagrams.

(ii) The coordinate (T, μ) in QCD at finite $\mu_B = 3\mu$ is outside the critical line of the phase-quenched QCD. (Note that it is *not* a physical phase transition in QCD at finite μ_B.)

The condition (i) is satisfied in the large-N_c QCD and the models in the mean-field approximation, but it could be violated in real QCD. In other words, if the QCD critical point exists in a region accessible by the present lattice QCD techniques, the condition (i) must be broken; it necessitates some kind of enhancement of the flavor-disconnected contribution, which should also be observed in the lattice QCD simulations (see section 3 for more details).

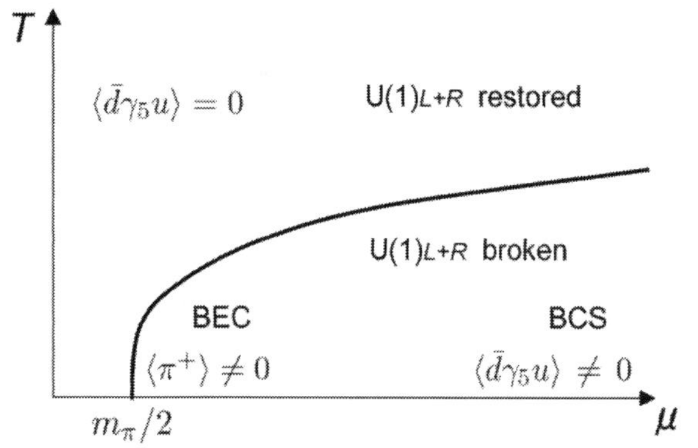

Figure 1. Phase diagram of phase-quenched QCD with μ or QCD with isospin chemical potential $\mu_I = 2\mu$ [7] (figure taken from [8]). See the text in section 2.3 for further detail.

2. Theorem on the QCD critical point

2.1. Sketch of the proof

The main observation to arrive at the conclusion above is that pions, which are the (pseudo-)Nambu-Goldstone modes associated with chiral symmetry breaking, are the lightest among all the mesonic modes (even in the medium) as long as the above conditions (i) and (ii) are met [15].[1] We empirically know that this is true in the QCD vacuum. If this is also true in some region at finite T and/or μ_B, the flavor-singlet mesonic mode σ, which has the quantum number $\bar{\psi}\psi$ with the mass m_σ, satisfies the relation:

$$m_\sigma \geq m_\pi > 0, \tag{1}$$

where m_π is the pion mass and the latter inequality follows due to the explicit breaking of chiral symmetry in the presence of the quark mass.

On the other hand, the definition of the QCD critical point is that the static correlation length ξ defined by

$$\langle \bar{\psi}\psi(\boldsymbol{x})\bar{\psi}\psi(\boldsymbol{y})\rangle \sim e^{-|\boldsymbol{x}-\boldsymbol{y}|/\xi}, \tag{2}$$

is divergent, or $m_\sigma = 1/\xi$ is vanishing at the point. However, from the bound (1) for m_σ, this is apparently impossible, and thus, the QCD critical point cannot exist in this region. Below we shall show, based on the rigorous QCD inequalities [17], that pions are indeed the lightest mesonic modes under the conditions (i) and (ii) given in section 1. We shall work in the Euclidean and two-flavor QCD with the degenerate quark mass m. (For the generalization to QCD with any number of flavors N_f, see [15].)

[1] It is interesting to note that the maximal severity of the sign problem inside the critical line of the phase-quenched QCD is also related to the lightness of pions [16] (see also [11]).

2.2. Proof at $\mu_B = 0$

For clarity, we first illustrate the essence of the idea at $\mu_B = 0$, and will generalize it to the case for $\mu_B \neq 0$. Consider the Dirac operator $\mathcal{D} = D + m$, where the operator $D = \gamma_\mu(\partial_\mu + igA_\mu)$. satisfies the anti-Hermiticity and chiral symmetry, $D^\dagger = -D$ and $\gamma_5 D \gamma_5 = -D$. From these two properties, we can easily show

$$\gamma_5 \mathcal{D} \gamma_5 = \mathcal{D}^\dagger, \tag{3}$$

and the positivity, $\det \mathcal{D} \geq 0$.

Let us now consider a correlation function for a generic flavor nonsinglet fermion bilinear $M_\Gamma = \bar{\psi}\Gamma\psi$,

$$C_\Gamma(x,y) \equiv \langle M_\Gamma(x) M_\Gamma^\dagger(y) \rangle_{\psi,A} = -\langle \mathrm{tr}[S_A(x,y)\Gamma S_A(y,x)\bar{\Gamma}] \rangle_A. \tag{4}$$

Here, $S_A(x,y) \equiv \langle x|\mathcal{D}^{-1}|y\rangle$ denotes a propagator from the point y to x in a background gauge field A, the symbols $\langle \,\cdot\, \rangle_{\psi,A}$ and $\langle \,\cdot\, \rangle_A$ denote the average over ψ, A and the average over A, respectively, and $\bar{\Gamma} \equiv \gamma_0 \Gamma^\dagger \gamma_0$. From (3) and the positivity, we have

$$C_\Gamma = \langle \mathrm{tr}[S_A(x,y)\Gamma i\gamma_5 S_A^\dagger(x,y) i\gamma_5 \bar{\Gamma}] \rangle_A \leq \langle \mathrm{tr}[S_A(x,y) S_A^\dagger(x,y)] \rangle_A. \tag{5}$$

Here we used the Cauchy-Schwarz inequality, which is saturated when $\Gamma = i\gamma_5 \tau_A$ with τ_A being the traceless flavor generators. The asymptotic behavior of C_Γ at large distance $|\boldsymbol{x} - \boldsymbol{y}|$ in the static limit can be written as

$$C_\Gamma \sim e^{-m_\Gamma |\boldsymbol{x}-\boldsymbol{y}|}, \tag{6}$$

where m_Γ is the mass of the lowest mesonic state in the channel Γ. Then the inequality (5) leads to

$$m_\Gamma \geq m_\pi. \tag{7}$$

In this derivation of the QCD inequalities, we need to assume that M_Γ is not flavor singlet, otherwise the contribution of flavor-disconnected diagrams $\langle \mathrm{tr}[\Gamma S_A(x,x)]\, \mathrm{tr}[\bar{\Gamma} S_A(y,y)] \rangle_A$ must also be taken into account. Phenomenologically disconnected diagrams may be suppressed compared with connected ones, as is known as the Okubo-Zweig-Iizuka (OZI) rule. Theoretically it can be justified in the 't Hooft large-N_c limit [18], since the flavor-disconnected diagrams are shown to be subleading in $1/N_c$ compared with the connected ones. Here we simply assume that the contribution of the flavor-disconnected diagrams is negligible [condition (i)], where (7) is also applicable to the flavor-singlet channel, $\Gamma = \sigma$.

This argument is valid for any temperature T. Therefore, we conclude that the QCD critical point cannot exist on the T-axis (at $\mu_B = 0$) under the condition (i).

2.3. Proof at $\mu_B \neq 0$

Unfortunately the above argument cannot be naively generalized to $\mu_B \neq 0$, because the Dirac operator $\mathcal{D}(\mu) = D + \mu\gamma_0 + m$ does no longer satisfy (3) and the positivity of the measure is lost. This is indeed the origin of the notorious sign problem in QCD at $\mu_B \neq 0$.

On the other hand, we can generalize it to the phase-quenched QCD, or equivalently, QCD at finite isospin chemical potential $\mu_I = 2\mu$. In this case, the Dirac operator $\mathcal{D}(\mu_I) = D + \mu_I \gamma_0 \tau_3/2 + m$ satisfies the relation $\tau_1 \gamma_5 \mathcal{D} \gamma_5 \tau_1 = \mathcal{D}^\dagger$ and $\det \mathcal{D}(\mu_I) \geq 0$ [7]; or complex Dirac eigenvalues for two flavors are conjugate with each other, and $\det \mathcal{D}(\mu_I) = |\det \mathcal{D}(\mu)| \geq 0$. Then the following inequality follows at *any* μ_I [7]:

$$C_\Gamma \leq \langle \mathrm{tr}[S_A(x,y) S_A^\dagger(x,y)] \rangle_A, \tag{8}$$

which is saturated when $\Gamma = i\gamma_5\tau_{1,2}$. This leads to the inequality, $m_\Gamma \geq m_{\pi\pm}$. Under the condition (i) this is again applicable to $\Gamma = \sigma$, and repeating a similar argument to QCD at $\mu_B = 0$, the critical point would be prohibited at $\mu_I \neq 0$.

However, one should note that this is not necessarily applicable to any coordinate (T, μ_I). Indeed when the pion condensation phase appears (where $m_\pm = 0$) as we will see shortly, $m_\sigma = 0$ is still possible and the QCD critical point is not ruled out.[2]

Let us look at the phase diagram of the phase-quenched QCD (figure 1), first discussed in [7]. As one increases μ_I at $T = 0$, π_+ meson is first excited (instead of a baryon) so that it exhibits the Bose-Einstein condensation (BEC) for $\mu > m_\pi/2$. On the other hand, at large μ_I where fundamental degrees of freedom are quarks and gluons, the attractive interaction between u and $\bar{d}$ near the Fermi surface leads to the Bardeen-Cooper-Schrieffer (BCS) pairing $\langle \bar{d}\gamma_5 u\rangle$ [7]. Because of the same symmetry breaking patterns $\mathrm{U}(1)_{L+R} \to \mathbb{Z}_2$, it is plausible that there is no phase transition between the two regimes. This is the so-called BEC-BCS crossover phenomena. Below we simply call it the pion condensation phase. Combining the arguments above altogether, it follows that the critical point cannot appear in QCD at $\mu_I \neq 0$ outside the pion condensation phase under the condition (i).

Now we can generalize this result to QCD at $\mu_B \neq 0$, taking into account the following property which holds under the conditions (i) and (ii):

$$\langle \bar{\psi}\psi\rangle_{\mu_B = 3\mu} = \langle \bar{\psi}\psi\rangle_{\mu_I = 2\mu}, \tag{9}$$

where the same (T, μ) is taken in both hand sides. This property can be understood as follows [19]: the contribution of u and d to the chiral condensate are decoupled from each other under the condition (i), and there is a charge conjugation symmetry which flips the sign of the chemical potential for u and d independently. As (μ_u, μ_d)=$(\mu, \pm\mu)$ correspond to the baryon and isospin chemical potentials respectively, the relation (9) follows. However, this is not true inside the pion condensation phase of the phase-quenched QCD where u and d are coupled in the ground state, and hence, the condition (ii) is necessary [8, 20]. Therefore the locations and the orders of the chiral transition must be the same between QCD at $\mu_B \neq 0$ and phase-quenched QCD outside the pion condensation phase of the latter. The exactness of the phase quenching can be actually shown for *any* observable to $O(N_f/N_c)$ [8, 20] using the large-N_c orbifold equivalence [21] between QCD at μ_B and QCD at μ_I (see also [22] for the orbifold equivalence between QCD and $\mathrm{SO}(2N_c)$ gauge theory at $\mu_B \neq 0$).

Since the QCD critical point is already shown to be excluded outside the pion condensation phase of the phase-quenched QCD, the same must be true for QCD at $\mu_B \neq 0$. This is our main result [15]. This also provides the underlying reason why the critical points are always observed in this region with maximal severity of the sign problem in the model calculations [12, 13, 14].

3. Discussion

So far we have assumed the condition (i), but how is this reasonable in real QCD? To understand it, first note that as far as we assume the condition (i), the chiral transition cannot be first order outside the pion condensation phase from our theorem. This is because, if it was first order, we could weaken the phase transition by gradually increase the quark mass m and we would eventually arrive at the critical mass $m = m_c$ where the chiral transition becomes second order; this is prohibited from our theorem. It means that the only thing that can make the chiral transition stronger to be first order is the effect of flavor-disconnected diagrams, including the effect of the axial anomaly [23]. So whether the QCD critical point can appear outside the pion

[2] The statement in [15] that the critical point cannot appear in QCD at any μ_I in the large-N_c limit was not precise. However, the conclusion that the QCD critical point is ruled out at finite μ_B under the conditions (i) and (ii) remains unchanged.

condensation phase is determined by the competition between the contribution of disconnected diagrams and the quark mass. We know from the lattice QCD simulation [24] that the latter overwhelms the former at $\mu_B = 0$ to make the chiral transition crossover at finite T, and the condition (i) is satisfied in this sense. Therefore, unless there is enhancement of the latter such that it overcomes the former at $\mu_B \neq 0$ for some reason, the critical point would not be accessible by the present lattice QCD techniques.

In addition, nuclear matter is shown to lie inside this pion condensation phase independently of the quark mass, at least $T = 0$ [25]. Most likely these physics at finite baryon density appear due to the subtle cancellation of the complex phase. It is certainly an important challenge for the theory to find out other ways to sample appropriate configurations in this region, which could enable us to probe the interesting physics in dense QCD including the color superconductivity. (For such recent attempts, see, e.g., [26, 27, 16, 28].)

Acknowledgments

Y.H. was supported by JSPS KAKENHI Grant Numbers 23340067 and 24740184. N.Y. is supported by JSPS Research Fellowships for Young Scientists.

References

[1] Stephanov M A 2004 *Prog. Theor. Phys. Suppl.* **153** 139 (*Preprint* hep-ph/0402115)
[2] Asakawa M and Yazaki K 1989 *Nucl. Phys. A* **504** 668; Barducci A, Casalbuoni R, Curtis S De, Gatto R and Pettini G 1989 *Phys. Lett. B* **231** 463
[3] Kitazawa M, Koide T, Kunihiro T and Nemoto Y 2002 *Prog. Theor. Phys.* **108** 929 (*Preprint* hep-ph/0207255)
[4] Hatsuda T, Tachibana M, Yamamoto N and Baym G 2006 *Phys. Rev. Lett.* **97** 122001 (*Preprint* hep-ph/0605018); 2007 *Phys. Rev. D* **76** 074001 (*Preprint* 0704.2654); Abuki H, Baym G, Hatsuda T and Yamamoto N 2010 *Phys. Rev. D* **81** 125010 (*Preprint* 1003.0408)
[5] Fodor Z and Katz S D 2002 *JHEP* **0203** 014 (*Preprint* hep-lat/0106002); 2004 *JHEP* **0404** 050 (*Preprint* hep-lat/0402006)
[6] Splittorff K 2006 *PoS LAT* **2006** 023 (*Preprint* hep-lat/0610072)
[7] Son D T and Stephanov M A 2001 *Phys. Rev. Lett.* **86** 592 (*Preprint* hep-ph/0005225)
[8] Hanada M and Yamamoto N 2012 *JHEP* **1202** 138 (*Preprint* 1103.5480)
[9] Ejiri S 2006 *Phys. Rev. D* **73** 054502 (*Preprint* hep-lat/0506023)
[10] Golterman M, Shamir Y and Svetitsky B 2006 *Phys. Rev. D* **74** 071501 (*Preprint* hep-lat/0602026)
[11] Splittorff K and Verbaarschot J J M 2007 *Phys. Rev. Lett.* **98** 031601 (*Preprint* hep-lat/0609076)
[12] Han J and Stephanov M A 2008 *Phys. Rev. D* **78** 054507 (*Preprint* 0805.1939)
[13] Andersen J O, Kyllingstad L T and Splittorff K 2010 *JHEP* **1001** 055 (*Preprint* 0909.2771)
[14] Sakai Y, Sasaki T, Kouno H and Yahiro M 2010 *Phys. Rev. D* **82** 096007 (*Preprint* 1005.0993)
[15] Hidaka Y and Yamamoto N 2012 *Phys. Rev. Lett.* **108** 121601 (*Preprint* 1110.3044)
[16] Grabowska D, Kaplan D B and Nicholson A N *Preprint* 1208.5760
[17] Weingarten D 1983 *Phys. Rev. Lett.* **51** 1830; Witten E 1983 *Phys. Rev. Lett.* **51** 2351; Nussinov S 1984 *Phys. Rev. Lett.* **52** 966; Espriu D, Gross M and Wheater J F 1984 *Phys. Lett.* **B146** 67
[18] 't Hooft G 1974 *Nucl. Phys.* **B72** 461; Witten E 1979 *Nucl. Phys.* **B160** 57
[19] Cohen T D 2004 *Phys. Rev. D* **70** 116009; (*Preprint* hep-ph/0410156); Toublan D 2005 *Phys. Lett. B* **621** 145 (*Preprint* hep-th/0501069)
[20] Hanada M, Matsuo Y and Yamamoto N 2012 *Phys. Rev. D* **86** 074510 (*Preprint* 1205.1030)
[21] Bershadsky M and Johansen A 1998 *Nucl. Phys.* **B536** 141 (*Preprint* hep-th/9803249); Schmaltz M 1999 *Phys. Rev. D* **59** 105018 (*Preprint* hep-th/9805218); Strassler M J *Preprint* hep-th/0104032; Kovtun P, Ünsal M and Yaffe L G 2005 *JHEP* **0507** 008 (*Preprint* hep-th/0411177)
[22] Cherman A, Hanada M and Robles-Llana D 2011 *Phys. Rev. Lett.* **106** 091603 (*Preprint* 1009.1623); Cherman A and Tiburzi B C 2011 *JHEP* **1106** 034 (*Preprint* 1103.1639)
[23] Pisarski R D and Wilczek F 1984 *Phys. Rev. D* **29** 338
[24] Aoki Y, Endrodi G, Fodor Z, Katz S D and Szabo K K 2006 *Nature* **443** 675 (*Preprint* hep-lat/0611014)
[25] Cohen T D 2003 *Phys. Rev. Lett.* **91** 032002 (*Preprint* hep-ph/0304024)
[26] Aarts G 2009 *Phys. Rev. Lett.* **102** 131601 (*Preprint* 0810.2089)
[27] Cristoforetti M *et al.* [AuroraScience Collaboration] 2012 *Phys. Rev. D* **86** 074506 (*Preprint* 1205.3996)
[28] Ohnishi A, Ichihara T and Nakano T Z *Preprint* 1211.2282

QCD Thermodynamics with Domain-Wall Fermions

Prasad Hegde

Dept. of High-Energy Physics, National Taiwan University, Taipei 10617, Taiwan.

E-mail: phegde@ntu.edu.tw

Abstract. The study of the QCD chiral phase transition on the lattice has been hampered in the past by the lack of a chiral-invariant fermion formulation. Only recently has it become possible to perform fully dynamical simulations with chiral formulations such as domain-wall and overlap fermions. Here we present results for various chiral observables at finite temperature that were obtained using domain-wall fermions. We also present results showing that the axial anomaly is still present even above the chiral phase transition. Lastly, we show our results for the Dirac spectrum at finite temperature and speculate on the mechanism of $U(1)_A$ breaking beyond the chiral phase transition.

1. Chiral Symmetry and the QCD Phase Diagram

The restoration of a spontaneously broken continuous global symmetry is the prototype of a phase transition in several quantum field theories. In QCD for e.g. while the Lagrangian for $N_f \geqslant 2$ massless quarks is formally invariant under separate $SU(N_f)$ rotations of the right- and left-handed quarks, interactions break this group down to its vector subgroup i.e. the familiar isospin symmetry. According to the conventional picture of Pisarski and Wilczek [1], the restoration of this chiral symmetry at high temperatures is expected to be second-order for two massless flavors. For massive quarks the transition is widely believed to be a crossover.

On the other hand, the singlet part of the axial group viz. $U(1)_A$, is broken by the famous axial anomaly [2, 3] at the perturbative level itself so that there is no question of its complete restoration at any finite temperature [4, 5, 6, 7]. However the configurations that contribute to the axial anomaly are Boltzmann-suppressed at high temperatures due to the screening of the coupling constant [8]. It is conceivable that this suppression is nearly complete by some temperature that is not too high. We may then speak of an *effective* restoration of the axial symmetry. If this temperature is close to the chiral phase transition temperature T_c, then the chiral phase transition can even be of first order [1].

1.1. Chiral Symmetry on the Lattice

Since chiral symmetry breaking is a non-perturbative effect, the only way of studying it from first principles is through lattice QCD. Unfortunately the study of chiral symmetry on the lattice is not straightforward owing to the well-known fermion doubling problem: Any discretization D of the Dirac operator which preserves exact chiral symmetry i.e. $\{\gamma_5, D\} = 0$, will necessarily describe more than one fermion species in the continuum[9, 10]. Traditionally, this problem has been circumvented through the use of either staggered or Wilson fermions. Wilson fermions break chiral symmetry explicitly by adding an irrelevant second derivative term, the Wilson term, which does not commute with γ_5. Although chiral symmetry is regained in the continuum, its loss at finite lattice spacing leads to problems such as unwanted additive renormalizations and operator mixing as well as exceptional configurations. Staggered fermions on the other hand preserve a $U(1)$ subgroup of the full chiral symmetry group. Unfortunately one does not get rid of all the doublers; four of these remain even in the continuum limit and introduce additional complications at finite lattice spacing [11, 12, 13]. From the point of view of the phase transition, the biggest problem in this case is that the chiral and continuum limits do not commmute: The continuum limit must be taken separately for each pion mass before a chiral extrapolation can be performed. Lastly, neither action satisfies an index theorem and therefore the famous relation between topology and the zero modes of the Dirac operator is regained only in the continuum.

All these problems may be avoided by the use of chiral fermions such as the overlap [14, 15], domain-wall [16, 17] or fixed-point [18]. For example, chiral fermions satisfy an exact, if altered, definition of chiral symmetry at finite lattice spacing a, namely [19]

$$\psi_x \to \left[1 + i\theta\gamma_5\left(1 - \frac{a}{2}D_{x\,x'}\right)\right]\psi_{x'}, \qquad \overline{\psi}_x \to \overline{\psi}_{x'}\left[1 + i\theta\left(1 - \frac{a}{2}D_{x'\,x}\right)\gamma_5\right]. \tag{1}$$

This works provided the Dirac matrix D satisfies the famous Ginsparg-Wilson relation [20]

$$\{\gamma_5, D\} = aD\gamma_5 D. \tag{2}$$

Chiral fermions avoid the doubling problem since $\{\gamma_5, D\} \neq 0$. They also satisfy the index theorem even at finite a. The downside is that all three actions are much more expensive to simulate than either staggered or Wilson fermions. Of the three, the overlap and fixed-point operators satisfy the Ginsparg-Wilson relation, Eq (2) exactly. By contrast, domain-wall fermions live in five dimensions and satisfy the Ginsparg-Wilson relation only when the fifth dimension becomes infinite, $L_s = \infty$ [21]. For $L_s < \infty$ the quark mass receives an additive correction $m_{\rm res}$ even when its input value is set to zero.

Both domain-wall and overlap fermions encounter difficulties during evolution due to dislocations. Dislocations occur whenever the system tunnels from one topological sector to another. In the case of overlap fermions, they significantly increase the cost of the HMC algorithm. As a result, overlap simulations are generally performed at fixed topology. In the case of domain-wall fermions, dislocations result in a large value of $m_{\rm res}$ [22, 23]. In this work we used the Dislocation Suppressing Determinant Ratio (DSDR) action which suppresses dislocations, while still allowing for topology change, by augmenting the gauge action with a ratio of Wilson determinants [24, 25, 26]. Our lattice sizes were $16^3 \times 8$, with $L_s = 32$ or 48. Our pion mass was held fixed at approximately 200 MeV.

2. Symmetry Restoration: $SU(2)_L \times SU(2)_R$ Versus $U(1)_A$

A convenient way of studying symmetry restoration on the lattice is by looking at the appropriate correlators of Dirac bilinears. For example, the vector (ρ) and axial vector (a_1) correlators become equal when chiral symmetry is restored. Similarly, the scalar (δ) and pseudoscalar (π) correlators are expected to become degenerate once $U(1)_A$ is restored. Fig. 1 presents our results for these correlators: We see that while the ρ and a_1 become degenerate around $T_c \approx 160$ MeV, the scalar (δ) and pseudoscalar (π) correlators remain distinct even at $T = 195$ MeV. This suggests that $U(1)_A$ stays broken at the chiral phase transition, although it must be remembered that we are still quite far from the chiral limit.

This difference may be quantified in terms of susceptibilities, obtained by integrating the appropriate correlators. Fig. 2 shows how the correlators in the scalar-pseudoscalar sector are related by chiral and axial symmetries. Chiral symmetry relates the flavored mesons π and δ to the flavor-singlets σ and η'. While the flavor-singlet correlators are hard to measure, their susceptibilities are somewhat easier since they are related to the fluctuations of $\langle\overline{\psi}\psi\rangle$ and

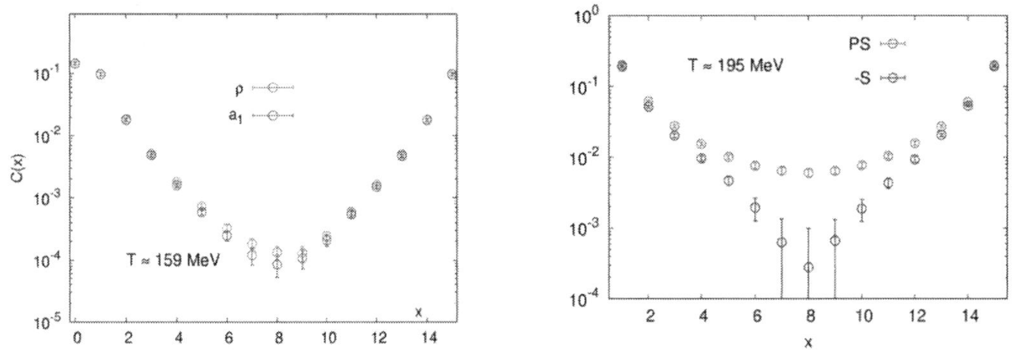

Figure 1. (Left) The vector (ρ) and axial vector (a_1) correlators for $T = 159$ MeV. (Right) The scalar (δ) and pseudoscalar (π) correlators for $T = 195$ MeV.

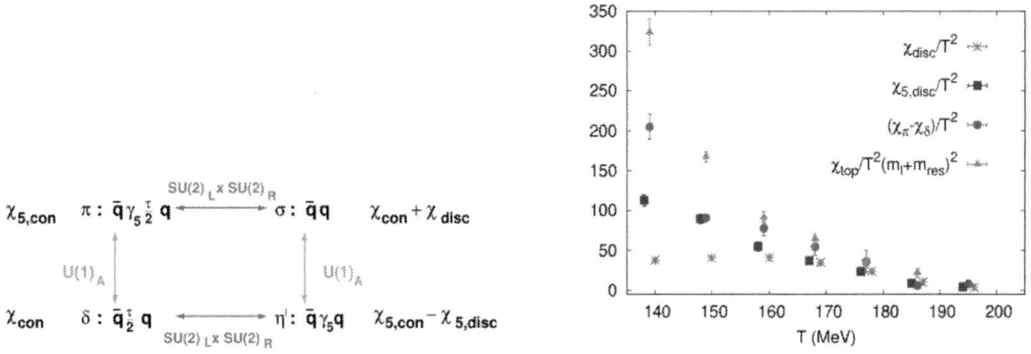

Figure 2. (Left) The expected relations between the different correlators whenever chiral/axial symmetry is restored. (Right) Our results for these susceptibilities as a function the temperatures. The relation to the topological susceptibility is expected on the basis of the Ward identity.

$\langle\overline{\psi}\gamma_5\psi\rangle$. From Fig. 2, one can derive relations among the susceptibilities that must hold once either symmetry is restored; the extent to which these are satisfied by our data is shown in the plot on the right. Although strictly speaking these relations are valid only in the chiral limit, they hold to a good degree for our current simulations.

3. Eigenvalue Distribution

In the domain-wall formalism, the chiral, four-dimensional fermion operator arises in the low-energy limit of the five-dimensional operator. As a result, the low-lying spectrum of both operators should be the same. Unfortunately lattice artifacts complicate this picture at finite lattice spacing. Furthermore although we may set the bare quark mass to zero, we cannot similarly eliminate the residual mass m_{res} from the four-dimensional theory. We could subtract m_{res} from each eigenvalue for each configuration, but it must be remembered that m_{res} is defined only as an ensemble average rather than on a configuration-by-configuration basis. However the spectral density $\rho(\lambda)$ is also an ensemble-averaged quantity and subtracting m_{res} from each eigenvalue is equivalent to simply shifting the origin of $\rho(\lambda)$.

We display our results for $\rho(\lambda)$ in Fig. 3. In each of these figures, $\lambda = \sqrt{\Lambda^2 - (m_l + m_{\mathrm{res}})^2}$, where Λ are the eigenvalues obtained directly from our lattice calculations. Lattice artifacts appear as eigenvalues $\Lambda < m_l + m_{\mathrm{res}}$; we have binned these eigenvalues in the unphysical region

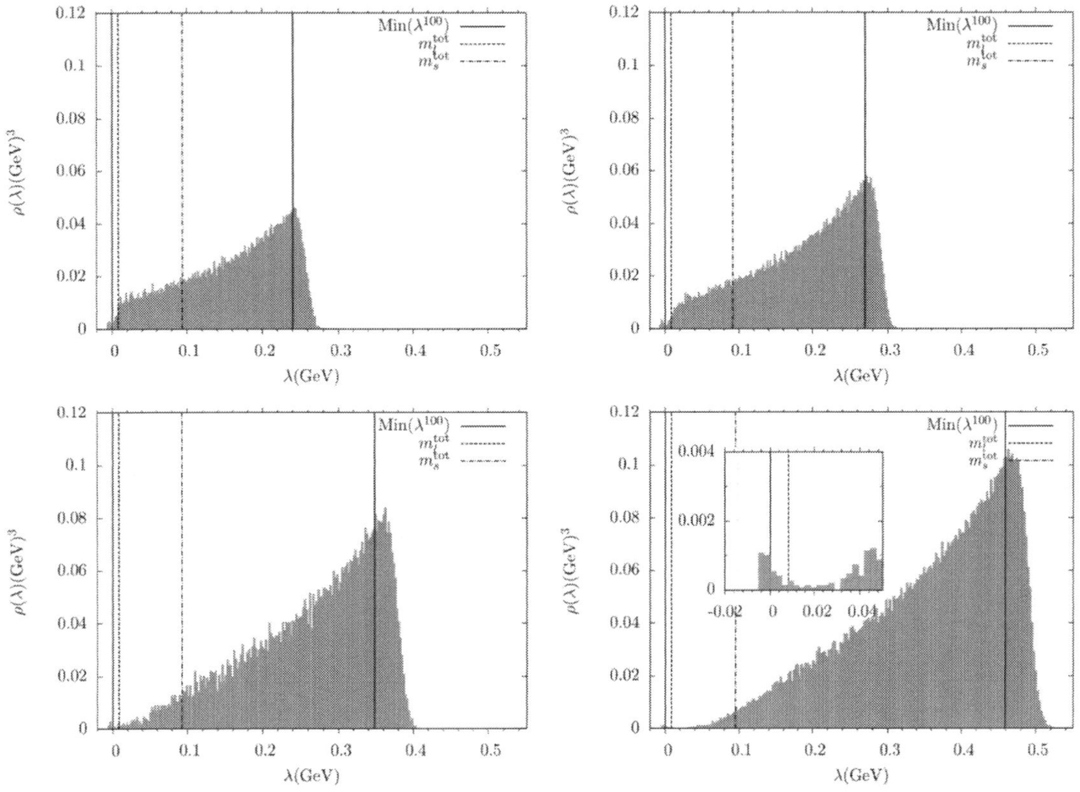

Figure 3. (From top left to bottom right) Renormalized Dirac spectrum for $T = 149$, 159, 177 and 195 MeV.

$\lambda < 0$; we refer the reader to our paper [27] for details.

We see that the eigenvalue density at the origin shrinks rapidly as the temperature is increased, as one would expect on the basis of the Banks-Casher relation [28]. Nevertheless, a small peak at the origin persists even at $T = 195$ MeV i.e. well above the chiral phase transition temperature [29]. This could be consistent with a spectral density of the type $\rho(\lambda) = m^2\delta(\lambda)$. Such a form is expected within the Dilute Instanton Gas framework, but it is not the only possibility. There are atleast two other possibilities, namely $\rho(\lambda) \propto \lambda$ and $\rho(\lambda) \propto m$. These yield qualitatively different predictions for thermodynamic observables (Table 1). Studies on larger volumes and at smaller quark masses should thus be useful in distinguishing between these possibilities. In this way, it is hoped that one may be eventually able to shed light on the mechanism of $U(1)_A$ breaking at high temperatures.

Acknowledgments
This work was carried out by the HOTQCD collaboration and I would like to thank its members for their help and guidance. The simulations were performed on the QCDOC computers of the Riken/BNL Research Center, the Bluegene/L and Bluegene/P computers at Lawrence Livermore National Laboratory and the New York Center for Computational Sciences (NYCCS) at Brookhaven National Laboratory. This work was supported in part by contract DE-AC02-98CH10886 with the U.S. Department of Energy.

Term	$\langle\bar{\psi}\psi\rangle$	χ_π	χ_δ	$\chi_\pi - \chi_\delta$	χ^{disc}
λ	$-2m\ln(m)$	$-2\ln(m)$	$-2\ln(m)$	2	0
m	πm	π	0	π	π
$m^2\delta(\lambda)$	m	1	-1	2	2

Table 1. Scaling forms for the chiral condensate and the various susceptibilities depending on the possible form of $\rho(\lambda, m)$ near the origin *i.e.* for $\lambda \to 0$, $m \to 0$. The leading m-dependence comes from dimensional analysis, however the additional logarithmic dependence can result in divergences in the chiral limit, such as in χ_π for $\rho(\lambda) \propto \lambda$.

[1] R. D. Pisarski, F. Wilczek, Phys. Rev. **D29**, 338-341 (1984).
[2] S. L. Adler, Phys. Rev. **177**, 2426-2438 (1969).
[3] J. S. Bell, R. Jackiw, Nuovo Cim. **A60**, 47-61 (1969).
[4] E. V. Shuryak, Comments Nucl. Part. Phys. **21**, 235 (1994) [hep-ph/9310253].
[5] T. D. Cohen, Phys. Rev. D **54**, 1867 (1996) [hep-ph/9601216].
[6] S. H. Lee and T. Hatsuda, Phys. Rev. D **54**, 1871 (1996) [hep-ph/9601373].
[7] N. J. Evans, S. D. H. Hsu and M. Schwetz, Phys. Lett. B **375**, 262 (1996) [hep-ph/9601361].
[8] D. J. Gross, R. D. Pisarski, L. G. Yaffe, Rev. Mod. Phys. **53**, 43 (1981).
[9] H. B. Nielsen and M. Ninomiya, Phys. Lett. B **105**, 219 (1981).
[10] L. H. Karsten and J. Smit, Nucl. Phys. B **183**, 103 (1981).
[11] G. P. Lepage, Phys. Rev. D **59**, 074502 (1999) [hep-lat/9809157].
[12] S. R. Sharpe, PoS LAT **2006**, 022 (2006) [hep-lat/0610094].
[13] M. Creutz, PoS LAT **2007**, 007 (2007) [arXiv:0708.1295 [hep-lat]].
[14] H. Neuberger, Phys. Lett. B **417**, 141 (1998) [hep-lat/9707022].
[15] R. Narayanan and H. Neuberger, Phys. Rev. Lett. **71**, 3251 (1993) [hep-lat/9308011].
[16] D. B. Kaplan, Phys. Lett. B **288**, 342 (1992) [hep-lat/9206013].
[17] Y. Shamir, Nucl. Phys. B **406**, 90 (1993) [hep-lat/9303005].
[18] P. Hasenfratz and F. Niedermayer, Nucl. Phys. B **414**, 785 (1994) [hep-lat/9308004].
[19] M. Luscher, Phys. Lett. B **428**, 342 (1998) [hep-lat/9802011].
[20] P. H. Ginsparg and K. G. Wilson, Phys. Rev. D **25**, 2649 (1982).
[21] H. Neuberger, Phys. Rev. D **57**, 5417 (1998) [hep-lat/9710089].
[22] Y. Aoki, T. Blum, N. Christ, C. Cristian, C. Dawson, T. Izubuchi, G. Liu, R. Mawhinney *et al.*, Phys. Rev. **D69**, 074504 (2004). [hep-lat/0211023].
[23] D. J. Antonio *et al.* [RBC and UKQCD Collaboration], Phys. Rev. D **75**, 114501 (2007) [hep-lat/0612005].
[24] P. M. Vranas, Phys. Rev. **D74**, 034512 (2006). [hep-lat/0606014].
[25] H. Fukaya, S. Hashimoto, K. I. Ishikawa, T. Kaneko, H. Matsufuru, T. Onogi and N. Yamada [JLQCD Collaboration], Phys. Rev. D **74**, 094505 (2006) [arXiv:hep-lat/0607020].
[26] D. Renfrew, T. Blum, N. Christ, R. Mawhinney, P. Vranas, PoS **LATTICE2008**, 048 (2008). [arXiv:0902.2587 [hep-lat]].
[27] A. Bazavov *et al.* [HotQCD Collaboration], arXiv:1205.3535 [hep-lat].
[28] T. Banks and A. Casher, Nucl. Phys. B **169**, 103 (1980).
[29] R. G. Edwards, U. M. Heller and R. Narayanan, Nucl. Phys. B **535**, 403 (1998) [hep-lat/9802016].

Exact Chiral Invariance at Finite Density on Lattice

Rajiv V. Gavai
Department of Theoretical Physics, Tata Institute of Fundamental Research, Homi Bhabha Road, Mumbai 400005, India
E-mail: gavai@tifr.res.in

Sayantan Sharma
Fakultät für Physik, Universität Bielefeld, D-33615 Bielefeld, Germany
E-mail: sayantan@physik.uni-bielefeld.de

Abstract. A new lattice action is proposed for the overlap Dirac matrix with nonzero chemical potential. It is shown to preserve the full chiral invariance for all values of lattice spacing exactly. It is further demonstrated to arise in the domain wall formalism by coupling the chemical potential only to the physically relevant wall modes.

1. Introduction

A fundamental aspect of the phase diagram of Quantum Chromo Dynamics (QCD) in the T-μ_B plane is the critical point, where T and μ_B denote the the temperature and baryonic chemical potential respectively. Theoretical [1, 2, 3] as well as experimental searches for locating it are currently going on [4]. Its discovery would be exciting in many ways. Apart from becoming a new milestone in our understanding of the nature of the strongly interacting matter, it would also be unique compared to the other known phase diagrams in the way theory and experiment compliment each other in locating the critical point in it. It is desirable to predict the existence of the QCD critical point (or rule it out) starting from first principles, employing only the QCD Hamiltonian. Lattice QCD is the best suited tool for such an exploratory exercise in view of its success in other non-perturbative aspects.

The prime reason for expecting the critical point are the chiral symmetries QCD has due to the quark mass spectra. Our world has two light quarks, having masses much smaller than Λ_{QCD}, and a moderately heavy quark, making it a '2+1' flavour QCD. All attempts to explore nonzero μ_B on the lattice use the staggered quarks. They have a $U(1)_V \times U(1)_A$ chiral symmetry on the lattice, and a corresponding order parameter, the chiral condensate. Staggered quarks, however, break the flavour and spin symmetry on the lattice, and have no flavour singlet axial anomaly. In the continuum limit of $a \to 0$, they have 4 quark 'flavours' of the same mass. The '2+1' flavour QCD is approximated by taking square root and fourth root of the staggered quark determinant. Whether this can restore the correct flavour symmetry as well as the chiral anomaly in the continuum limit has been a controversial issue. Moreover, it has been argued the 'rooting problem' becomes even worse [5] at nonzero μ_B. Since the QCD critical point is expected [6] to exist if only if one has two light flavours, and the flavour singlet anomaly is mildly temperature dependent [7], as in our world, it appears desirable to improve upon them. The overlap Dirac fermions [8], or the closely related domain wall fermions [9], offer such a possibility to improve. Indeed, the overlap quarks have all the symmetries of the continuum QCD and also have an index theorem [10] as well, raising the hope that even the anomaly effects could be well treated. Unfortunately, adding the chemical potential turns out to be nontrivial for them. Bloch and Wettig [11] made a proposal to do so but it violates [12] the exact chiral invariance on the lattice as does the simple addition [13] of a baryon number term. In this talk I report on our alternative action [14] which does have exact and continuum-like chiral invariance on lattice for any value of the lattice spacing and any chemical potential. Its chiral order parameter permits, in principle, the task of mapping out the T-μ_B phase diagram, assuming that the algorithmic developments can handle the fermion sign problem well.

2. Exact Chiral Invariance for $\mu \neq 0$

We begin by noting that the massless continuum QCD action can be written in a form where the chiral symmetry is manifest in terms of the fields appearing in the action and ask [14] if this can be imitated on lattice. The continuum QCD action at finite density in an explicit chiral invariant form is

$$S_{QCD} = \int d^3x \, d\tau [\bar{\psi}_L(\not{D} + \mu\gamma^4)\psi_L + \bar{\psi}_R(\not{D} + \mu\gamma^4)\psi_R - F^{\mu\nu}F_{\mu\nu}/4] \ , \tag{1}$$

where $\psi_L = (1-\gamma_5)\psi/2$ and $\psi_R = (1+\gamma_5)\psi/2$ with $\bar{\psi}_L = \bar{\psi}(1+\gamma_5)/2$ and $\bar{\psi}_R = \bar{\psi}(1-\gamma_5)/2$. The last term in the action is for gluons and will play no role in our discussion below. We assume that some usual convenient lattice form for it has been chosen and focus on the lattice fermionic action only. We propose that addition of chemical potential on the lattice be done in such an explicit chiral symmetry preserving manner as well, opting therefore for the overlap quarks on the lattice.

The overlap quarks have all the symmetries of the continuum QCD but also have a nonlocal action:

$$S_F = \sum_{n,m} \bar{\psi}_n \, aD_{ov,nm} \, \psi_m \ , \tag{2}$$

where the sum over n and m runs over all the space-time lattice sites, a is the lattice spacing, and the overlap Dirac matrix D_{ov} is defined by $aD_{ov} = 1 + \gamma_5 sgn(\gamma_5 D_W)$. sgn denotes the sign function. D_W is the standard Wilson-Dirac matrix on the lattice but with a negative mass term $M \in (0, 2)$:

$$aD_W(x,y) = (4-M)\delta_{x,y} - \sum_{i=1}^{4}[U_i^\dagger(x-\hat{i})\delta_{x-\hat{i},y}\frac{1+\gamma_i}{2} + \frac{1-\gamma_i}{2}U_i(x)\delta_{x+\hat{i},y}] \ . \tag{3}$$

The overlap Dirac matrix satisfies the Ginsparg-Wilson relation [15], $\{\gamma_5, D\} = aD\gamma_5 D$ and has exact chiral symmetry on lattice. The corresponding infinitesimal chiral transformations [16] are

$$\delta\psi = i\alpha\gamma_5(1 - aD_{ov})\psi \quad \text{and} \quad \delta\bar{\psi} = i\alpha\bar{\psi}\gamma_5 \ . \tag{4}$$

The generators of the transformation in Eq.(4) satisfy $\gamma_5^2 = 1$ and $\hat{\gamma}_5^2 \equiv [\gamma_5(1 - aD_{ov})]^2 = 1$. One therefore defines the left-right projections for quark fields as $\psi_L = (1 - \hat{\gamma}_5)\psi/2$ and $\psi_R = (1 + \hat{\gamma}_5)\psi/2$, leaving the antiquark field decomposition as in the continuum. Such a decomposition is commonly done for writing chiral gauge theories [17] on the lattice and is possible since ψ and $\bar{\psi}$ are independent fields in the Euclidean field theory. In analogy with the Eq.(1) of continuum QCD, the action for the overlap quarks in presence of nonzero chemical potential may now be written down as

$$S = \sum_{n,m} [\bar{\psi}_{n,L}(aD_{ov} + a\mu\gamma^4)_{nm}\psi_{m,L} + \bar{\psi}_{n,R}(aD_{ov} + a\mu\gamma^4)_{nm}\psi_{m,R}] \tag{5}$$

$$= \sum_{n,m} \bar{\psi}_n[(1 - a\mu\gamma^4/2)aD_{ov} + a\mu\gamma^4]_{nm}\psi_m . \tag{6}$$

It is easy to verify that this action is invariant under the chiral transformation Eq.(4) for all values of $a\mu$ and a and it reproduces the continuum action in the limit of $a \to 0$ with $\mu \to \mu/M$ scaling. In order to obtain the order parameter for checking if the symmetry is spontaneously broken, one usually adds a linear symmetry breaking term. Adding a quark mass term, $am(\bar{\psi}_R\psi_L + \bar{\psi}_L\psi_R)$, one obtains the order parameter valid for all T and μ on the lattice by taking a derivative of the log of the partition function with respect to am as,

$$\langle\bar{\psi}\psi\rangle = \lim_{am\to 0}\lim_{V\to\infty} \langle \mathrm{Tr}\, \frac{1}{aK_{ov} + am + a\mu\gamma^4}\rangle . \tag{7}$$

where $K_{ov} = D_{ov}(1 - aD_{ov}/2)^{-1}$, such that $\{\gamma_5, K_{ov}\} = 0$. Although the discussion above is for a single flavour of quark, i.e, $U(1)_L \times U(1)_R$ symmetry, its generalization to N_f flavours is straightforward. Indeed, since it relies only on the spin-structure, the flavour index as well as the corresponding generator matrices just carry through.

3. Physical Picture

Domain wall fermions are known to be akin to the overlap fermions, and may provide a physical picture for introduction of μ in the way above. Moreover, full QCD simulations with them are easier, so it is also practically useful to check how the above exact chiral invariance can be preserved. The domain wall action of [18] for $\mu = 0$, is

$$\begin{aligned} S = \sum_{x,x'}\sum_{s,s'=1}^{N_5} a^4\bar{\psi}(x,s)\,\big[am\;\delta_{x,x'}\left(\delta_{s,1}\delta_{s',N_5}P_+ + P_-\delta_{s,N_5}\delta_{s',1}\right) \\ -\delta_{x,x'}\left(P_-\delta_{s',s+1} + P_+\delta_{s',s-1}\right) + \left(a_5D_W(x,x') + \delta_{x,x'}\right)\delta_{s,s'}\big]\,\psi(x',s') , \end{aligned} \tag{8}$$

where $P_\pm = (1 \pm \gamma_5)/2$ and N_5, a_5 are the number of sites and the lattice spacing in the fifth direction respectively. The physically relevant 4D fermion field is identified with the fermion fields at the boundaries of the fifth dimension as,

$$\Psi(x) = P_-\psi(x,1) + P_+\psi(x,N_5) \quad , \quad \bar{\Psi}(x) = \bar{\psi}(x,1)P_+ + \bar{\psi}(x,N_5)P_- . \tag{9}$$

These are the lowest energy field configurations of the five dimensional theory and these correspond to the four dimensional chiral fermions in the limit $m \to 0$ and $N_5 \to \infty$. In order to get a four dimensional fermion determinant representing the dynamics of the physical quark fields, one has to integrate out all the fermion degrees of freedom from the five dimensional action and remove the contribution of the higher energy configurations. The latter are the bulk modes since these are delocalized in the fifth dimension even when $N_5 \to \infty$. Integrating out the fermion degrees of freedom is conveniently done by recasting the domain wall action in terms of the fields η_i localized on four dimensional branes existing at each site i along the fifth dimension [19]. These are related to the 5D fermion fields in an asymmetric way. The neighbouring pairs of η are related by a transfer matrix, $T = (1 + a_5H_WP_+)^{-1}(1 - a_5H_WP_-)$ with $H_W = \gamma_5 D_W$. Integrating over the η fields successively one obtains a determinant which is a functional of this transfer matrix:

$$D^{(5)}(ma) = \det\left[P_- - maP_+ - T^{-N_5}\left(P_+ - maP_-\right)\right] . \tag{10}$$

It has the contribution of all the fermions in the five dimensional theory. Choosing $ma = 1$, one obtains contribution from all but the lowest energy configurations which is divided out to obtain the physical quark determinant. Taking first $a_5 \to 0$ limit and then $N_5 \to \infty$, the ratio of determinants turns out to be the determinant of the traditional overlap Dirac matrix.

To get correct physics at finite μ one needs to construct the number density corresponding to the physical degrees of freedom. We proposed [14] that the following term be added to the domain wall action in Eq. (8) for nonzero density:

$$\sum_{x,x'} \sum_{s,s'=1}^{N_5} a^4 \bar{\psi}(x,s) a\mu \, \gamma_4 \, \delta_{x,x'} \left(\delta_{s,1}\delta_{s',1} P_- + P_+ \delta_{s,N_5}\delta_{s',N_5}\right) \psi(x',s').$$

We then follow the same procedure outlined above for the $\mu = 0$ case and rewrite the full action in terms of the η-fields. The μ-dependent term then becomes as below:

$$\mu a \left[\bar{\eta}_1 (a_5 H_W P_- - 1)^{-1} \gamma_4 P_- \eta_1 - \bar{\eta}_{N_5} (a_5 H_W P_- - 1)^{-1} \gamma_4 P_+ \eta_1\right] \quad (11)$$

It is important to note that unlike the Bloch-Wettig case [20], the chemical potential in our proposal is coupled only to the physical fermions which are strongly localized on the boundaries of the fifth dimension, hence our transfer matrix continues to remain the same, i.e., is μ-independent. Moreover since we did not couple μ to the number density of the bulk modes, the pseudo-fermion action remains the same as in the $\mu = 0$ case. Integrating over the η fields one obtains the physical fermion determinant in a ratio form similar to that for $\mu = 0$ with the $\det D^{(5)}(ma)$ in the numerator generalised to,

$$\begin{aligned} \det D^{(5)}(ma, \mu a) &= \det \Big[P_- - maP_+ + a\mu \left(a_5 H_W P_- - 1\right)^{-1} \gamma_4 P_- \\ &\quad - T^{-N_5} \Big(P_+ - maP_- - a\mu \left(a_5 H_W P_+ + 1\right)^{-1} \gamma_4 P_+ \Big) \Big] . \end{aligned} \quad (12)$$

and the pseudo-fermion determinant in the denominator remaining unchanged. On taking the limits $a_5 \to 0$ and $N_5 \to \infty$, the ratio of determinant simplifies to $\det[D_{ov} + (1 - D_{ov}/2)(ma + a\mu\gamma^4)]$, where both the dimensional parameters μ and m have been scaled by a factor of $1/M$ due to the tree level wavefunction renormalization [21]. A little algebra shows that γ^4 can be commuted through in the determinant above to yield the same overlap matrix of Eq. (6) with exact chiral symmetry on the lattice.

The action in Eq.(6) leads to an overlap fermion determinant which is identical to that in the recent work [22] with fermionic sources in the overlap formalism of [23]. The main difference is, however, in the necessity of sources in [22] to define the chiral symmetry. Indeed, the chiral symmetry transformation there is local, defined as rotation of the sources, while our Eq.(4) is nonlocal, defined as the rotation of quark fields. The left-right symmetry is in-built in the formalism there whereas we needed to introduce the left-right projections in form of L- and R-fields to do so.

It is easy to verify that our $D_{ov}(a\mu)$, defined above, is not γ_5-hermitian. In general, therefore, it is not clear whether it may be diagonalizable. Noting that an M-scaling was essential in the continuum limit of our D_{ov} with $a\mu/M$ governing the density dependent term, it is easily seen that for any arbitrary value of μ the density term can be made small by choosing small enough a. Hence it suffices to study the chiral anomaly at $\mathcal{O}((a\mu/M)^2)$. The leading term in such an expansion in $a\mu/M$, is $D_{ov}(0)$ which is γ_5-hermitian. Its eigenvalues come in complex pairs (λ, λ^*), with the corresponding eigenvectors related by a γ_5-rotation. Using these properties, we showed that the chiral anomaly arises as usual, and is governed by the number of zero modes of $D_{ov}(0)$ even on the lattice. In particular, the contribution of the eigenvectors for the eigenvalue two, i.e, the 'doublers', to the chiral anomaly is identically zero while that of the pairs (λ, λ^*) cancels mutually. Using first order perturbation theory in $a\mu/M$, it is easy to show that the latter statement remains unchanged even for nonzero μ on a fine enough lattice for our proposal while the former is true for any $a\mu/M$. Since zero modes continue to be eigenvalues of γ_5, all one needs to do is to compute the trace of γ_5 in a new basis, rotated by a unitary transformation. After some algebra, one can show that the chiral anomaly remains unchanged on the lattice for arbitrary μ using our proposal to introduce a chemical potential for the overlap quarks, exactly as in the continuum.

4. Summary

In conclusion, we used the analogy with continuum QCD to demand a manifest chiral invariance in terms of L and R-fields on the lattice to obtain $aD_{ov} + a\mu\gamma^4(1 - aD_{ov}/2)$ as the exact chiral invariant form of the overlap Dirac matrix for any μ and/or lattice spacing. We wrote down the exact chiral order parameter on the lattice for chiral symmetry restoration in the entire $T - \mu_B$ plane. The chiral anomaly remains unaffected for our proposal for small enough lattice spacing a, just as in continuum. We provided a physically appealing interpretation of our proposal by explicitly writing down the corresponding domain wall fermion action at nonzero chemical potential and taking the limit of infinite fifth dimension.

[1] Z Fodor and S D Katz 2004 JHEP **0404** 050.
[2] R V Gavai and S Gupta 2008 Phys. Rev. D **78** 114503.
[3] C Schmidt 2010 Prog. Theor. Phys. Suppl. **186** 563.
[4] B Mohanty for STAR collaboration 2011 J. Phys. G **38** 124023.
[5] M Golterman, Y Shamir and B Svetitsky 2006 Phys. Rev. D **74** 071501.
[6] K Rajagopal and F Wilczek, Chapter 35 in the Festschrift in honor of B.L. Ioffe, "At the Frontier of Particle Physics / Handbook of QCD", M. Shifman, ed., (World Scientific).
[7] R D Pisarski and F Wilczek 1984 Phys. Rev. D **29** 338.
[8] H Neuberger 1998 Phys. Lett. B **417** 141.
[9] D B Kaplan 1992 Phys. Lett. B **288** 342.
[10] P Hasenfratz, V Laliena and F Neidermeyer 1998 Phys. Lett. B **427** 125.
[11] J Bloch and T Wettig 2006 Phys. Rev. Lett. **97** 012003.
[12] D Banerjee, R V Gavai and S Sharma 2008 Phys. Rev. D **78** 014506.
[13] R V Gavai and S Sharma 2010 Phys. Rev. D **81** 034501.
[14] R V Gavai and S Sharma 2012 Phys. Lett. **B 716** 446.
[15] P H Ginsparg and K G Wilson 1982 Phys. Rev. D **25** 2649.
[16] M Luscher 1998 Phys. Lett. B **428** 342.
[17] M Luscher 1999 Nucl. Phys. B **549** 295; F Niedermayer 1999 Nucl. Phys. B(PS) **73** 105.
[18] Y Shamir 1993 Nucl. Phys. B **406** 90.
[19] R G Edwards and U M Heller 2001 Phys. Rev. D **63** 094505.
[20] J Bloch and T Wettig 2007 Phys. Rev. D **76** 114511.
[21] R G Edwards, U M Heller and R Narayanan 1999 Phys. Rev. D **59** 094570.
[22] R Narayanan and S Sharma 2011 JHEP **1110** 151.
[23] R Narayanan and H Neuberger 1995 Nucl. Phys. B **443** 305.

Journal of Physics: Conference Series **432** (2013) 012020 doi:10.1088/1742-6596/432/1/012020

The Phase Diagram of Two Color QCD

Simon Hands[1], Seamus Cotter[2], Pietro Giudice[1] and Jon-Ivar Skullerud[2]

[1] Physics Department, College of Science, Swansea University, Singleton Park, Swansea SA2 8PP, UK

[2] Department of Mathematical Physics, National University of Ireland Maynooth, Maynooth, County Kildare, Ireland

E-mail: s.hands@swan.ac.uk

Abstract. I present recent results from lattice simulations of SU(2) gauge theory with $N_f = 2$ Wilson quark flavors, at non-zero quark chemical potential μ. The thermodynamic equation of state is discussed along with the nature of the high density matter which forms. It is conjectured that deconfinement may mean different things for bulk and Fermi surface phenomena.

1. Why Two Colors?

In this talk I will give an update on our project to simulate SU(2) gauge theory with $N_f = 2$ quark flavors on cold lattices with quark chemical potential $\mu \neq 0$ [1, 2, 3]. Because of the Sign Problem, QC_2D offers currently the best prospect of using lattice simulations to study gauge theories at non-zero baryon charge density. In particular, we can explore the systematics of the lattice approach in this unexplored physical regime. There is good news: unlike the case of hot QCD, it is possible to perform a scan in μ at fixed cutoff, and the primary thermodynamic observable, the quark number density $n_q \equiv -\partial f/\partial\mu$, as a component of a conserved current has no quantum corrections. The bad news is that both UV and IR artifacts are significant and complicated; only recently have we begun to make progress in disentangling them [3].

A second, more theoretical motivation is that QC_2D with $\mu \neq 0$ offers the chance to explore color deconfinement in a new physical régime, complementary to the more usually studied transition at $T > 0$. There are interesting differences which will challenge us to refine the language used to discuss this essential feature of non-abelian gauge theories.

2. The Simulation

We define our lattice theory with the Wilson plaquette action and unimproved Wilson fermions with hopping parameter κ. Chemical potential is introduced via the standard prescription of weighting forward/backward temporal hops by $e^{\pm\mu a}$. Reality of the fermion measure follows from $\det M = \det\tau_2 M^* \tau_2$ where the Pauli matrix τ_2 acts on color. Our only innovation [1] is the introduction of a scalar, isoscalar and gauge-invariant diquark source term $jqq \equiv j\kappa(\psi_2^{tr} C\gamma_5\tau_2\psi_1 - \bar\psi_1 C\gamma_5\tau_2\bar\psi_2^{tr})$, where subscripts denote flavor. Setting $j \neq 0$ mitigates the IR fluctuations due to Goldstone modes associated with superfluidity at non-zero quark density. Indeed, j is nothing but a Majorana mass for the quarks. Fig. 1 shows how the computational effort, measured by both the number of `congrad` iterations and the HMC timestep dt required

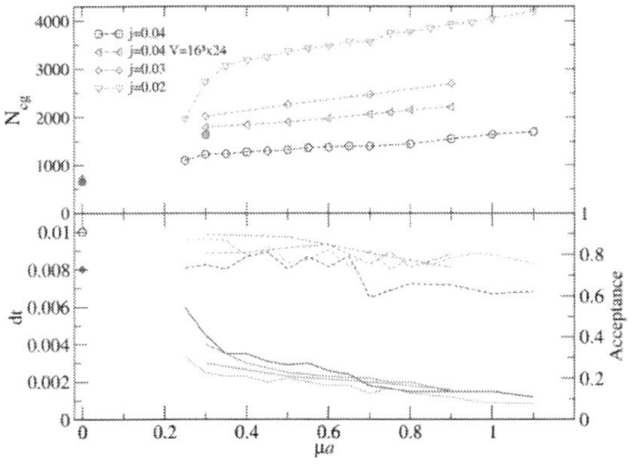

Figure 1. Computational effort needed for QC_2D.

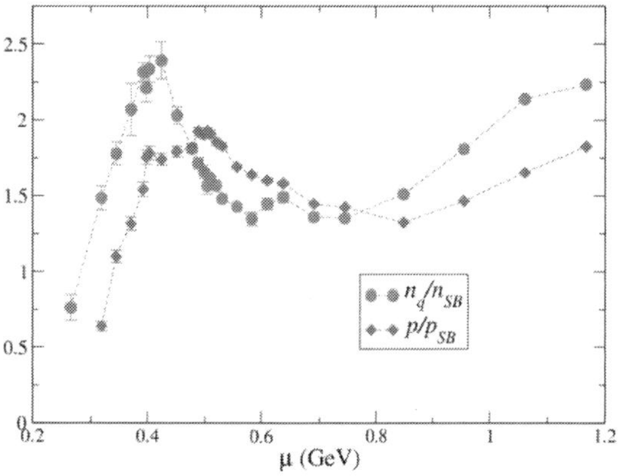

Figure 2. n_q/n_q^{SB} and p/p^{SB} for $ja = 0.04$ on $12^3 \times 24$.

for decent acceptance varies with μ and j. Despite the absence of a Sign Problem, studies of dense matter still require an order of magnitude more cpu than that needed for vacuum QCD.

By now we have accumulated matched ensembles on a coarse $8^3 \times 16$ lattice with $a = 0.229(3)$fm [1] and a finer $12^3 \times 24$ lattice with $a = 0.178(6)$fm [2], with scale set by assuming the two color string tension is $(440\text{MeV})^2$. Both lattices have temperature $T \approx 50$MeV, making them "cold" in QCD terms. The quark mass is relatively heavy, corresponding to $m_\pi/m_\rho \approx 0.8$ — up to now we have not considered the chiral limit to be so important, because effects due to $m \neq 0$ could naively be expected to lie deep at the bottom of the Fermi Sea. In our most recent simulations [3] we have explored the systematics both of varying the diquark source $ja = 0.04, \ldots, 0.02$ and varying T, with $N_\tau = 24, \ldots, 8$ corresponding to $T = 47, \ldots, 141$MeV.

3. Equation of State

Fig. 2 plots n_q/n_q^{SB} as a function of μa, calculated on the fine lattice with $ja = 0.04$ [2]. Here n_q^{SB} denotes the "Stefan-Boltzmann" (SB) result evaluated for free Wilson fermions on the same volume; taking the ratio is a first step towards correcting for discretisation and finite volume artifacts. These are considerable, as shown by the departures from unity in Fig. 5. For μa greater than unity, the dominant error is due to UV artifacts, but for smaller μ the disparity between the two curves points to the IR; indeed, the oscillatory behaviour is due to departures of the Fermi surface from sphericity as $T \to 0$ due to the discretisation of momentum space [4]. Fig. 3 illustrates the difficulty of disentangling these two effects, updating Fig. 2 for a range of temperatures with the limit $j \to 0$ taken. Fig. 4 presents the same data normalised by the continuum SB result $N_f N_c(\mu^3/\pi^2 + \mu T^2)/3$. The apparent peak at $\mu a \approx 0.4$, previously identified as a signal of Bose-Einstein condensation (BEC) of tightly bound diquarks [1, 2], is most likely an IR artifact due to the dip in the dashed red free fermion curve of Fig. 5. On the other hand the difference in vertical scales shows that UV artifacts still need to be corrected.

Another previously neglected factor is the $j \to 0$ limit, which appears to be a much more important effect for interacting quarks; setting $j \neq 0$ enhances their tendency to form pairs, in effect promoting BEC formation which could deform the results. Taking all factors into account, we conclude the approximate plateau in the range $0.4 < \mu a < 0.7$ is consistent with $n_q/n_q^{SB} \approx 1$, rather than 1.4 as suggested by Fig. 2. The same gross features are also apparent in a calculation of the pressure $p = \int^\mu n_q d\mu$; Fig. 6 shows the result of three different approaches to correcting for lattice artifacts at $T = 47$MeV [3].

Figs. 3,4 and 6 support an interpretation in terms of four distinct regimes separated by three

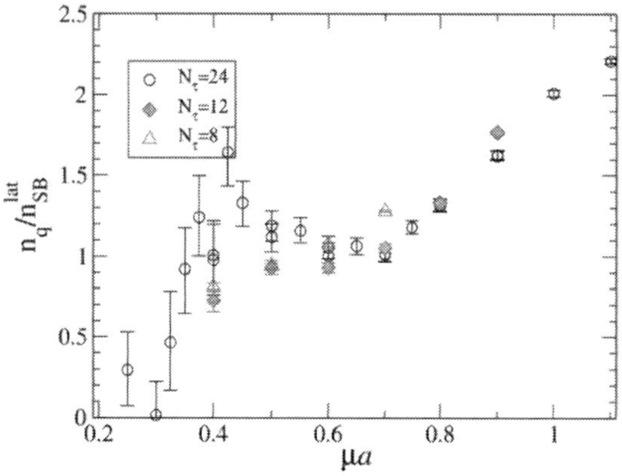

Figure 3. n_q/n_q^{SB} for $j \to 0$ using lattice free fermions.

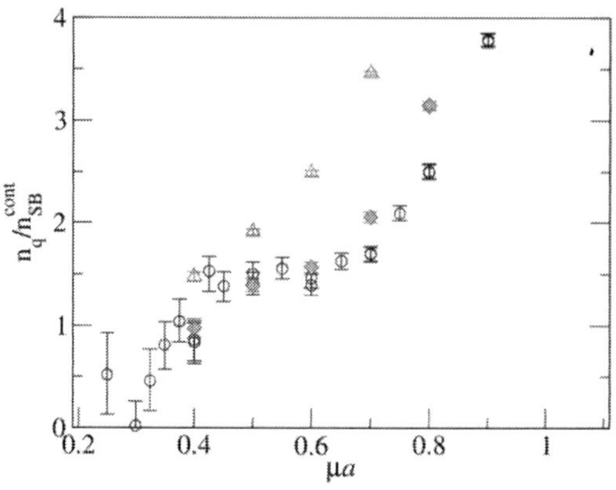

Figure 4. n_q/n_q^{SB} for $j \to 0$ using continuum free fermions.

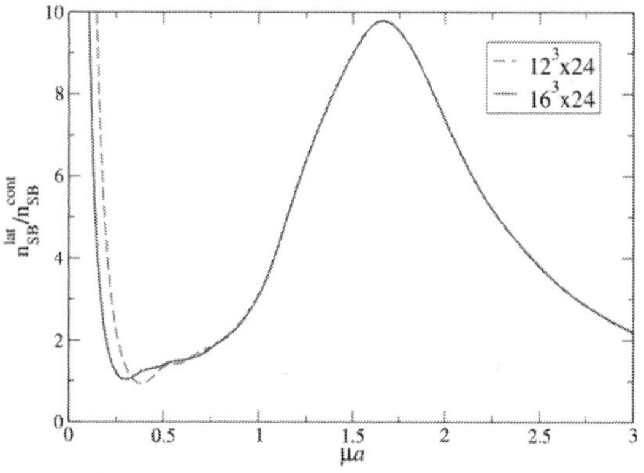

Figure 5. $3\pi^2 n_q^{SB}/N_f N_c \mu^3$ vs. μa on two different lattice volumes.

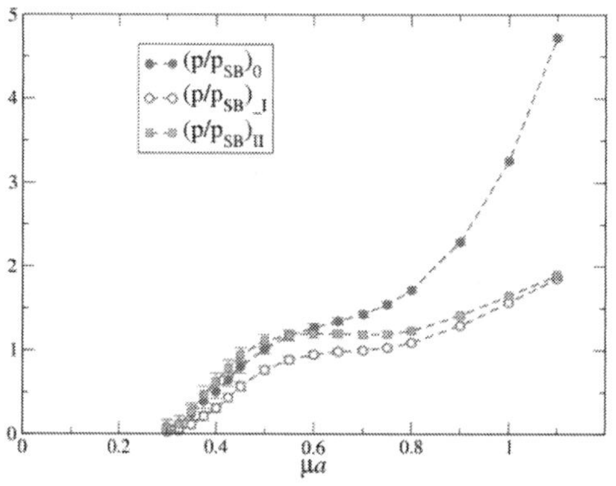

Figure 6. p/p^{SB} on $12^3 \times 24$ for $j \to 0$ using three different artifact correction schemes.

thresholds. The first is the onset threshold, taking place at $\mu_o = m_\pi/2 \approx 0.32a^{-1} \approx 360$MeV, separating the vacuum state from a non-zero quark density $n_q > 0$. In mean field theory this is predicted to be a second order phase transition [5], unlike the first order transition expected in QCD. The non-monotonic behaviour of n_q/n_q^{SB} above onset due to a BEC, predicted in mean field-theory, and reported in [1, 2] is, however, no longer a robust feature of the numerical results; it may well be that simulations far closer to the chiral limit are needed to expose this behaviour. Next, at $\mu_Q \approx 0.5a^{-1} \approx 530$MeV is a crossover to a régime where both n_q and p are approximately equal to their SB values, suggestive that a degenerate system of quarks with a well-defined Fermi momentum $k_F \propto n_q^{1/3}$ and Fermi energy $E_F \approx \mu$ has formed. For reasons that will be elaborated we refer to this as the *quarkyonic* régime. Finally at $\mu_d \approx 0.8a^{-1} \approx 850$MeV both n_q and p start to climb above the SB values. If for the moment we assume the degenerate quark description still holds, then for $\mu_Q < \mu < \mu_d$ we have $E_F \approx k_F$ as expected for weakly interacting massless quarks, but $E_F < k_F$ for $\mu > \mu_d$, implying that the quark matter at very large densities has a large negative correction to the kinetic energy, ie. it is strongly self-bound.

4. Order Parameters and Phase Diagram

In order to characterise the high density system we have also examined two "order parameters", one exact, the other approximate. Fig. 7 shows the superfluid order parameter $\langle qq \rangle$ as a

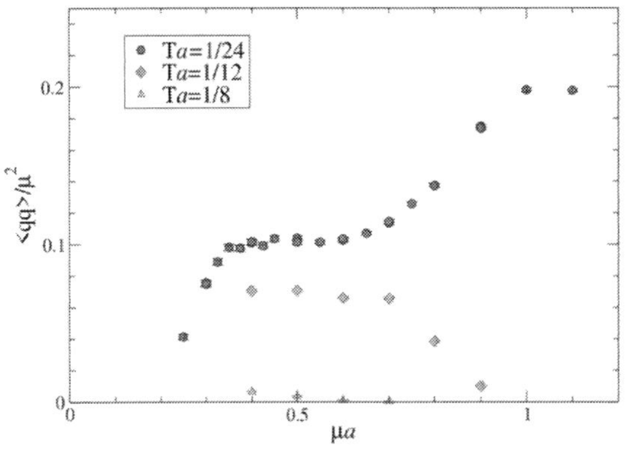

Figure 7. $\langle qq\rangle/\mu^2$ for $j \to 0$ for various T.

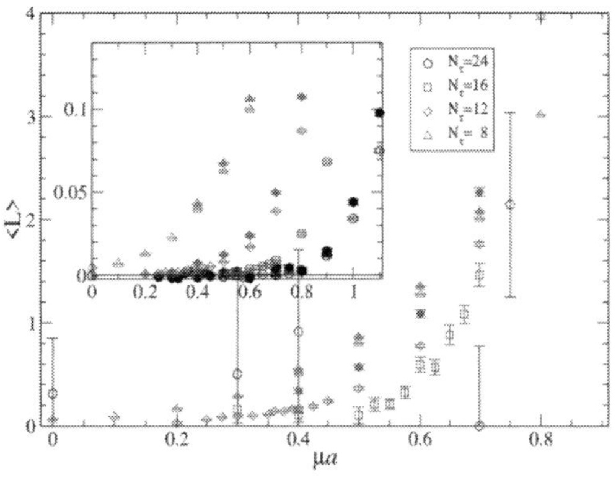

Figure 8. Polyakov line $L(\mu)$ for various T.

function of μ, extrapolated to $j = 0$, for three different temperatures. For a degenerate system superfluidity arises through Cooper pair condensation of diquark pairs at the Fermi surface; hence $\langle qq\rangle$ should scale as the area of the Fermi surface k_F^2. The plot shows $\langle qq\rangle/\mu^2$, and indeed it is remarkably constant within the quarkyonic régime, although clearly T-sensitive. The temperature sensitivity becomes more marked once $\mu > \mu_d$, and at the highest temperature studied the order parameter vanishes everywhere, implying restoration of the "normal" phase.

We also studied the Polyakov loop L as a function of both μ and T. Since this involves comparison of data with different N_τ, it has been necessary to renormalise L via multiplication by a factor $Z_L^{N_\tau}$ [6], determined at $\mu = 0$ and normalised so that $L(T = 1/4a) \equiv 1$. The results are shown in Fig. 8; the inset shows the unrenormalised data. First consider the data from the lowest temperature with $N_\tau = 24$. Although in the presence of fundamental matter L is not an exact order parameter for global Z_{N_c} center symmetry, its behaviour strongly suggests that the transition at μ_d is for all practical purposes identical with deconfinement, ie. the free energy for a static fundamental color source becomes finite once $\mu > \mu_d$. The quark density at deconfinement is $16 - 32\ \mathrm{fm}^{-3}$ (the uncertainty arises from the difficulty in handling lattice artifacts discussed in the previous section), some $30 - 60$ times that of nuclear matter.

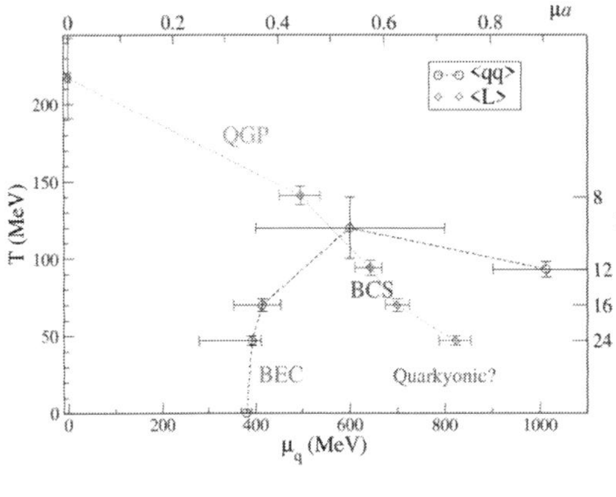

Figure 9. Tentative QC$_2$D phase diagram.

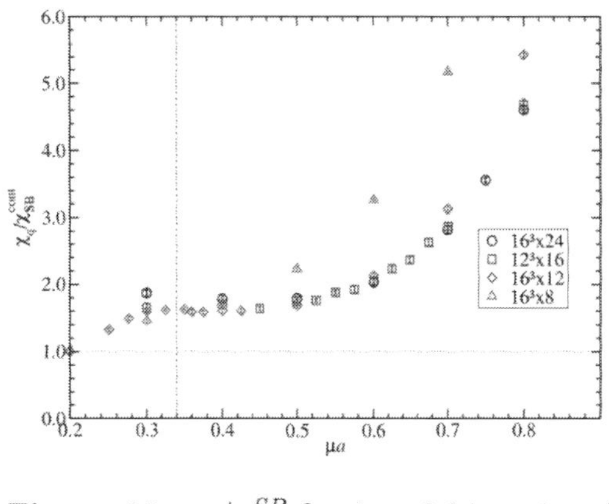

Figure 10. χ_q/χ_q^{SB} for $ja = 0.04$, various T.

It's then interesting that μ_d defined via L falls rapidly as T rises. A pragmatic definition of $\mu_d(T)$ is the value at which $L(\mu, T) \approx L(0, T_d)$. The resulting tentative phase diagram is shown in Fig. 9. There are at least three distinct regions/phases: a normal hadronic phase with

$\langle qq\rangle = 0$, $L \approx 0$ at low T and μ; the quarkyonic region with $\langle qq\rangle \propto \mu^2$ and $L \approx 0$ at low T and intermediate μ; and a deconfined, normal phase with $\langle qq\rangle = 0$, $L > 0$ at large T and/or large μ. At this stage we cannot exclude a deconfined superfluid phase at large μ and small/intermediate T. The "quarkyonic" nomenclature is now justified; this phase has the thermodynamic bulk scaling of a weakly-interacting degenerate quarks, but remains confined. It thus has two of the features of dense baryonic matter originally proposed on the basis of large-N_c arguments [7].

5. Quark Number Susceptibility and Deconfinement

We have also recently calculated the quark number susceptibility $\chi_q \equiv \partial n_q/\partial\mu$ [8, 3]. Whilst naively χ_q is expected to reflect local fluctuations of the n_q operator, it turns out that at large μ the dominant term comes from a connected "hairpin" diagram. Fig. 10 shows χ_q divided by the continuum SB result. Once again, we note a large range over which the ratio is approximately constant; indeed it is compatible with one as $j \to 0$, though more sensitive to the quark mass value used for the free fermions than other bulk observables [3].

The most striking feature of Fig. 10, however, is the absence of T-dependence at all but the highest temperature studied (141 MeV), is in stark contrast to the behaviour of L in Fig. 8. Despite our intuition from the thermal transition in QCD [9], χ_q cannot be regarded as a proxy for L once $\mu/T \gg 1$. We conjecture that the change in behaviour of the bulk thermodynamic quantities n_q, p and χ_q observed at μ_d is a transition from short-ranged binary confining interactions to longer-ranged interactions among several quarks within the medium, and that this corresponds with the transition from weak to strong self-binding noted in Sec. 3. The T-dependent behaviour of L, in contrast, must be due to degrees of freedom close to the Fermi surface which can be thermally excited; a correlation between gapless excitations and $L > 0$ has also been noted in analytical and numerical studies on small cold volumes [10]. The same phenomenon should inform a theory of transport properties.

6. Summary

QC_2D offers an accessible theoretical laboratory for the study of dense baryonic matter. For low T at least three physical regions can be identified: the vacuum for $\mu < \mu_o$; a confined quarkyonic superfluid for $\mu_Q < \mu < \mu_d$; and a deconfined phase for $\mu > \mu_d$. The most recent simulations have only made our findings in the quarkyonic régime more robust. Not discussed here are new results for renormalised energy density, conformal anomaly, and chiral symmetry restoration [3].

Acknowledgments

This work used the DiRAC Facility jointly funded by STFC, the Large Facilities Capital Fund of BIS and Swansea University. We thank the DEISA Consortium (www.deisa.eu), funded through the EU FP7 project RI-222919, for support within the DEISA Extreme Computing Initiative.

References

[1] S. Hands, S. Kim and J.-I. Skullerud, *Eur. Phys. J. C* **48** (2006) 193.
[2] S. Hands, S. Kim and J.-I. Skullerud, *Phys. Rev. D* **81** (2010) 091502(R).
[3] S. Cotter, P. Giudice, S. Hands and J.-I. Skullerud, *Preprint* arXiv:1210.4496 [hep-lat].
[4] S. Hands and D.N. Walters, *Phys. Lett. B* **548** (2002) 196.
[5] J.B. Kogut, M.A. Stephanov, D. Toublan, J.J.M. Verbaarschot and A. Zhitnitsky, *Nucl. Phys. B* **582** (2000) 477.
[6] S. Borsanyi *et al.*, *Phys. Lett. B* **713** (2012) 342.
[7] L. McLerran and R.D. Pisarski, *Nucl. Phys. A* **796** (2007) 83.
[8] P. Giudice, S. Hands and J.-I. Skullerud, *PoS LATTICE* **2011** (2011) 193.
[9] Y. Aoki, Z. Fodor, S.D. Katz and K.K. Szabo, *Phys. Lett. B* **643** (2006) 46; A. Bazavov *et al.*, *Phys. Rev. D* **80** (2009) 014504.
[10] S. Hands, T.J. Hollowood and J.C. Myers, *JHEP* **1007** (2010) 086; *JHEP* **1012** (2010) 057.

Photon azimuthal anizotropy and magnetic field in heavy-ion collisions

Vladimir Skokov

Department of Physics, Brookhaven National Laboratory, Upton, NY 11973, USA

E-mail: VSkokov@bnl.gov

Abstract. Recent measurements of the azimuthal anisotropy of direct photons in heavy-ion collisions at the energies of RHIC showed that it is of the same order as the hadronic one. This finding appears to contradict the expected dominance of photon production from a quark-gluon plasma at an early stage of a heavy-ion collision. A possible explanation of the strong azimuthal anisotropy of the photons, given recently, is based on the presence of a large magnetic field in the early phase of a collision. In this talk, we consider a novel photon production mechanism stemming from the conformal anomaly of QCDxQED and the existence of strong (electro)magnetic fields in heavy ion collisions. Using the hydrodynamical description of the bulk modes of QCD plasma, we show that this mechanism leads to the photon production yield that is comparable to the yield from conventional sources. The comparison of the results to the data from the PHENIX collaboration, show that this mechanism might be the most responsible for the observed azimuthal anisotropy of photons.

1. Introduction

The "direct" photons leave the fireball almost without interacting with the medium, and thus they may play an important role in unraveling the properties of hot and dense matter. According to expectations, based on the large yield of thermal photons, photon production is believed to be dominated by hot quark-gluon plasma at the early stage of a collision at the top energies in RHIC (Brookhaven) and the LHC (CERN). Indeed, measuremants from the PHENIX collaboration showed that the observed temperature of photon radiation at energy $\sqrt{s} = 200$ GeV in heavy-ion collisions is about $T_{\text{ave}} \approx 220$ MeV [1]. One can argue that this value can be considered as an average over the entire evolution of the matter created in heavy-ion collisions. It is higher than the temperature of the phase transition and thus it supports the picture of photon production from the early stage in the collision. An alternative explanation of the apparent high temperature of the photon source is the formation of prethermal glasma shining photons early in a collision [2]. In both scenarios, we would expect that the photons' azimuthal anisotropy, characterized by the second Fourier component $v_2^\gamma(p_t) = \int d\phi \cos(2\phi) \frac{dN^\gamma}{dp_t d\phi} / \frac{dN^\gamma}{dp_t}$, is small, because, at the early QGP stage, the evolving medium does not develop an appreciable amount of azimuthal anisotropy. Others studied this phenomenon in hydrodynamic calculations that well describe hadron production and hadron azimuthal anisotropy. The hydrodynamic calculations [3], as expected, demonstrated that the photon v_2^γ is almost an order of magnitude smaller than the one of hadrons. However, recent measurements by the PHENIX collaboration (RHIC) [4] and preliminary results by ALICE (LHC) revealed that the anisotropy of produced photons is very close to that of hadrons.

Journal of Physics: Conference Series **432** (2013) 012021 doi:10.1088/1742-6596/432/1/012021

In this talk, we review a recently proposed novel mechanism which is based on the presence of a large highly anisotropic magnetic field in heavy-ion collisions that could serve as a natural source of the photons' anisotropy. The magnetic field is maximal at the early stage of a collision, thereby allowing us to reconcile the large thermal yield of photons with significant azimuthal anisotropy.

2. Magnetic field in heavy-ion collisions

The results of Refs. [5, 6, 7] showed that in non-central heavy-ion collisions the amplitude of magnetic field can reach very high values up to a few $m_\pi^2 \approx 10^{18}$ Gauss owing to a fast movement of spectators. This field is large at the time of the collision and decreases inversely proportional to the square of time. The magnetic field essentially is anisotropic and, on average, points in the direction perpendicular to the reaction plane. It is important to remember that this estimates of the magnetic field have been established without taking into account effects related to induced magnetic field in quark-gluon plasma. Depending on the magnitude of the electric conductivity, the induced magnetic field may live much longer then the one generated by spectators. Presently, we do not know neither precise values for the electric conductivity, nor we are able to perform quantitative calculations for the induced magnetic field because this requires to perform calculations in relativistic magneto-hydrodynamics. Consequently, in further estimates we will consider the magnetic field from spectators only.

Different microscopic mechanisms of the photon emission were proposed recently, including the synchrotron radiation of photons from moving charged quarks [8], and that reflecting the scale anomaly of QCD×QED [9] and owing to the axial anomaly [10]. It is important to mention that the last two mechanisms generate photons even in case of a vanishing quark number, which is expected to be rather low at the time scales of order R/γ [1], when the magnetic field is the highest and that in all the above mechanisms photons are emitted in the direction perpendicular to the magnetic field. As we show below photon production from the conformal anomaly results in a considerable contribution to v_2^γ [9] and potentially can describe the measured data.

In what follows, we neglect the spatial gradients of magnetic field and estimate the time dependence in the eikonal approximation taking into account only the (leading at large times) contribution from spectators:

$$eB_{x,y}(t) \simeq \frac{eB^0_{x,y}}{1+(t/t_B)^2}, \tag{1}$$

where eB^0_i it the magnitude of the i-th component of the magnetic field at $t=0$ and t_B is the characteristic decay time. The x-component of magnetic field at $t=0$, B^0_x, is approximately independent of the impact parameter b, while the y-component is linear in b. Both components $B^0_{x,y}$ are linear as a function of the collision energy, $\sqrt{s}$; the typical decay time is inversely proportional to $\sqrt{s}$.

3. Conformal anomaly as a source of photons

In this talk, we investigate photon production mechanism stemming from the *conformal SVV* anomaly that involves a dilatation current S_μ and two vector currents V_μ and reflects the violation of conformal invariance of QCD by quantum effects. The conformal anomaly results from the running of the coupling constant in QCD and expresses the non-conservation of the dilatation current S^μ, so that the trace of the energy-momentum tensor θ^μ_μ does not vanish even in the chiral limit of massless quarks: $\partial^\mu S_\mu = \theta^\mu_\mu = \frac{\beta(g)}{2g} G^{\mu\nu a} G_{\mu\nu a} + \sum_q m_q \left[1+\gamma_m(g)\right] \bar{q}q$.

The quarks carry both color and electric charges, so when QCD is coupled to electromagnetism, the quark triangle diagram induces an anomalous coupling of the trace of the

[1] Here, R is the radius of colliding nuclei and γ is the Lorentz factor.

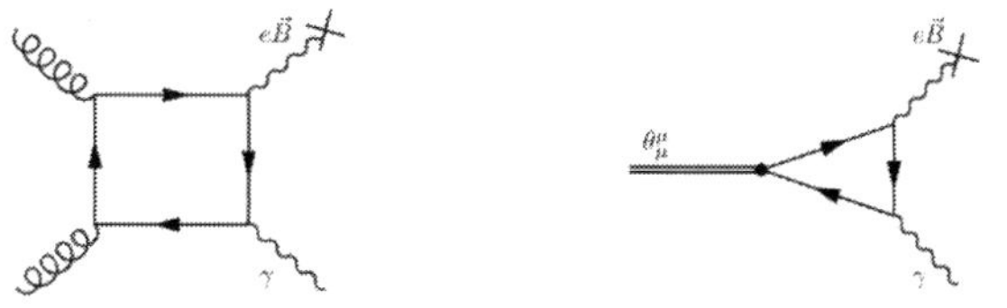

Figure 1. The coupling of the conformal anomaly to the external magnetic field resulting in photon production.

energy-momentum tensor to photons [11, 12, 13]. The trace of the energy-momentum tensor in hydrodynamics excites the bulk modes of the fluid that are abundant in (non-conformal) quark-gluon plasma [14, 15, 16, 17]. As a result, the conformal anomaly coupled to the magnetic field acts as a source of photon production that is powered by the energy of the bulk hydrodynamical modes in the plasma (see Fig. 1, left panel). This is the mechanism of photon production that will be discussed in detail in Ref. [9]. Note that while in what follows the hydrodynamics will be used to describe the bulk modes in the plasma, the deviation from equilibrium in general need not be small for this mechanism to operate. For example, the non-equilibrated Bose-Einstein condensate of gluons [18] may be even more effective in producing photons, see right panel in Fig. 1. Note that unlike in the conventional scenario, the quarks in this case appear only in the triangle loop that receives contributions from the virtual UV modes – so the production of real on-shell quarks is not required, and the mechanism can operate even at very early times.

In vacuum, the coupling of QCD scale anomaly to electromagnetism can be defined by the coupling of the scalar meson to photons. This coupling is described by the triangle quark diagram, see Figure 1, and leads to the following effective interaction [11, 12, 13]: $\mathcal{L}_{\sigma\gamma\gamma} = g_{\sigma\gamma\gamma}\ \sigma\ F_{\mu\nu}F^{\mu\nu}$,where numerical value of the coupling $g_{\sigma\gamma\gamma}$ can be fixed using the parameter of the lightest scalar meson. We obtain $g_{\sigma\gamma\gamma} \simeq 0.02\ \text{GeV}^{-1}$ (see Ref. [9] for details).

To compute the photon production rate from the diagram of Fig. 1, we evaluate the imaginary part for the photon self-energy, see [19, 20]. A straightforward calculation yields for the production rate at mid-rapidity ($q_z = 0$) the following expression:

$$q_0\frac{d\Gamma_B}{d^3q} \;=\; C\frac{(B_y^2 - B_x^2)q_x^2 + q_\perp^2 B_x^2}{\exp(\beta q_0) - 1}\rho_\theta(q_0 = |\vec{q}|), \tag{2}$$

where the proportionality factor C is defined by the properties of the lowest mass scalar meson and the couplig $g_{\sigma\gamma\gamma}$; ρ_θ is the spectral function for the operator θ. Since we consider production of photons in the QCD plasma, it is appropriate to use the hydrodynamic spectral function of the bulk mode θ [21, 22]: $\rho_\theta(q_0, \vec{q}) = 9q_0\frac{\zeta}{\pi} + \rho_{\text{sound}}$, where ζ is the bulk viscosity. The second term describes the sound peak at $q_0 = c_s|\vec{q}|$. The sound mode does not contribute to the production of real photons since the width of the sound peak is not large enough to reach the null dispersion of photons. Therefore the photon production is dominated by the bulk viscosity ζ.

In deriving Eq. (2) we neglected the z-component of the magnetic field, because it is expected to be an order of magnitude smaller than B_x and B_y ($B_z \sim B_{x,y}/\gamma$); we also neglect the contribution of the electric field.

In what follows we will compare our result with the baseline provided by the conventional thermal photon production rate. Note that this conventional mechanism is expected to be the dominant one for low transverse momentum, $p_\perp$, photons. For photons with $p_\perp \sim 2$ GeV and above there will be additional contributions to the rate which can be calculated perturbatively. However we did not include these additional contributions as we are mainly interested in low $p_\perp$ photons.

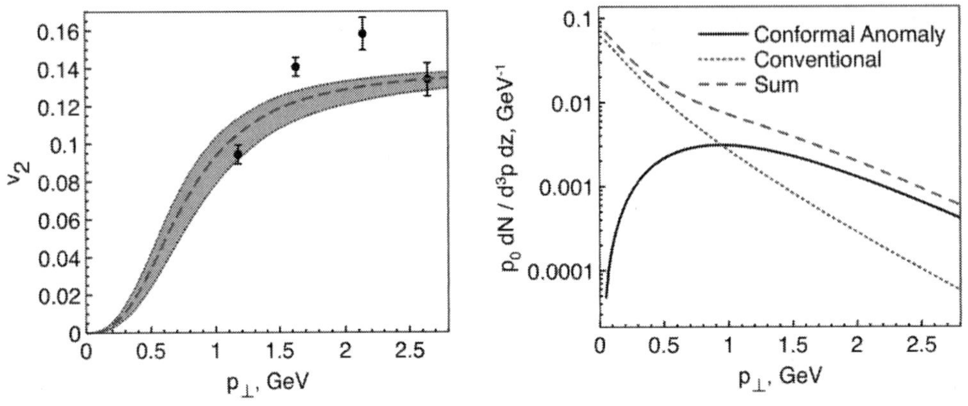

Figure 2. Left panel: The azimuthal anisotropy v_2 of the direct photons for different values of bulk viscosity corresponding to C_ζ in the range of $2.5 \div 5$ calculated for minimum bias Au-Au collisions. The dashed line represents the results with $C_\zeta = 4$. The black dotes are the data from the PHENIX collaboration [4] for minimum bias Au-Au collisions at $\sqrt{s} = 200$ GeV. Right panel: The transverse momentum spectra of the produced direct photons for $C_\zeta = 2.5$ calculated for minimum bias Au-Au collsisions, see text for details.

The spectral function for θ and the bulk viscosity was calculated in lattice QCD [16, 22]. However the extraction of bulk viscosity from the lattice data is notoriously difficult. To get an independent estimate of the bulk viscosity we thus follow [23, 24] and assume that $\frac{\zeta}{\eta} = C_\zeta \left(\frac{1}{3} - c_s^2\right)^2$. Thus the bulk viscosity vanishes in the conformal limit, $c_s^2 = 1/3$. In the relaxation time approximation, this expression is obtained in the kinetic theory with $C_\zeta = 15$ (see e.g. [25]). The paper [25] contains also a phenomenological estimate of the value of bulk viscosity inferred from the comparison of viscous hydrodynamical computations with the data on the elliptic flow of mesons and baryons. The resulting estimate is $\zeta/s = 0.005$ [25]. Using the lattice data for the speed of sound in the freeze-out temperature range from Ref. [26], $c_s^2 = 0.175 \div 0.221$, we infer for the bulk viscosity $C_\zeta = 2.5 \div 5$. The leading log calculations in SU(3) Yang Mills theory results in a much larger value $C_\zeta \simeq 48$, see Ref. [25]. In our calculations, we choose the lowest value available in the literature, $C_\zeta = 2.5 \div 5$, with an assumption $\eta/s = 1/4\pi$.

Here we neglect the transverse expansion of the fireball and approximate the time evolution of the temperature at early times using the Bjorken hydrodynamics $T/T_0 = (\tau_0/\tau)^{1/3}$, where T_0 is the initial temperature and τ_0 is the initial time (given by the characteristic thermalization time of the gluons) that can be estimated in terms of the saturation scale, Q_s, and the coupling constant, α_s, see e.g. Ref. [18]. For Au-Au collisions at $\sqrt{s} = 200$ GeV we use $\tau_0 = 0.1$ fm/c.

The results for the azimuthal anisotropy of photons calculated using both conventional production mechanism and the one from the conformal anomaly are shown in Fig. 2 for the minimum bias Au-Au collisions at $\sqrt{s} = 200$ GeV. In our approximation (no transverse flow), the conventional mechanism does not give any contribution to the azimuthal anisotropy. The comparison with the experimental data from PHENIX [4] indicates that conformal anomaly could account for a large fraction of the observed photon anisotropy.

In Fig. 2 we show our result for the transverse momentum spectrum of direct photons. Due to the factor of q^2 in the production rate (2), the spectrum of photons produced due to conformal anomaly is enhanced in comparison to the conventional one at transverse momenta $k_\perp > 1$ GeV. The factor of q^2 in the rate hardens the transverse momentum spectrum, and magnetic field grows with the impact parameter of the collision; these two effects thus conspire in mimicking both the elliptic and radial flow of photons in non-central collisions.

4. Experimental signatures

In this section we discuss the experimental signatures of the photon produced in a magnetic field. Firstly, because the anisotropic contribution to the photon production rate is proportional to q_x^2, the fourth harmonics v_4^γ will be significantly suppressed, violating the established scaling for hadrons $v_4^{\rm hadr}/v_2^{\rm hadr} \sim 1$. Secondly, by properly selecting specific system geometries, we could diminish either the magnetic field, or the hadronic elliptic flow. The former possibility can be experimentally realized by performing central collisions with deformed nuclei, e.g. U-U collisions. The magnetic field is defined by the centrality, while the hadronic flow by the initial eccentricity. Consequently, if our mechanism gives dominant contribution to the photon azimuthal anisotropy, the central U-U collisions would show small photon v_2^γ compared to the hadronic one. The later possibility can be achieved in non-central Au-Au collision of a given centrality class. Since the initial eccentricity and, consequently, the hadronic $v_2^{\rm hadr}$ fluctuates at given centrality, it is possible to select non-central events with zero $v_2^{\rm hadr}$. Again if the proposed mechanism gives the dominant contribution to the photon production, v_2^γ will be much larger then $v_2^{\rm hadr}$.

5. Acknowledgement

I thank my co-authors, A. Bzdak, G. Bazar and D. Kharzeev, with whom I have been working on the subject of this talk. I also grateful to L. McLerran and E. Shuryak for interesting discussions. This work was supported by Contract No. DE-AC02-98CH10886 with the U.S. Department of Energy. I thank ECT* for support and hospitality during the workshop "QCD in strong magnetic field", which gave me inspiration to work on these proceedings.

[1] A. Adare *et al.* [PHENIX Collaboration], Phys. Rev. Lett. **104**, 132301 (2010).
[2] M. Chiu, T. K. Hemmick, V. Khachatryan, A. Leonidov, J. Liao and L. McLerran, arXiv:1202.3679 [nucl-th].
[3] M. Dion, J. F. Paquet, B. Schenke, C. Young, S. Jeon and C. Gale, Phys. Rev. C **84**, 064901 (2011).
[4] A. Adare *et al.* [PHENIX Collaboration], Phys. Rev. Lett. **109**, 122302 (2012).
[5] D. E. Kharzeev, L. D. McLerran and H. J. Warringa, Nucl. Phys. A **803**, 227 (2008).
[6] V. Skokov, A. Y. Illarionov and V. Toneev, Int. J. Mod. Phys. A **24**, 5925 (2009).
[7] A. Bzdak and V. Skokov, Phys. Lett. B **710**, 171 (2012).
[8] K. Tuchin, arXiv:1206.0485 [hep-ph].
[9] G. Basar, D. Kharzeev, D. Kharzeev and V. Skokov, arXiv:1206.1334 [hep-ph].
[10] K. Fukushima and K. Mameda, Phys. Rev. D **86**, 071501 (2012).
[11] J. R. Ellis and J. Lanik, Phys. Lett. B **150**, 289 (1985).
[12] R. J. Crewther, Phys. Rev. Lett. **28**, 1421 (1972).
[13] M. S. Chanowitz and J. R. Ellis, Phys. Lett. B **40**, 397 (1972).
[14] D. Kharzeev and K. Tuchin, JHEP **0809**, 093 (2008).
[15] F. Karsch, D. Kharzeev and K. Tuchin, Phys. Lett. B **663**, 217 (2008).
[16] H. B. Meyer, Phys. Rev. Lett. **100**, 162001 (2008).
[17] P. Romatschke and D. T. Son, Phys. Rev. D **80**, 065021 (2009).
[18] J. P. Blaizot, F. Gelis, J. F. Liao, L. McLerran and R. Venugopalan, Nucl. Phys. A **873**, 68 (2012).
[19] H. A. Weldon, Phys. Rev. D **28**, 2007 (1983).
[20] L. D. McLerran and T. Toimela, Phys. Rev. D **31**, 545 (1985).
[21] J. Hong and D. Teaney, Phys. Rev. C **82**, 044908 (2010).
[22] H. B. Meyer, JHEP **1004**, 099 (2010).
[23] S. Weinberg, Astrophys. J. **168**, 175 (1971).
[24] P. B. Arnold, C. Dogan and G. D. Moore, Phys. Rev. D **74**, 085021 (2006).
[25] K. Dusling and T. Schafer, Phys. Rev. C **85**, 044909 (2012).
[26] S. Borsanyi *et al.*, JHEP **1011**, 077 (2010).

QCD with many fermions and QCD topology

Edward Shuryak

Department of Physics and Astronomy, Stony Brook University, Stony Brook, NY 11794 USA

E-mail: `Edward.Shuryak@stonybrook.edu`

Abstract. Major nonperturbative phenomena in QCD – confinement and chiral symmetry breaking – are known to be related with certain topological objects. Recent lattice advances into the domain of many $N_f = O(10)$ fermion flavors have shown that both phase transitions had shifted in this case to much stronger coupling. We discuss confinement in terms of monopole Bose condensation, and discuss how it is affected by fermions "riding" on the monopoles, ending with the N_f dependence of the critical line. Chiral symmetry breaking is discussed in terms of the (anti)selfdual dyons, the instanton constituents. The fermionic zero modes of those have a different meaning and lead to strong interaction between dyons and antidyons. We report some qualitative consequences of this theory and also some information about our first direct numerical study of the dyonic ensemble, in respect to both chiral symmetry breaking and confinement (via back reaction to the holonomy potential).

1. Lattice advances into the domain of multi-flavor QCD

Let us start by reviewing recent advances in lattice data using our own version of the phase diagram, see Fig.1 . We plot the "critical lattice coupling at scale T_c" $\beta_c(T_c) = 2N_c/g^2(T_c)$ as a function of N_f. The "bare" coupling values reported in lattice works (defined at the lattice UV scale a) are evolved (by the two-loop beta function) by a factor N_t (the number of points in the time direction) to the physical scale $N_t a = 1/T_c$. The near overlapping points in the figure are from different lattice N_t: their spread is a measure of the inaccuracy of the two-loop beta function used.

The lines at the right show trajectory of the infrared fixed point in various perturbative approximations: but since the bottom of the plot corresponds to strong coupling, the real curve there can deviate from the perturbative trend.

The open boxes correspond to lattice data from Ref. [1] : they show the critical line of the chiral restoration. For completeness we've also displayed (by red diamonds) results from few other lattice groups: a spread is due to different lattice actions used. The qualitative trend with N_f is clear: the curve is moving downward with N_f, which means the phase transitions is happening at *stronger gauge coupling*. Thus going to $N_f \sim 10$ produces "the most strongly coupled QGP", which is by itself a very interesting phenomenon. Why is it happening?

Let us add that the situation at $N_f = 12$ is special. Ref. [2] argued that chiral symmetry remains unbroken. More recent paper [3] have studied the region of even stronger coupling $6/g^2 \sim 1$, and perhaps clarified the picture. They do observe two distinct transitions, with chiral condensate disappearing at stronger coupling than deconfinement, with a novel intermediate phase in between. Furthermore, those phases are separated by a bulk transition from weaker coupling domain, in which there seem to exist a conformal (infrared fixed point) behavior. If so, the two phase transitions must be below (on the other side of) the true trajectory of the fixed point.

2. Confinement and unusual confinement

Pioneering works by 't Hooft and Mandelstam proposed the so called "dual superconductor" model, which attributed it to Bose-Einstein condensation (BEC) of color magnetic monopoles. Seiberg and Witten [4] had shown how it works in the $\mathcal{N}$=2 supersymmetric theories. Monopoles in certain gauges were identified on the lattice, and the so called "Pisa operator" inserting them into vacuum has been shown to have a nonzero VEV at $T < T_c$ [5]. Approaching T_c from above, one had observed k-clusters of identical monopoles exchanging places [6], as all bosons do near a BEC transition. Thus there is little doubt in my mind that the "dual superconductor" picture is correct. (As a digression let me mention here that even up to $T \sim 2 - 3T_c$ studies in heavy ion collisions at RHIC/LHC monopoles remain important component of the plasma [7], even significantly contributing to its record-small viscosity [8].)

The main issue to be addressed below is why fermions apparently affect the deconfinement transition line so strongly. But before we go into it in more detail, let me also mention another general question I raised in a note [9], if *"unusual confinement phases"* are possible, in which BEC of magnetic objects with quantum numbers different from that of a single monopole can be formed instead. (In such case one would of course expect the phase transitions between different confining phases, if their order parameters do not have the same quantum numbers.)

In fact such examples are known in supersymmetric QCD literature. Seiberg-Witten $\mathcal{N}$=2 QCD with $N_f = 3$ had found two possible confinements: one for (i) a quartet of states with magnetic charge $n_m = 1$ and electric charge $n_e = 0$; and (ii) a singlet with charges $n_m = 2, n_e = 1$. The former automatically leads to confinement *with* the chiral symmetry breaking, and the latter has confinement *without* it. Dynamics of the latter case is especially interesting, as it is related to BEC of a two-monopole composite bound by fermions. Note further

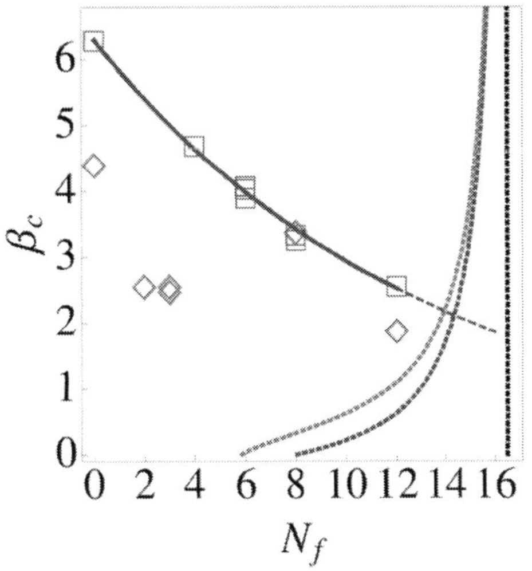

Figure 1. Dependence of the critical lattice coupling β_c at scale T_c versus the number of fundamental quark flavors N_f in QCD-like theories. Blue boxes are from [1] with near-coincident boxes being lattice data for the same N_f with different N_τ which demonstrate lattice spacing consistency. Red diamonds are from various other literature. The thick blue line is the fitting curve, extended as dashed blue line beyond $N_f = 12$. The black/purple/red dashed curves on the right are lines for vanishing beta function at 1,2,3-loop levels.

that $N_f = 3$ is the last value before the conformal window (starting in this case at $N_f = 4$): perhaps it is a hint that something like that should be looked at right before the conformal window also in theories without supersymmetry?

3. Monopoles with fermions

How can the light fermions affect the monopoles? It is known that the light fermions can become attached to them. The so called "fermionic zero modes" are specific bound states in which positive kinetic energy of localization exactly cancels the magnetic-moment-field interaction. These zero modes are known explicitly for 't Hooft-Polyakov monopoles which are present in gauge theories with adjoint scalars, such as in $\mathcal{N} = 2$ supersymmetric gauge theories. Since flipping the charge and the spin leads to the same Dirac equation, antiquarks also have the same zero modes. Furthermore, for each of these zero mode states, it can be either populated or not, so the number of totally available states grows exponentially $\sim 2^{2N_f N_M}$. Spectroscopy of those states in the supersymmetric setting was developed in 1990's. [1] There is one zero mode for the fundamental while two for the adjoint (Dirac) fermions: so these monopole-single-fermion states case into spin 0 and $\frac{1}{2}$ objects, respectively. While in the static case zero mode states are degenerate with the pure monopole, it is not so for the dynamical lattice monopoles with non-static paths.

Now back to confinement/BEC transition. The new idea, from Ref[11], is that if fermions can "ride" on the monopoles, they are giving them flavor indices and thus are making them "distinguishable", unsuitable for Bose-Einstein Condensation! Such dilution of *"unoccupied"* monopoles need to be compensated somehow for BEC to happen, and this can only be done by *a shift stronger coupling* that makes monopoles lighter and more numerous.

A very interesting BEC criterium for an interacting boson ensemble was proposed by Feynman in 1950's (originally for liquid ^4He) and was recently revived/generalized to interacting systems by Cristoforetti and Shuryak [10]. In the finite-T description with periodic paths there appear "k-clusters" of bosons interchanging their initial (at Matsubara time 0) and final (at Matsubara time $\beta = 1/T$) positions. Those clusters are "Feynman polygons" with k points. Their probability depends on two competing factors, suffering from suppression due to the extra action $exp(-kS_{ex})$ while benefiting from large combinatory number of k-polygons. The balance point marks the onset of condensation by divergence of the sum over the k-clusters. According to Feynman, quantitative BEC condition is a *universal* critical value of the extra action per particle, which in three spatial dimensions is given by $S_{ex} \leq S_c \approx 1.655$ Upon fulfilling it, infinitely long sequences of "hopping" bosons will occur, creating macroscopic "supercurrents". Let us now apply the above criteria to the monopole condensation in QCD-like theories. The minimal exchange action S_{ex} for two nearest-neighbor bosons that are separated by a typical distance $d = n^{-1/3}$ (with n the number density) during the Matsubara time from $\tau = 0$ to $\tau = \beta = 1/T$ could be estimated as

$$S_{ex} = m^* \sqrt{\beta^2 + d^2} - m^* \beta + S_V \tag{1}$$

with an explicitly written kinetic term, containing an effective mass m^*, and the implicit potential term S_V. When close to condensation, the monopoles are very dense, with typical spatial separation d comparable or smaller than the inverse temperature β, therefore justifying a further approximation of the above expression: $S_{ex} \approx \frac{1}{2} m^* T d^2 + S_V = \frac{1}{2}\left(\frac{m^*}{T}\right)\left(\frac{n}{T^3}\right)^{-\frac{2}{3}} + S_V$.

[1] "Magnetic supermultiplets" have been constructed and for two famous conformal theories, the $\mathcal{N}$=4 SYM and the $\mathcal{N}$=2 SQCD with $N_f = 4$ one finds exactly the same set of spins/multiplicities as that in the original electric one. Since those theories are electric-magnetic self-dual, with couplings related by $g \leftrightarrow 1/g$, their coupling cannot run at all!

The term $S_V = \int_0^{1/T} V(r(\tau))d\tau$ is basically the ratio of the inter-monopole interaction potential and T, also known as classical plasma coupling $\Gamma_M \sim< V > /T \sim S_V$. As shown in Fig.3b of [7], at high T where the "magnetic scaling" $d \sim 1/(g^2T)$ and $g_{magnetic} \sim 1/g$ works, this ratio Γ_M does not depend on the coupling or T and is a constant ≈ 5, while close to condensation it decreases to a value of about 1 as $T \to T_c$ for $N_f = 0$. The following onset condition will then be applied for the monopole condensation in QCD-like theories for the rest of our analysis:

$$\left(\frac{m^*}{T}\right)\left(\frac{n}{T^3}\right)^{-\frac{2}{3}} \leq \tilde{S}_c \equiv 2\left(S_c - S_V\right) \tag{2}$$

The constant $\tilde{S}_c$ is of order one and its precise value is not needed as long as its N_f-dependence is negligible. Now, how would the transition get affected by adding light fermions? As already pointed out, the fermions can be attached to some of the monopoles via zero modes and effectively reduce the number of identical monopoles. Consider a monopole with one flavor of light quark added: for each of its allowed zero modes there is a probability for it to be occupied by a fermion or not. Let us assume the ratio of the probabilities (occupied/not-occupied) to be f (a kind of zero-mode fugacity), we then see that the overall probability for a monopole (with N_M number of zero modes for *each fermion flavor*) to stay as a "pure" monopole is simply $1/(1+f)^{2N_M}$ (with the factor 2 accounting for both quark and anti-quark contributions for Dirac fermions). So effectively the available density for BEC condensation will be $n/(1+f)^{2N_f N_M}$. Combined with the BEC condition in Eq.(2) we obtain

$$\left(\frac{m^*}{T}\right)\frac{(1+f)^{4N_f N_M/3}}{(n/T^3)^{2/3}} \leq \tilde{S}_c \tag{3}$$

This implies that with increasing N_f, the density n has to increase and mass m^* to decrease correspondingly so as to reach the same condensation condition. This pushes the transition to stronger coupling, therefore explaining the N_f-dependence of the critical coupling in Fig.1.

To make a semi-quantitative estimate, we use the following magnetic scaling relations that connect the monopole mass and density with gauge coupling: $m^*/T \sim 1/g$ and $n^{1/3}/T \sim g^2$ [7]. Combined with the above condition, we obtain the critical gauge coupling for monopole condensation to be: $g(N_f) = g_0\,(1+f)^{4N_f N_M/15}$ where g_0 is the corresponding critical coupling for pure gauge case. This can be further converted to the lattice coupling $\beta = 2N_c/g^2$:

$$\beta_c(N_f) = \beta_0(1+f)^{-8N_f N_M/15} \tag{4}$$

For fundamental fermions (with $N_M = 1$) the critical (lattice) gauge coupling, $\beta_c \equiv \beta(T = T_c) = 2N_c/g^2(T_c)$ as a function of N_f, the resulting curve is shown in fig.1. We've used the above Eq.(4) to make a fit for these data (blue boxes) and obtained the optimal value $f \approx 0.154$: the fitting curve is shown as the thick blue line in Fig.1. Our model formula in Eq.(4) with one parameter nicely describes all the data points in [1] from $N_f = 0$ to $N_f = 12$! The suppression factor f of monopole-quark as compared with pure monopole may be understood as follows: the monopole-quark has color-electric charge, and in the near-T_c plasma it was known from previous studies of the "electric mass" on the lattice, telling us that the electric particles are heavier than the magnetic particles by roughly $\Delta M \sim 2T_c$, thus leading to a suppression factor $\sim e^{-\Delta M/T} \sim 0.135$ which is fairly close to $f \approx 0.154$.

It is well known that the $N_c = 2$ theory is a very special case, with extra symmetry between quarks-antiquarks and mesons-diquarks. It also allows the finite density lattice simulations without the "sign problem". Lattice study of this theory was recently extended to the low-T finite-μ region with $N_f = 2, 4$ quarks by Hands, et al [12]. The quark density (per flavor) shown in their Fig.3 displays (i) a structure at $\mu \approx m_\pi/2$ due to rotation from the $\bar\psi\psi$ to diquark condensate; (ii) the usual quark Fermi sphere at higher μ; and (iii) an unexpected growth of quark density at still higher μ. The deconfinement as per the Polyakov loop appears concurrent with (iii). We now propose that this extra quark density (iii) appears due to the condensate of the *monopole-quark* states. At low T of this simulation the dominant monopole-single-quark objects, being bosonic, would appear mostly as a condensate, like the diquarks. Assuming standard effective potential with a repulsive binary interaction $V_{int} = \lambda n^2/4$, one gets the condensate density growing linearly with μ, $n_{BEC} \sim (\mu - m)/\lambda$ at $\mu > m$, which is consistent with observations. Furthermore, a similar density per flavor for both $N_f = 2$ and $N_f = 4$ is consistent with our view that these objects are dominantly states with only a *single* quark per monopole. This proposal should and can be checked in many ways.

4. Chiral symmetry breaking and Zero Mode Zone

Spontaneous breaking of $SU(N_f)$ chiral symmetry has been modeled by a hypothetical Nambu-Jona-Lasinio 4-fermion interaction since 1960's, and attributed to instanton-induced 't Hooft interactions in 1980's. Its detailed studies in 1990's the context of the so called Instanton Liquid Models (ILM) are reviewed in [13]. Chiral restoration at $T > T_c$ in this approach can be viewed as a structural phase transition in the instanton ensemble, from a random plasma at low T into a gas of strongly correlated $\bar{I}I$ instanton-antiinstanton pairs [19]. The pairing mechanism is due to the fermion exchange, thus it gets stronger as N_f grows. Therefore studies covered in [13] were not able to confirm existence of chiral symmetry breaking for $N_f > 5$.

The nonzero holonomy (VEV of Polyakov line different from 1) becomes important at $T < 2T_c$ and, in particular, it modifies the instantons. Classical solutions with nonvanishing fields at spatial infinity has been constructed [14, 16]: they revealed "instanton quarks", or (anti)self dual constituent dyons, which in a set of N_c form an instanton. The names and quantum numbers of those (for the simplest $SU(2)$ gauge group we will discuss below) are given in the Table 1. The *redesign of the Instanton Liquid in terms of these dyons* is the main topic of the rest of the talk.

name	E	M	mass
M	+	+	v
$\bar{M}$	+	-	v
L	-	-	$2\pi T - v$
$\bar{L}$	-	+	$2\pi T - v$

Table 1. The charges and the mass (in units of $8\pi^2/g^2T$) for 4 SU(2) dyons.

The main consequence of the ILM is the formation of the so called "Zero Mode Zone" (ZMZ) of fermionic Dirac eigenvalues inside some narrow strip

$$|\lambda| < \sigma \sim < T_{IA} > \sim \frac{\rho^2}{R^3} \sim 20\, MeV \tag{5}$$

where we used the typical instanton size and separation $\rho \approx 1/3\,fm, R \approx 1fm$ [20]. Alternatively, one can estimate this width from requirements that their total number in a give lattice volume is about the number of instantons/antiinstantons V_4/R^4 since those are supposed to be made of their *collectivized* zero modes. ZMZ states are qualitatively different from perturbative fermionic plane waves and concentrate in them the chiral symmetry breaking. ILM studies had shown that the lowest hadrons, pions especially, are "made of" the ZMZ states. This has also been confirmed in lattice studies in which only the ZMZ contribution has been kept in quark propagators. Recent lattice work [21] had demonstrated what happens if one *removes*

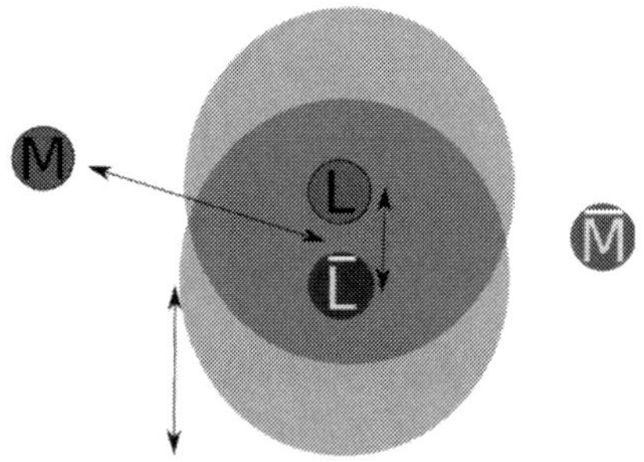

Figure 2. The schematic picture of the dyonic molecule, for 2 colors and large N_f.

the ZMZ states from propagators: the hadronic spectroscopy changes drastically, with chiral partners (A_1 for ρ, $N*(1/2^-)$ for the nucleon) getting much lighter and near-degenerate with their partners. Thus a tiny fraction $\sim 10^{-3}$ of all states inflict $O(1)$ changes of the hadronic masses!

Quasizero or ZMZ fermionic states are also responsible for main fluctuations in the Dirac matrix inversion and bad convergence of chiral perturbation theory on the lattice, as fermionic masses used are not small compared to its width: so even from practical perspective of effective usage of computer time one should study and understand them better than other states. Looking at the eigenvectors $\psi_\lambda(x)$ one tried to understand their nature. Early studies which used a bit "cooled" lattice configurations had identified instantons, with ILM size and density. Yet uncooled configurations had shown structures more consistent with lower dimensional topology [22, 23]. How well the "dyonic vacuum" we study now can explain those observations remains to be studied.

5. Statistical mechanics of the selfdual dyons

Here I report two papers in which this issue is addressed. The first [24] deals with a number of qualitative observations made in various lattice works, which can be understood directly from the main properties of the dyonic zero modes. For example, one can see why there is such huge difference between periodic or anti periodic fermions, or between different Z_N sectors (fermions couple to different dyons of different mass).

Due to a very limited space, we only discuss one effect, related with large N_f behavior of the chiral symmetry boundary. Depending which kind of fermions the theory has, their zero modes fall on different dyons. For physical antiperiodic fermions those mode exist for (twisted) or $L, \bar{L}$ dyons. As the number of fundamental fermions N_f in the theory increases, they bind this dyon-antidyon pair into tight $L\bar{L}$ "clusters", which play a role of the nucleus of these molecules, see Fig.5. Chiral symmetry is unbroken and the lowest Dirac eigenstates "at the gap" are due to independent $L\bar{L}$ clusters,

The standard Abelian electric charges of both L and $\bar{L}$ are equal to -1, so the cluster nuclei of such molecule has a charge -2, so it looks like anti-He, with $M, \bar{M}$ as "positrons" around it. (This sign of a charge does not violate C parity, of course, because the Abelian projected fields are color-projected to color direction of the holonomy field $\sim \mathrm{Tr}(F_{\mu\nu}A_4)$ and while both non-Abelian field in it are C-odd, the product is C-even. So these signs is just matter of the definitions already used in the field.)

One qualitative difference due to the nonzero holonomy is that the fermions are "Higgsed", with nonzero masses (in SU(2)) $m_f = \pm gv/2$. Therefore zero modes at large distances are exponentially decreasing. The fermion-induced binding of the dyons is of course further enhanced by the number N_m of fermionic zero modes. Defining potential via $e^{-V} \sim \det T \sim e^{-N_m m_f r}$ one finds a linear confining potential for the $L\bar{L}$ dyons, producing "clusters" of the size $< r >\sim (N_m m_f)^{-1}$, decreasing at large N_f. A dilute gas made of small $L\bar{L}$ clusters does not break chiral symmetry: it will only happen if the inter cluster distances are similarly small. The only way to do so is to move into stronger coupling and generate much higher dyon density. In any case, the expression (5) for the width of the ZMZ is modified, substituted by the exponentially small value of "hopping amplitudes" $< T >\sim exp(-N_m M_f R)$ between the L and $\bar{L}$ dyons.

The second paper [25] formulates the partition function and perform the first Monte-Carlo simulations of the ensemble. We are not presenting the simulation results at this point yet, but outline the structure of the partition function. The bosonic interactions of the dyons is separated into the so called *moduli space metric*, associated with bosonic zero-modes, and *electric screening* associated with the nonzero bosonic modes. The third essential ingredient of the partition function is the *fermionic determinant*, so schematically it is written as

$$dZ \sim \{dX_i\} e^{-S_c} \,\mathrm{det}G \,\mathrm{det}F_{zm} \frac{\mathrm{det}' F_{nzm}}{\sqrt{\mathrm{det}' B}} \tag{6}$$

where the first factor is the product of the differentials of all collective variables, the exponent contains the classical actions of all dyons, two subsequent determinants of matrices G and F_{zm} are related to bosonic and fermionic zero-modes, respectively. Two remaining (primed) determinants correspond to nonzero-modes of the one-loop bosonic and fermionic operators describing small linearized perturbations of the classical solution.

The moduli spaces metric is described by the matrix G in the approximation proposed by Diakonov [18]. For arbitrary $SU(N_c)$ group it is

$$[\hat{G}]_{mn} = \delta_{mn}(4\pi\nu_m + \frac{1}{T|x_m - x_{m-1}|} + \frac{1}{T|x_m - x_{m+1}|}) - \frac{\delta_{mn-1}}{T|x_m - x_{m+1}|} - \frac{\delta_{mn+1}}{T|x_m - x_{m-1}|}$$

where the indexes m, n run over the different types of dyons. When the system is dilute, these expressions reduce to Coulombic classical plasma, with electric and magnetic charges. Its behavior at higher densities is non-trivial. For two identical monopoles the shape/metric is known as the famous Atiyah-Hitchin manifold: and we need to simulate the anybody version of it. The obvious condition is that the simulation is restricted to positive positive eigenvalues of the matrix G. We also found that Coulombic terms in it need to be regulated at small distances.

The so called screening term (from non-zero mode determinant) produces effective linear $V \sim r$ potential between dyons, with sign-altering coefficient proportional to the Debye mass. This as well as the Coulomb terms require simulation of only charge-neutral systems: this forces us to have equal number of all $L, M, \bar{L}, \bar{M}$ dyons. We also decided not to use transitional 4-torus geometry, and use S^3 sphere instead. Current study is done with 64 dyons.

The fermionic zero modes of the self dual (antisefdual) dyons are treated as the zero modes of individual instantons/antiinstantons in the ILM. One writes the Dirac operator as a matrix of the type

$$i\not{D} = \begin{pmatrix} im & T_{SA} \\ T_{AS} & im \end{pmatrix}, \tag{7}$$

with sermonic masses on the diagonal and the overlap sub-matrices T_{SA} in the other two blocks: those are defined by

$$T_{SA} = \int d^4x\, \psi^\dagger_{0,Selfdual}(x-z_I) i\not{D} \psi_{0,Antiselfdual}(x-z_A) \tag{8}$$

Here S, A are understood as indices which run over all selfdual and antiselfdual dyons, and ψ_0 are their fermionic zero modes. The individual matrix elements have the meaning of a hopping amplitude for a quark from one pseudo-particle to another, and the determinant of this matrix is nothing else but the sum over the loop diagrams in which quarks visit each relevant dyon once.

At high temperature – low dyon density – the approximate factorization of the Dirac matrix into independent 2×2 boxes (separate dyon-antidyon clusters) explain disappearance of the near-zero eigenvalues and the so called spectral gap. As T goes down and effective coupling constant g grows, the dyons become lighter and thus more numerous: eventually they form dense enough ensemble to develop ZMZ with a nonzero density at $\lambda = 0$, breaking spontaneously the chiral symmetry. We had studied about two orders of magnitude variation of the dyon density, with the number of light fermions $N_f = 0, 1, 2, 4$.

The simulation determines correlation functions of the dyons and the chiral phase transition line, as a function of the dyon density. We also calculate the free energy of the system, relative to that of noninteracting dyon gas. Its magnitude explains the strength of the back reaction, the dyonic contribution to the effective holonomy potential. Where this potential becomes comparable to perturbative one, it cancels it and widely fluctuating holonomy matrix leads to confinement.

At the moment we are not calculating the fermionic propagators and hadron correlation functions, but it can be done in a way similar to that in ILM. Another set of questions for future studies is whether the resulting ZMZ states do or do not agree with those seen on the lattice.

Acknowledgements

I am thankful to organizers for invitation to speak at this meeting, especially valuable as I was not attended lattice-related conferences for quite some time. Needless to say, I am much indebted to my collaborators Jinfeng Liao, Pietro Faccioli and Tin Sulejmanpasic.

[1] K. Miura, M. P. Lombardo and E. Pallante, arXiv:1110.3152 [hep-lat].
[2] A. Deuzeman, M. P. Lombardo and E. Pallante, arXiv:1012.6023 [hep-lat].
[3] A. Cheng, A. Hasenfratz and D. Schaich, arXiv:1111.2317 [hep-lat].
[4] N. Seiberg, E. Witten, Nucl. Phys. **B426**, 19-52 (1994). [hep-th/9407087].
[5] C. Bonati, G. Cossu, M. D'Elia and A. Di Giacomo, Phys. Rev. D **85**, 065001 (2012) [arXiv:1111.1541 [hep-lat]].
[6] A. D'Alessandro, M. D'Elia and E. V. Shuryak, Phys. Rev. D **81**, 094501 (2010) [arXiv:1002.4161 [hep-lat]].
[7] J. Liao and E. Shuryak, Phys. Rev. C **75**, 054907 (2007) [arXiv:hep-ph/0611131].
[8] C. Ratti and E. Shuryak, Phys. Rev. D **80**, 034004 (2009) [arXiv:0811.4174 [hep-ph]].
[9] E. Shuryak, arXiv:1204.2806 [hep-lat].
[10] M. Cristoforetti, E. Shuryak, Phys. Rev. **D80**, 054013 (2009).
[11] J. Liao and E. Shuryak, PRL in press, arXiv:1206.3989 [hep-ph].
[12] S. Hands and P. Kenny, Phys. Lett. B **701**, 373 (2011) [arXiv:1104.0522 [hep-lat]].
[13] T. Schafer and E. V. Shuryak, Rev. Mod. Phys. **70**, 323 (1998) [arXiv:hep-ph/9610451].
[14] K. -M. Lee and C. -h. Lu, Phys. Rev. D **58**, 025011 (1998) [hep-th/9802108].
[15] K. M. Lee and P. Yi, Phys. Rev. D **56**, 3711 (1997) [arXiv:hep-th/9702107].
[16] T. C. Kraan and P. van Baal, Phys. Lett. B **435**, 389 (1998) [arXiv:hep-th/9806034].
[17] F. Bruckmann, D. Nogradi and P. van Baal, Acta Phys. Polon. B **34**, 5717 (2003) [arXiv:hep-th/0309008].
[18] D. Diakonov, *Topology and Confinement*, arxiv:0906.2456v1 (2009)
[19] E. M. Ilgenfritz and E. V. Shuryak, Nucl. Phys. B **319**, 511 (1989).
[20] E. V. Shuryak, Nucl. Phys. B **203**, 93 (1982).

[21] L. Y .Glozman, C. B. Lang and M. Schrock, Phys. Rev. D **86**, 014507 (2012) [arXiv:1205.4887 [hep-lat]].
[22] I. Horvath, A. Alexandru, J. B. Zhang, Y. Chen, S. J. Dong, T. Draper, F. X. Lee and K. F. Liu *et al.*, Phys. Lett. B **612**, 21 (2005) [hep-lat/0501025].
[23] P. V. Buividovich, T. Kalaydzhyan and M. I. Polikarpov, arXiv:1111.6733 [hep-lat].
[24] E. Shuryak and T. Sulejmanpasic, Phys. Rev. D **86**, 036001 (2012) [arXiv:1201.5624 [hep-ph]].
[25] P. Faccioli and E. Shuryak, arXiv:1301.2523 [hep-ph].

Journal of Physics: Conference Series **432** (2013) 012023 doi:10.1088/1742-6596/432/1/012023

A subset solution to the sign problem in simulations at non-zero chemical potential

Jacques Bloch
Institute for Theoretical Physics, University of Regensburg, 93040 Regensburg, Germany
E-mail: jacques.bloch@physik.uni-regensburg.de

Abstract. At nonzero quark chemical potential lattice QCD simulations with dynamical quarks are hampered by the sign problem caused by the complex fermion determinant. In this talk we discuss the relation between QCD and random matrix theory and present a novel solution to the sign problem in dynamical random matrix simulations. The sign problem is solved by gathering matrices in subsets, whose sums of complex determinants are real and positive. Markov chains of relevant subsets were generated in Monte Carlo simulations and various observables, like the chiral condensate and quark number, were accurately computed. The subset formula, which is at the basis of the positivity proof for the subset weights, also provides a simple solution to the Silver Blaze puzzle.

1. Introduction

The main motivation for this work is the study of the phases of QCD as a function of temperature and chemical potential. Although the phase diagram is believed to resemble Fig. 1, QCD has in fact only been studied extensively at nonzero temperature T for zero chemical potential μ while very little is known about its behavior when μ is nonzero. It is therefore of utmost importance to make progress in the study of QCD at nonzero chemical potential.

The major reason for this deficit is that dynamical lattice simulations of QCD at nonzero chemical potential are prohibitively expensive due to the *sign problem* caused by the complex fermion determinant [1]. This complex nature prohibits the inclusion of the fermion determinant

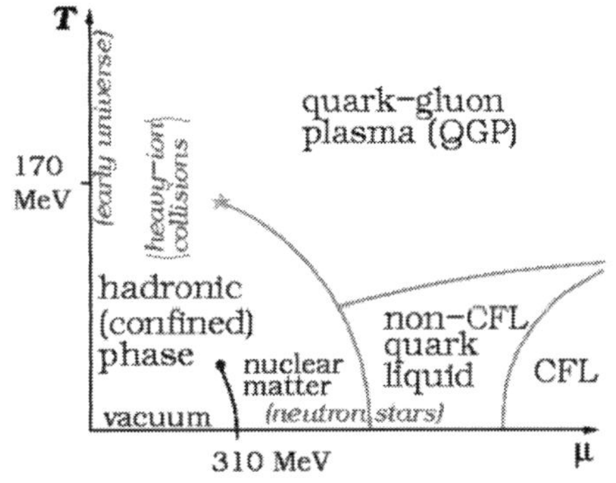

Figure 1. Conjectured QCD phase diagram.

in the importance sampling weight of Monte Carlo simulations. The methods developed so far to circumvent this problem, e.g., reweighting using auxiliary ensembles, continuation from imaginary μ and Taylor series expansion around $\mu = 0$, typically require an amount of work that increases exponentially with growing volume and chemical potential in order to produce reliable results, and are therefore only usable at small μ.

Note that the sign problem also occurs in theories other than QCD, and for some of these simpler models alternative methods, like complex Langevin or world-line and loop formulations, were able to simulate the model without encountering a sign problem [2, 3].

In this talk we will present a subset solution to the sign problem in a random matrix model, which is equivalent to QCD in the ε-regime [4, 5]. In Sec. 2 we introduce QCD at nonzero chemical potential and its relation to random matrix theory. In Sec. 3 we explain the subset method for dynamical simulations of the random matrix model. In Sec. 4 we present various numerical results computed with the method. Finally, we conclude with a summary in Sec. 5.

2. QCD and random matrix theory

2.1. QCD at nonzero quark chemical potential

After integration over the fermion fields the QCD partition function for N_f dynamical fermions is given by

$$Z_{\rm QCD} = \int \mathcal{D}A_\mu \, \underbrace{e^{-S_G} \prod_{f=1}^{N_f} \det[D(m_f, \mu)]}_{\text{MCMC weight function ?}} . \tag{1}$$

with gauge fields A_μ, gauge action S_G and Dirac operator $D(m, \mu) = m + \gamma_\mu \partial^\mu_{\rm cov} + \mu\gamma_4$ for a quark of mass m with quark chemical potential μ. Because of the hermiticity and chirality properties of the Dirac operator its determinant satisfies

$$\det[D(m_f, \mu)] \text{ is } \begin{cases} \text{real and} \geq 0 & \text{for } \mu = 0 \\ \text{complex} & \text{for } \mu \neq 0 \end{cases} . \tag{2}$$

At nonzero chemical potential QCD has a sign problem as the fermion determinant cannot be interpreted as a probabilistic weight in Monte Carlo simulations.

2.2. The random matrix model

To investigate the sign problem in QCD one can make use of the equivalence between QCD in the ε-regime ($m_\pi, \mu \ll \frac{1}{L} \ll \Lambda$) and random matrix theory (RMT), which was derived from chiral perturbation theory (χPT) and holds both at zero and nonzero chemical potential.

One matrix model which was shown to satisfy this equivalence in its universal limit is the two-matrix model of Osborn [6]. In this model, ϕ_1 and ϕ_2 are complex $(N + \nu) \times N$ matrices distributed according to partition function

$$Z_\nu^{N_f}(\mu; \{m_f\}) = \int d\phi_1 d\phi_2 w(\phi_1)\, w(\phi_2) \prod_{f=1}^{N_f} \det D_{\mu, m_f}(\phi_1, \phi_2), \tag{3}$$

with Gaussian weights $w(\phi) \propto \exp(-N \operatorname{tr} \phi^\dagger \phi)$ and Dirac matrix (with ν zero modes):

$$D_{\mu,m}(\phi_1, \phi_2) = \begin{pmatrix} m & i\phi_1 + \mu\phi_2 \\ i\phi_1^\dagger + \mu\phi_2^\dagger & m \end{pmatrix} . \tag{4}$$

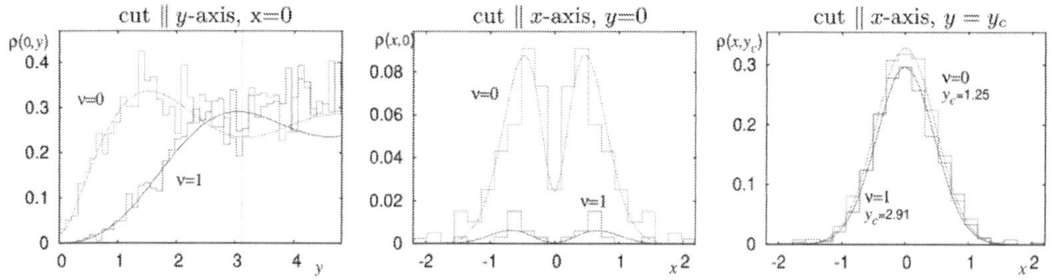

Figure 2. Three cuts of the microscopic spectral density over the overlap operator at chemical potential $\mu = 0.1$ on a 4^4 lattice using 8703 quenched configurations at $\beta = 5.1$. The curves clearly distinguish between topological sectors $\nu = 0$ and 1 [18].

The equivalence with QCD holds in the microscopic limit of RMT, where $N \to \infty$ while $\hat{m} = 2Nm = \Sigma V m_{\rm QCD}$ and $\hat{\mu}^2 = 2N\mu^2 = F^2 V \mu_{\rm QCD}^2$ are kept fixed.

Similarly to QCD, the Dirac matrix in the matrix model becomes non-Hermitian for $\mu \neq 0$. Its massless spectrum no longer lies on the imaginary axis, but spreads in the complex plane and is contained in a strip along the imaginary axis with width $\approx 4\mu^2$. Its determinant is generically complex and dynamical simulations of the random matrix model also suffer from a sign problem.

Much information was gathered about the QCD Dirac operator and the sign problem using RMT. There are analytical solutions for the microscopic spectral densities [6, 7] and individual eigenvalue distributions [8]. There are also analytical results for the average phase of the fermion determinant [9–13], which characterizes the severity of the sign problem.

At finite μ the Banks-Casher relation ($\Sigma \propto \rho(0)$) is no longer valid and a new mechanism was uncovered relating the strong oscillations in the Dirac spectrum when $\mu > m_\pi/2$ with the discontinuity of Σ at $m = 0$ [14, 15]. This mechanism also explains the *Silver Blaze puzzle*, i.e. the fact that the thermodynamical quantities are independent of μ up to $m_{\rm N}/3$ in the full theory, while the quenched and phase-quenched approximations violate the Silver Blaze for $\mu > m_\pi/2$. An analogous scenario was recently found explaining the Silver Blaze for the quark number density [16].

2.3. Matching QCD to RMT

The equivalence between RMT and QCD in the ε-regime can be used to determine the low energy constants Σ and F of χPT by fitting Dirac spectra of lattice QCD to RMT predictions. Because of the sign problem in dynamical simulations of QCD, RMT predictions with real chemical potential μ were only verified for quenched QCD. The agreement of the microscopic spectral density was successfully verified for staggered fermions [17] and for overlap fermions [8,18], which have the additional peculiarity of distinguishing between different topological sectors. A typical result is shown in Fig. 2. In a study of the overlap operator with imaginary chemical potential realistic values of Σ and F were computed [19].

2.4. Dynamical simulations and sign problem in RMT

As the Dirac operator $D_{\mu,m}$ is non-Hermitian for $\mu \neq 0$ its determinant $\det[D_{\mu,m}] \equiv Re^{i\theta}$ is generically complex. The average phase factor $\langle e^{2i\theta}\rangle$ reflects the fluctuations of the fermion determinant and characterizes the strength of the sign problem in dynamical simulations:

$$\langle e^{2i\theta}\rangle_{N_f} = \left\langle \frac{\det[D_{\mu,m}]}{\det[D^\dagger_{\mu,m}]} \right\rangle_{N_f} = \frac{Z_\nu^{N_f+1|1^*}(\mu, m, m_f)}{Z_\nu^{N_f}(\mu, m_f)}, \quad (5)$$

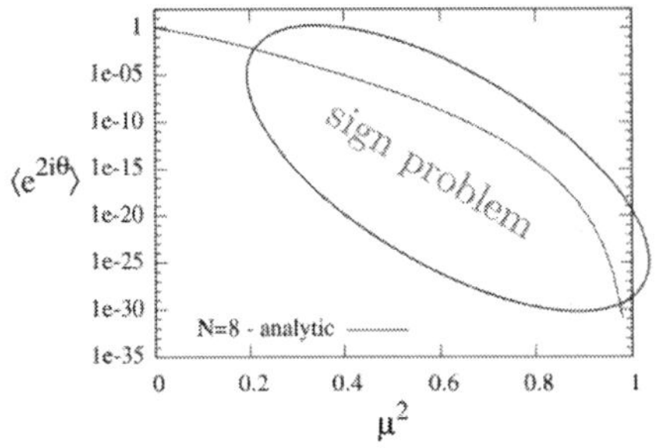

Figure 3. Average phase factor $\langle e^{2i\theta}\rangle$ vs μ^2 in the phase-quenched RMT ensemble for $N = 8$.

where the numerator in the right-hand side is the partition function with $N_f + 1$ quarks and one conjugate bosonic quark. The dependence of the average phase on the chemical potential is shown in Fig. 3. In dynamical simulations the sign problem will show up when the statistical sampling is flooded by the strong fluctuations of the phase.

Dynamical simulations of the random matrix partition function at nonzero μ can be used to probe the sign problem numerically and can serve as a playground for algorithmic developments. Standard reweighting methods only work at small μ. In the next section we will introduce a subset method which performs partial resummations of configurations to overcome the sign problem.

3. Subset method

In the subset method [4, 5] the random matrix configurations are gathered in subsets using the following principle: For of any given configuration $\Phi = (\phi_1, \phi_2)$ consider the set of matrices

$$\Omega(\Phi) = \left\{\Psi(\Phi;\theta_n) : \theta_n = \frac{\pi n}{N_s} \wedge n = 0, \ldots, N_s - 1\right\}, \tag{6}$$

containing N_s *orthogonal rotations* $\Psi = (\psi_1, \psi_2)$ of Φ defined by:

$$\begin{pmatrix}\psi_1\\ \psi_2\end{pmatrix} = \begin{pmatrix}\cos\theta & \sin\theta\\ -\sin\theta & \cos\theta\end{pmatrix}\begin{pmatrix}\phi_1\\ \phi_2\end{pmatrix}. \tag{7}$$

The Gaussian weights $W(\Omega) \equiv w(\psi_1)w(\psi_2)$ in the partition function are independent of θ such that all $\Psi(\Phi;\theta)$ in a subset Ω belong to the RMT ensemble and have the same Gaussian weights. If we factorize out the Gaussian weight, the remaining weight for the subset is given by the sum of fermion determinants, which we denote as fermionic subset weight:

$$\sigma_\Omega(\mu, m) = \sum_{n=0}^{N_s-1} \det{}^{N_f} D_{\mu,m}[\Psi(\Phi;\theta_n)]. \tag{8}$$

It is easy to see that the integral over all subsets Ω,

$$Z = \int d\Omega\, W(\Omega)\, \sigma_\Omega(\mu, m), \tag{9}$$

is an N_s-fold covering of the original RMT partition function such that both formulations are equivalent. In the subset formulation, the observables of the original RMT ensemble are computed as:

$$\langle O\rangle = \frac{1}{Z}\int d\Omega\, W(\Omega)\, \sigma_\Omega(\mu, m)\, \langle O\rangle_\Omega \tag{10}$$

with *subset measurements*

$$\langle O \rangle_\Omega = \frac{1}{\sigma_\Omega} \sum_{n=0}^{N_s-1} \det{}^{N_f} D(\Psi_n)\, O(\Psi_n), \qquad \Psi_n \in \Omega. \tag{11}$$

Of course, rewriting the partition function in terms of subsets does not a priori improve on the sign problem. What does cure the sign problem is that the special construction (6) of the subsets satisfies the following *positivity theorem:* For any such subset Ω the fermionic subset weight $\sigma_\Omega(\mu, m)$ is *real* and *positive* if $N_s > N_f N$ (for arbitrary m and $\mu < 1$). This theorem immediately follows from the subset formula:

$$\sigma_\Omega(\mu, m) = (1-\mu^2)^{N_f(N+\frac{\nu}{2})} \sigma_\Omega\left(0, \frac{m}{\sqrt{1-\mu^2}}\right) \qquad \text{if} \quad N_s > N_f N, \tag{12}$$

which was proven in Ref. [5]. As $\sigma_\Omega(0, m)$ is real and positive for any $m \in \mathbb{R}$, the left-hand side $\sigma_\Omega(\mu, m)$ is real and positive for $\mu < 1$. Note that for $m = 0$ and $\mu = 1$ we have $\sigma_\Omega(1, 0) = 0$, i.e., the sum of determinants is exactly zero for all subsets and the system is maximally non-Hermitian.

4. Simulations and results

We use the positive subset weights $W(\Omega)\sigma_\Omega$ to sample the ensemble of subsets using the Metropolis algorithm and generate a Markov chain with N_{MC} *relevant* subsets of random matrix configurations. The subset size is set to its minimum value required to ensure real positivity, i.e., $N_s = N_f N + 1$. Fermionic subset weights are computed by numerical evaluation of their N_s constituent determinants using an LU decomposition. The sample average $\overline{O}$ of an observable O in the RMT ensemble is computed as:

$$\overline{O} = \frac{1}{N_{\mathrm{MC}}} \sum_{k=1}^{N_{\mathrm{MC}}} \langle O \rangle_{\Omega_k}, \tag{13}$$

with the subset measurements $\langle O \rangle_{\Omega_k}$ defined in Eq. (11).

4.1. Chiral condensate and quark number density

We used the subset method to compute the chiral condensate

$$\Sigma = \frac{1}{2N} \frac{1}{Z} \frac{dZ}{dm} = \left\langle \frac{1}{2N} \operatorname{tr} D^{-1} \right\rangle \tag{14}$$

and the quark number density

$$n = \frac{1}{2N} \frac{1}{Z} \frac{dZ}{d\mu} = \left\langle \frac{1}{2N} \operatorname{tr} \left[\begin{pmatrix} 0 & \phi_2 \\ \phi_2^\dagger & 0 \end{pmatrix} D^{-1} \right] \right\rangle, \tag{15}$$

and compare our results with those obtained with standard reweighting methods. For the latter we generate $N_{MC} \times N_s$ random matrices so that we use the same total number of matrices as in subset method. We also compare the results with known analytical expressions.

We performed simulations for matrix sizes $N = 2, \ldots, 34$ with $N_f = 1$ and $m = 0.1/2N$, i.e., the mass is small with respect to the magnitude of the smallest eigenvalue. The number of Monte Carlo samples was $N_{MC} = 100,000$. The statistical errors displayed in the figures take into account the autocorrelations.

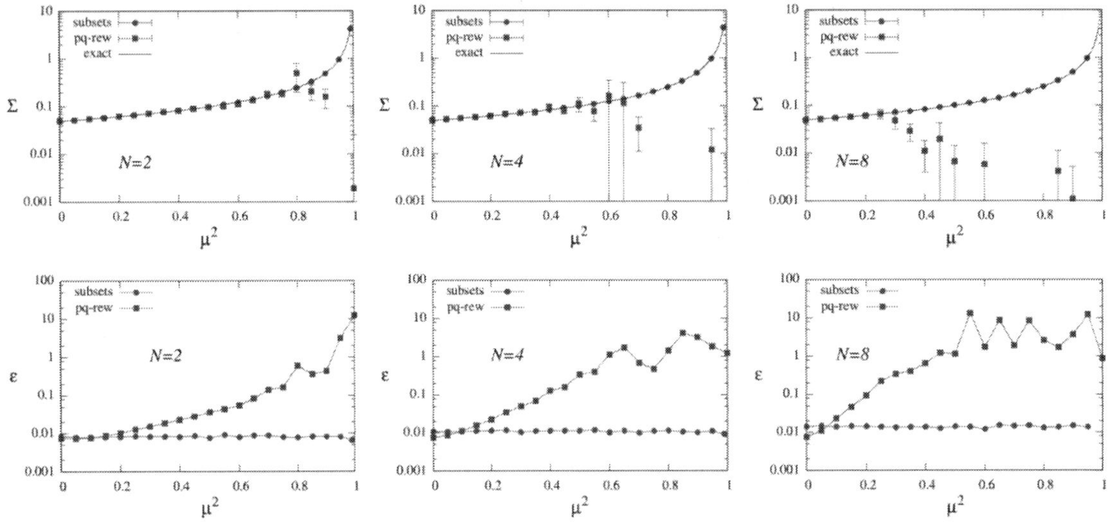

Figure 4. Chiral condensate Σ vs μ^2 for $N = 2, 4, 8$. The top row displays the data for the subset method (blue bullets), the phase-quenched reweighting method (red squares) and the analytic curve (solid line). The bottom row displays the relative statistical error ε on the data.

In Fig. 4 we show the chiral condensate (14) as a function of the chemical potential for $N = 2, 4, 8$. The results computed with the phase-quenched reweighting method clearly suffer from the sign problem, which sets in for smaller and smaller μ as N is increased. In contrast, the results of the subset method agree with the analytical curve from Ref. [15] over the complete μ-range. This is confirmed by the statistical error which is stable for all values of μ.

Similarly, Fig. 5 shows the results for the average quark number density (15). Again, the subset method agrees extremely well with the analytical curve, while the reweighting data are clearly plagued by the sign problem. Note that the relative error for the subset method is extremely small in this case; a peculiarity which can be understood from Eq. (21).

As the sign problem gets stronger with increasing volume, it is also interesting to look at the behavior of the error as a function of N; see Fig. 6 for the chiral condensate. For the subset method the N dependence is mild and almost independent of μ, however, for the reweighting method the error increases very rapidly with N until the method breaks down altogether.

Figures 4-6 very much show the power of the subset method to solve the sign problem in these dynamical random matrix simulations.

4.2. Understanding the chiral condensate

At $\mu = 0$ the Banks-Casher formula relates the nonzero chiral condensate to the spectral density of the Dirac operator at the origin. At $\mu \neq 0$ this relation no longer holds and an alternative mechanism was uncovered by Osborn, Splittorff and Verbaarschot [14], which relates Σ to extreme, microscopic oscillations in the eigenvalue spectrum of the Dirac operator when $m < 2\mu^2$ (see Fig. 7). As this region of parameter space exactly coincides with the sign problem the mechanism cannot be studied numerically with standard simulation methods. However, as can be seen in Fig. 8, the subset method seems to be insensitive to the large spectral oscillations and very nicely reproduces the analytical predictions for the dynamical chiral condensate when the quark mass moves into the cloud of eigenvalues: the reliability of the method remains unchanged when m crosses the $2\mu^2$ threshold.

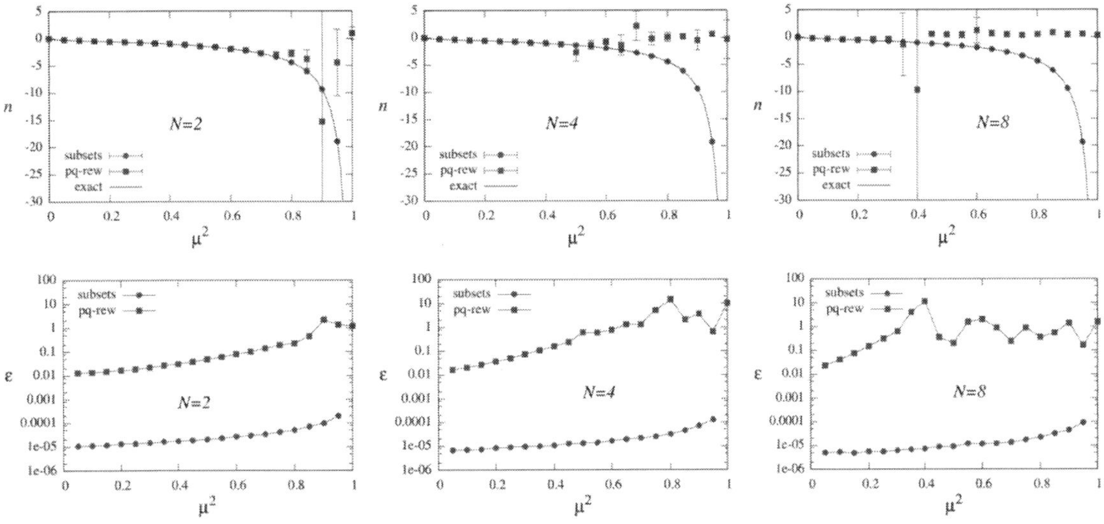

Figure 5. Quark number density n vs μ^2 for $N = 2, 4, 8$. The top row displays the data for the subset method (blue bullets), the phase-quenched reweighting method (red squares) and the analytic curve (solid line). The bottom row displays the relative statistical error ε on the data.

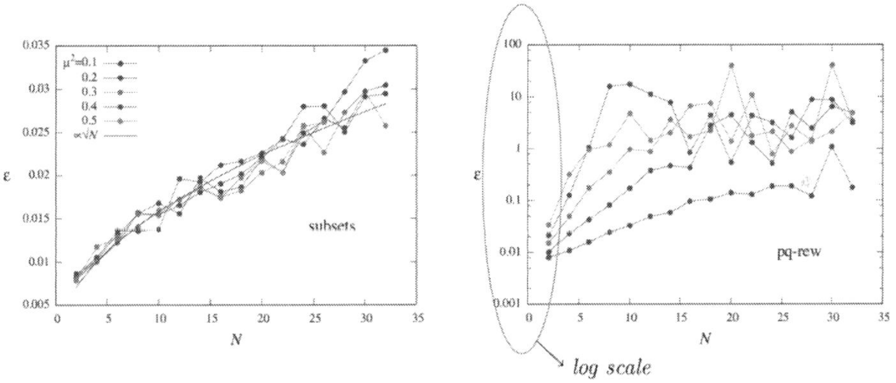

Figure 6. Relative statistical error on the chiral condensate as a function of the matrix size N for the subset method (left) and for phase-quenched reweighting (right) for chemical potential $\mu^2 = 0.1, 0.2, 0.3, 0.4, 0.5$.

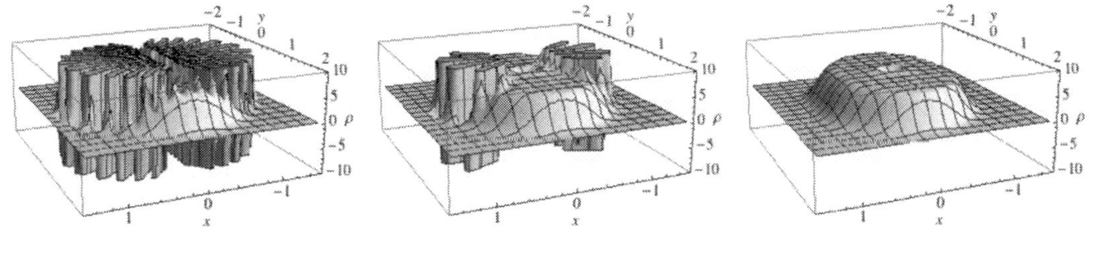

Figure 7. Spectral density $\rho(x + iy)$ of the Dirac operator for $N = 16$ and $\mu^2 = 0.6$. From left to right the quark mass is $m = 0.1, 0.4, 1.2$.

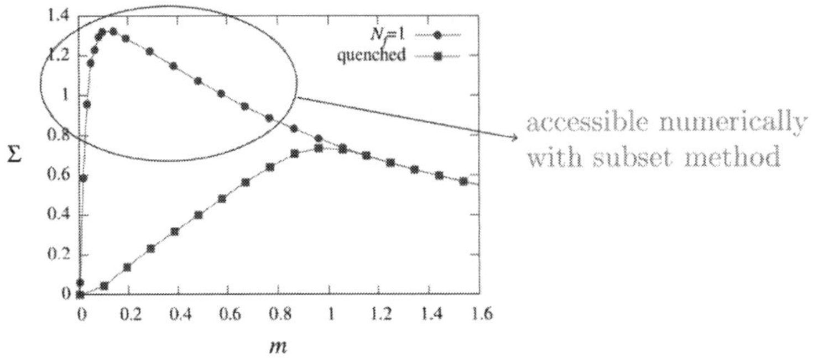

Figure 8. Dynamical and quenched chiral condensate as a function of the quark mass for $\mu^2 = 0.6$ and $N = 16$. The dynamical condensate was measured using the subset method.

4.3. Derivates from the subset weight formula

4.3.1. At the ensemble level Using the subset formula (12) the partition function and thermodynamic observables at chemical potential μ can be related to those at $\mu = 0$. For the $N_f = 1$ partition function we find

$$Z_{N_f=1}(\mu; m) = (1-\mu^2)^{N+\frac{\nu}{2}}\, Z_{N_f=1}\left(0; \frac{m}{\sqrt{1-\mu^2}}\right). \tag{16}$$

The derivatives with respect to the mass and chemical potential yield relations for the chiral condensate:

$$\Sigma(\mu; m) = \frac{1}{\sqrt{1-\mu^2}}\, \Sigma\left(0; \frac{m}{\sqrt{1-\mu^2}}\right), \tag{17}$$

and the quark number density:

$$n(\mu; m) = -\frac{\mu}{1-\mu^2}\left[\underbrace{1+\frac{\nu}{2N}}_{\text{main contribution}} - m\Sigma(\mu; m)\right]. \tag{18}$$

These relations immediately show that the Silver Blaze is satisfied in the microscopic limit, as

$$\Sigma(\hat{\mu}, \hat{m}) \to \Sigma(0; \hat{m}), \quad n(\hat{\mu}, \hat{m}) \to 0. \tag{19}$$

4.3.2. At the subset level The subset formula also allows us to relate the subset measurements of the thermodynamical observables at μ and $\mu = 0$ on individual subsets. The chiral condensate on a subset Ω obeys

$$\Sigma_\Omega(\mu; m) = \frac{1}{\sqrt{1-\mu^2}}\, \Sigma_\Omega\left(0; \frac{m}{\sqrt{1-\mu^2}}\right), \tag{20}$$

and the quark number density on a subset Ω satisfies

$$n_\Omega(\mu; m) = -\frac{\mu}{1-\mu^2}\left[\underbrace{1+\frac{\nu}{2N}}_{\text{main contribution}} - m\Sigma_\Omega(\mu; m)\right]. \tag{21}$$

Note that the main contribution is exactly reproduced by each subset, which explains why the relative error on n is so small (see Fig. 5).

It is interesting to note that the Silver Blaze is already solved on each individual subset:

$$\Sigma_\Omega(\hat{\mu}, \hat{m}) \to \Sigma_\Omega(0; \hat{m}), \quad n_\Omega(\hat{\mu}, \hat{m}) \to 0. \tag{22}$$

4.4. Sign problem in reweighting methods

In this section we take a closer look at the sign problem in reweighting methods. The ensemble average of y in an ensemble with weight w is given by

$$\langle y \rangle_w = \frac{\int dx\, w(x) y(x)}{\int dx\, w(x)}. \tag{23}$$

In reweighting an auxiliary ensemble with positive weight w_0 is introduced and the ensemble average is rewritten as

$$\langle y \rangle_w = \frac{\int dx\, w_0(x) \left[\frac{w(x)}{w_0(x)} y(x)\right]}{\int dx\, w_0(x) \left[\frac{w(x)}{w_0(x)}\right]} = \frac{\left\langle \frac{w}{w_0} y \right\rangle_{w_0}}{\left\langle \frac{w}{w_0} \right\rangle_{w_0}}. \tag{24}$$

The auxiliary distribution w_0 can be sampled using importance sampling, and the observables in the target ensemble can be evaluated using Eq. (24), even when w is complex. However, the expectation values in the numerator and denominator of (24) both involve cancellations that grow exponentially with the volume, and as they are computed by statistical sampling the work needed to make reliable measurements grows exponentially too and the sign problem sets in.

The denominator, denoted as the average reweighting factor r, is independent of the observable and its relative error can be used as a measure of the sign problem. Because of the large fluctuations of the complex weights, which are exactly at the origin of the sign problem, the reweighting factor can not be computed accurately in direct simulations of the auxiliary ensemble w_0. However, using the identity

$$r \equiv \left\langle \frac{w}{w_0} \right\rangle_{w_0} = \frac{\int dx\, w(x)}{\int dx\, w_0(x)} = \left[\left\langle \frac{w_0}{w} \right\rangle_w\right]^{-1}, \tag{25}$$

it can be computed indirectly to very high accuracy using the subset method as the subset measurements

$$\left\langle \frac{w_0}{w} \right\rangle_\Omega = \frac{1}{\sigma_\Omega(\mu, m)} \sum_{n=0}^{N_s-1} w_0(\Psi_n) \tag{26}$$

are all real and positive. The sign problem is caused by the relative error on the average reweighting factor, which is given by

$$\varepsilon_r = \sqrt{\frac{2\tau}{N_{\rm MC}}} \frac{\sigma_r}{r}, \tag{27}$$

where $N_{\rm MC}/2\tau$ is the number of independent measurements and σ_r is the variance of the reweighting factor in the auxiliary ensemble. We computed the reweighting factors and their relative standard deviations σ_r/r, which is proportional to the standard error ε_r, for the following reweighting schemes: quenched (q), phase-quenched (pq), μ-quenched (μq) and sign-quenched (sq). A comparison of the results for these schemes is shown in Fig. 9. The rapid increase of the relative standard deviation of the reweighting factors clearly shows that the sign problem quickly becomes unmanageable, independently of the chosen scheme, as $N_{\rm MC}$ would have to be increased accordingly to keep the error under control. It is noteworthy that the severity of the sign problem in reweighting methods can be measured accurately with the subset method.

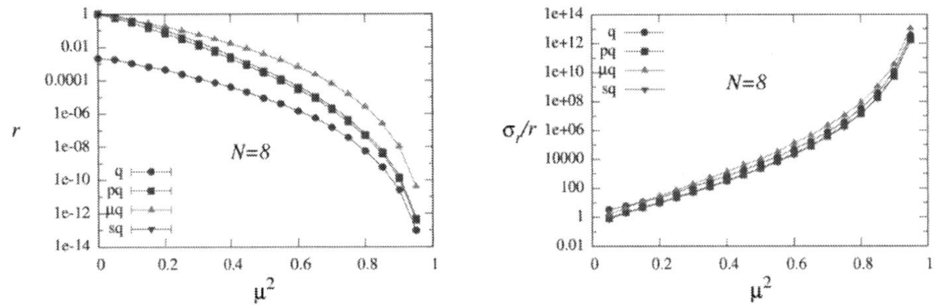

Figure 9. Comparison of the $N_f = 1$ reweighting factors (left) and their relative standard deviations σ_r/r (right) as a function of μ^2 for $N = 8$, for different reweighting schemes. These data are genuine predictions as there are no analytic results available for these quantities.

5. Summary

In this talk we presented a subset method that solves the sign problem in dynamical simulations of a two-matrix model. The matrices are gathered into subsets with real and positive weights, which can be simulated using importance sampling methods. In contrast to standard reweighting methods, where the work grows exponentially with N and μ due to the sign problem, the subset method is free of sign problem: Its results are reliable up to much larger values of μ^2 while the work only shows a mild dependence on N. The crucial feature of the subset method is that the cancellations, necessary to get the correct dynamical behavior, no longer happen stochastically through statistical sampling but deterministically inside subsets of size $\mathcal{O}(N)$.

To make further progress it will be useful to uncover the mechanism underpinning the subset method in order to extend these ideas to other models or even to QCD itself.[1]

References

[1] de Forcrand P 2009 *PoS* **LAT2009** 010 (*Preprint* 1005.0539)
[2] Aarts G 2009 *Phys. Rev. Lett.* **102** 131601 (*Preprint* 0810.2089)
[3] Chandrasekharan S 2008 *PoS* **LATTICE2008** 003 (*Preprint* 0810.2419)
[4] Bloch J 2011 *Phys. Rev. Lett.* **107** 132002 (*Preprint* 1103.3467)
[5] Bloch J 2012 *Phys. Rev.* **D86** 074505 (*Preprint* 1205.5500)
[6] Osborn J C 2004 *Phys. Rev. Lett.* **93** 222001 (*Preprint* hep-th/0403131)
[7] Akemann G, Osborn J C, Splittorff K and Verbaarschot J J M 2005 *Nucl. Phys.* **B712** 287–324 (*Preprint* hep-th/0411030)
[8] Akemann G, Bloch J C R, Shifrin L and Wettig T 2008 *Phys. Rev. Lett.* **100** 032002 (*Preprint* 0710.2865)
[9] Splittorff K and Verbaarschot J J M 2007 *Phys. Rev. Lett.* **98** 031601 (*Preprint* hep-lat/0609076)
[10] Osborn J, Splittorff K and Verbaarschot J 2008 *Phys.Rev.* **D78** 105006 (*Preprint* 0807.4584)
[11] Lombardo M P, Splittorff K and Verbaarschot J J M 2009 *Phys. Rev.* **D80** 054509 (*Preprint* 0904.2122)
[12] Bloch J and Wettig T 2009 *JHEP* **03** 100 (*Preprint* 0812.0324)
[13] Bloch J and Wettig T 2011 *JHEP* **05** 048 (*Preprint* 1102.3715)
[14] Osborn J C, Splittorff K and Verbaarschot J J M 2005 *Phys. Rev. Lett.* **94** 202001 (*Preprint* hep-th/0501210)
[15] Osborn J C, Splittorff K and Verbaarschot J J M 2008 *Phys. Rev.* **D78** 065029 (*Preprint* 0805.1303)
[16] Ipsen J and Splittorff K 2012 *Phys. Rev.* **D86** 014508 (*Preprint* 1205.3093)
[17] Akemann G and Wettig T 2004 *Phys. Rev. Lett.* **92** 102002 (*Preprint* hep-lat/0308003)
[18] Bloch J C R and Wettig T 2006 *Phys. Rev. Lett.* **97** 012003 (*Preprint* hep-lat/0604020)
[19] Lehner C, Bloch J, Hashimoto S and Wettig T 2011 *JHEP* **05** 115 (*Preprint* 1101.5576)
[20] Bloch J, Bruckmann F, Kieburg M, Splittorff K and Verbaarschot J 2012 (*Preprint* 1211.3990)

[1] After the conference new results revealed a connection between the subset method and a projection on canonical determinants [20].

Fermion bag solutions to some sign problems in four-fermion field theories

Anyi Li
Institute for Nuclear Theory, University of Washington, Seattle, 98195, USA
E-mail: anyili@uw.edu

Abstract. Lattice four-fermion models containing N flavors of staggered fermions, that are invariant under Z_2 and $U(1)$ chiral symmetries, are known to suffer from sign problems when formulated using the auxiliary field approach. Although these problems have been ignored in previous studies, they can be severe. In this talk, we show that the sign problems disappear when the models are formulated in the fermion bag approach, allowing us to solve them rigorously for the first time.

1. Introduction

Four-fermion field theories are interesting in both condensed matter and particle physics. The well known Hubbard model and its variants are often used in studying cuprate superconductors [1], antiferromagnets [2] and more recently graphene [3]. Low energy nuclear physics is also studied with four-fermion couplings in the effective field theory framework [4, 5]. In the context of more fundamental theories like QCD, four-fermion field theories offer a simpler setting to study phenomena like fermion mass generation and chiral symmetry breaking [6]. It has been suggested recently that quantum critical phenomena in graphene can be studied with four-fermion field theories [7, 8]. It has also been found there is a QCD-like sign problem in the four-fermion field theory [9]. Despite the wide interest, strongly coupled four-fermion field theories remain poorly understood as compared to their bosonic counterparts due to computational difficulties.

The only available method to compute quantities in a strongly interacting field theory with no small parameter is the Monte Carlo (MC) method. Due to the quantum nature of a fermion there are no natural fermion configurations with positive weights that can be used for importance sampling. In two space-time dimensions fermions can often be bosonized and models can we written in terms of world line configurations with positive weights. This fact can be used to design powerful MC methods [10, 11, 12]. In higher dimensions, the traditional MC approach is to integrate the fermions out in favor of a determinant of a large fermion matrix. Whenever this determinant is positive a non-local probability distribution emerges, which can be used to construct a MC method. The most popular is the Hybrid Monte Carlo (HMC) method [13, 14] which has continued to evolve in many ways since its discovery [15]. Unfortunately, small eigenvalues of the fermion matrix which naturally arise in the presence of massless fermions can cause singularities in the HMC approach. This makes it difficult to study quantum critical phenomena containing massless fermions. While other determinantal MC methods do not encounter such problems, they scale poorly with system size [16]. In cases where the determinant of the fermion matrix is not positive, the original theory is said to suffer from a sign problem and the traditional approach is not useful. The repulsive Hubbard model away from half filling is a classic example where progress has been limited due to sign problems. Other relativistic four-fermion field theories like the Gross-Neveu (GN) models and Nambu-Jona-Lasinio (NJL) models are also known to suffer from sign problems in three or more space-time dimensions [17].

Recently a new approach called the fermion bag approach was proposed to solve some four-fermion field theories [18, 19, 20, 21, 22]. It is an extension of the meron cluster idea proposed some time ago [23]. The idea behind the fermion bag is to identify fermion degrees of freedom that cause sign problems and collect them in a bag and sum only over them. This is in contrast to traditional approaches where all fermion degrees of freedom in the entire thermodynamic volume are summed to solve the sign problem. When the fermion bag contains only a small fraction of all the degrees of freedom and the summation can be performed quickly, the fermion bag approach can be used to design powerful MC methods. Sometimes, the bag splits into many disconnected pieces further simplifying the calculation. The fermion bag approach has three main advantages: (a) Due to a duality, fermion bag sizes are small both at weak and strong couplings, (b) Singularities in the massless limit can be tackled without a problem, (c) Some sign problems that haunt traditional approaches are naturally solved. While the first two advantages have been demonstrated, the third advantage is not so clear from previous work. Here we show how solutions to some unsolved sign problems in four-fermion models also emerge naturally in the fermion bag approach.

It is useful to clarify some confusions that may arise about what we mean by a sign problem and thus a solution to the sign problem. If one can write the partition of a quantum statistical mechanics system as a sum over configurations whose Boltzmann weights are all positive and if the cost of computation of the Boltzmann weights only scales as a polynomial in system size, then we say the model does not suffer from a sign problem. However, as already stated above, in fermionic systems there are no natural configurations where the Boltzmann weights are positive. The conventional method is to use the auxiliary field approach to expand the partition function as a sum of bosonic configurations where the fermion determinant is taken as part of the Boltzmann weight. If this weight can be negative one often says the model suffers from a sign problem. However, if an alternate approach can be found where the sign problem disappears, one can of course say the problem never suffered from a sign problem to begin with. On the other hand, if this alternate approach was not known earlier, the new approach can be considered as a solution to the sign problem present in the other method. This is what we mean when we say "solutions to unsolved sign problems". It must be noted that all sign problems are problems in exactly this sense. Once a solution is found there is no longer a problem. It is of course likely that some problems may remain unsolved [24].

We consider lattice GN models containing N flavors of massless staggered fermions with either a Z_2 or a $U(1)$ chiral symmetry [17]. While we work in three space-time dimensions, our results can easily be extended to higher dimensions. Although in three dimensions the symmetries we refer to are a part of a flavor symmetry, they are often loosely called chiral symmetries in the literature. The Z_2 models with odd N and all the $U(1)$ models are known to suffer from a sign problem when formulated in the traditional auxiliary field approach. Here we show that the sign problems disappear in the fermion bag approach. Our paper is organized as follows. In section 2 we review the auxiliary field approach to lattice GN models with both Z_2 and $U(1)$ chiral symmetries and discuss how the sign problems arise. In section 3 we discuss the severity of the sign problems. In section 4 we discuss the fermion bag approach and show that sign problems do not arise. Section 5 contains our conclusions.

2. Auxiliary field approach

Lattice GN models are formulated in the auxiliary field approach through the action

$$S_{GN} = \sum_{x,y,i} \overline{\chi}_i(x)(D[\vec{\phi}])_{x,y}\chi_i(y) + S_{AF} \tag{1}$$

where $\overline{\chi}_i(x), \chi_i(x)$ denote the Grassmann valued fermion fields of flavor $i = 1, 2.., N$ at the lattice site x. The explicit form of the auxiliary field action S_{AF} depends on the GN model and will be discussed below. The matrix $D[\bar{\phi}]$ is defined by

$$\left(D[\bar{\phi}]\right)_{xy} = D_{xy} + \delta_{xy}\,\bar{\phi}(x), \tag{2}$$

where $\bar{\phi}(x)$ is a function of the auxiliary fields as defined below and $D_{x,y}$ is the free staggered fermion matrix [25, 26, 27],

$$D_{x,y} = m\delta_{x,y} + \sum_{\alpha=1,2,3} \frac{\eta_{x,\alpha}}{2}\left[\delta_{x+\alpha,y} - \delta_{x,y+\alpha}\right]. \tag{3}$$

Since we work in three dimensions, α labels the three directions, $\eta_{x,\alpha} = e^{(i\pi\zeta_\alpha\cdot x)}, \zeta_1 = (0,0,0)$, $\zeta_2 = (1,0,0)$, $\zeta_3 = (1,1,0)$ are the staggered fermion phase factors and m is the bare fermion mass. We assume anti-periodic boundary conditions in all directions and denote the lattice volume by $V = L^3$.

Following [17], we define the auxiliary fields on dual sites $\tilde{x}$. The model with a Z_2 chiral symmetry is defined through a single real auxiliary field $\sigma(\tilde{x})$, such that

$$S_{AF}[\sigma] = \frac{N}{2g^2}\sum_{\tilde{x}} \sigma^2(\tilde{x}), \tag{4}$$

$$\bar{\phi}(x) = \frac{1}{8}\sum_{\langle\tilde{x},x\rangle} \sigma(\tilde{x}) \tag{5}$$

while the model with a $U(1)$ chiral symmetry requires two real auxiliary fields $\sigma(\tilde{x})$ and $\pi(\tilde{x})$, such that

$$S_{AF}[\sigma,\pi] = \frac{N}{4g^2}\sum_{\tilde{x}} \left(\sigma^2(\tilde{x}) + \pi^2(\tilde{x})\right), \tag{6}$$

$$\bar{\phi}(x) = \frac{1}{8}\sum_{\langle\tilde{x},x\rangle} \Big(\sigma(\tilde{x}) + i\varepsilon(x)\pi(\tilde{x})\Big), \tag{7}$$

where $\varepsilon(x)$ is the parity of a lattice site (1 on even sites and -1 on odd sites). In the above expressions, the set of nearest dual sites $\tilde{x}$ surrounding the fixed lattice site x is denoted as $\langle\tilde{x},x\rangle$ (see Fig. 1). In this work we only consider these two classes of models.

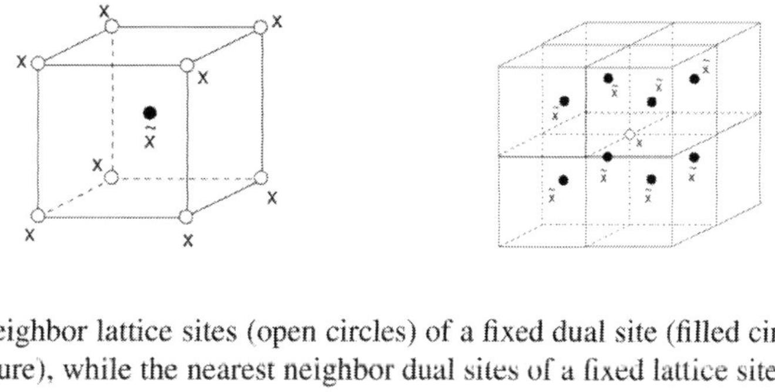

Figure 1. Nearest neighbor lattice sites (open circles) of a fixed dual site (filled circle) $\tilde{x}$ is represented by $[x,\tilde{x}]$ (see left figure), while the nearest neighbor dual sites of a fixed lattice site is denoted by $\langle\tilde{x},x\rangle$ (see right figure).

It is easy to verify that S_{GN} is invariant under $U(N)$ flavor transformations. When $m = 0$, additional chiral symmetries emerge. The Z_2 model is invariant under $\chi_i(x) \to \varepsilon(x)\chi_i(x)$, $\overline{\chi}_i(x) \to -\overline{\chi}_i(x)\varepsilon(x)$, $\sigma(\tilde{x}) \to -\sigma(\tilde{x})$ while the $U(1)$ model is invariant under the additional $U(1)$ chiral

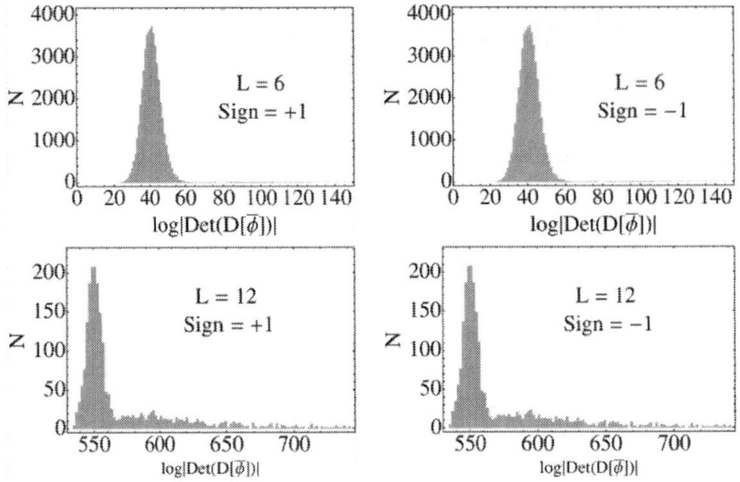

Figure 2. Distributions of positive (left graphs) and negative (right graphs) weight configurations as a function of $\log|\mathrm{Det}(D[\bar{\phi}])|$. One million configurations and 5000 configurations were generated at 6^3 and 12^3 lattices respectively. The distribution of positive configurations is almost identical to the one with negative configurations suggesting a severe sign problem.

symmetry $\chi_i(x) \to e^{i\varepsilon(x)\theta/2}\chi_i(x)$, $\overline{\chi}_i(x) \to \overline{\chi}_i(x)e^{i\varepsilon(x)\theta/2}$, $\sigma(\tilde{x}) \to \sigma(\tilde{x})\cos\theta + \pi(\tilde{x})\sin\theta$, $\pi(\tilde{x}) \to \pi(\tilde{x})\cos\theta - \sigma(\tilde{x})\sin\theta$. The models contain a quantum critical point (QCP) separating a chirally symmetric phase (at small couplings) from a phase where the chiral symmetry is spontaneously broken (at large couplings). The symmetries that govern the QCP need proper analysis due to fermion doubling. Without such an analysis it is difficult to establish the continuum field theory that emerges at the critical point [20].

In the traditional MC approach, one integrates over the Grassmann fields and writes the partition function of the GN models as

$$Z_{Z_2} = \int [\mathcal{D}\sigma]\, e^{-S_{AF}[\sigma]} \left\{ \mathrm{Det} D([\bar{\phi}]) \right\}^N, \tag{8}$$

$$Z_{U(1)} = \int [\mathcal{D}\sigma\mathcal{D}\pi]\, e^{-S_{AF}[\sigma,\pi]} \left\{ \mathrm{Det} D([\bar{\phi}]) \right\}^N, \tag{9}$$

In order to design a MC method the determinant terms in the above expressions have to be real and positive. In the Z_2 model since $\bar{\phi}$ is real, the matrix elements of $D[\bar{\phi}]$ are real. Hence, the determinant is real but not necessarily positive. In the case of the $U(1)$ model, $\bar{\phi}$ is complex and so the matrix elements of $D[\bar{\phi}]$ and its determinant can be complex. Hence, the Z_2 model as formulated in Eq. (8) suffers from a sign problem for all odd values of N, while the $U(1)$ model as formulated through Eq. (9) suffers from a sign problem for all values of N.

3. Severity of the sign problem

Earlier calculations in the Z_2 and $U(1)$ GN models have all been performed in the auxiliary field approach [28, 29, 30]. The essential focus has been to understand the quantum phase transition and compute the critical exponents. These calculations have circumvented the sign problem by studying even N in the Z_2 case or by introducing conjugate fermions with an opposite chiral charge in the $U(1)$.

Inclusion of conjugate fermions changes the partition function from Eq. (9) to

$$Z^c_{U(1)} = \int [\mathcal{D}\sigma][\mathcal{D}\pi]\ e^{-S_{AF}} \left| \mathrm{Det} D([\phi]) \right|^{2N}, \tag{10}$$

and changes the flavor symmetries to $U(N) \times U(N)$ while the chiral symmetry remains unchanged.

The $N = 1$ model with Z_2 chiral symmetry was also studied in the auxiliary field approach using the HMC algorithm [31]. Strangely, in this study the sign of the determinant was never discussed and seems to have been ignored. Since the results of the quantum critical behavior were in quantitative agreement with large N results (improved with Padé-approximations), it may have been assumed that the sign problem was mild. If this is indeed true then statistically, positive sign configurations should dominate over negative sign configurations. The Z_2 model studied in [31] is slightly different from the model studied here. The auxiliary fields $\sigma(x)$ also live on the main lattice site and the field $\bar{\phi}$ appearing in the Dirac operator $D[\bar{\phi}]$ of Eq. (2), is defined as $\bar{\phi}(x) = \frac{1}{6}\sum_{\langle z,x\rangle} \sigma(z)$, where now $\langle z, x\rangle$ refers to the six nearest neighbor sites z for a given site x. In order to study the sign problem, we generated several Gaussian random auxiliary field configurations according to the distribution

$$P(\sigma(x)) = \exp\left(-\sum_x \left\{ \sigma^2(x) - \frac{1}{2}\log(\pi) \right\} \right) \tag{11}$$

and computed $\mathrm{Det}(D[\bar{\phi}])$ for each of these configurations. We then separated the configurations into those with a positive determinant and those with a negative determinant. In Fig. 2 we plot the distribution of configurations with positive and negative determinants as a function of $\log|\mathrm{Det}(D[\bar{\phi}])|$ for 6^3 and 12^3 lattices. As can be seen, the distribution of configurations with positive and negative weights are almost identical suggesting a severe sign problem rather than a mild one! Although we are not performing important sampling, our results clearly show that the sign problem must be studied carefully.

An important question to study is whether the HMC algorithm is getting trapped in the sector of configurations with positive weights (or negative weights). Note that, in the Z_2 models the only way to move from a positive weight sector to the negative weight sector is to pass through configurations which have almost zero weight assuming the step size in the HMC algorithm is small. Perhaps the suppression of the tunneling between the two sectors leads to long auto-correlation times or even lack of ergodicity. This argument also applies to Z_2 models with even N [1].

4. Fermion bag approach

We will now show that the sign problems in both the Z_2 and the $U(1)$ models discussed in section 2, disappear in the fermion bag approach. The proof relies on the fact that any k_i-point correlation function involving the ith flavor of staggered fermions defined through

$$C_i(x_{i_1}, ..., x_{i_{k_i}}) = \int [d\overline{\chi}_i d\chi_i] \mathrm{e}^{-\sum_{x,y} \overline{\chi}_i(x)\, D_{xy}\, \chi_i(y)} \overline{\chi}_i(x_{i_1})\chi_i(x_{i_1}) \,...\, \overline{\chi}_i(x_{i_{k_i}})\chi_i(x_{i_{k_i}}) \tag{12}$$

is positive semi-definite. This is due to the special properties of the free staggered fermion matrix. Indeed, using the ideas developed in the fermion bag approach [20], we can write

$$C_i(x_{i_1}, .., x_{i_{k_i}}) = \mathrm{Det}(D)\ \mathrm{Det}(G[\{x\}_i]) = \mathrm{Det}(W[\{x\}_i]) \tag{13}$$

where $G[\{x\}_i]$ is the $k_i \times k_i$ matrix of propagators between the k_i sites in the set $\{x\}_i \equiv x_{i_p}$, $p = 1, .., k_i$ whose matrix elements are $G_{x_p,x_q} = D^{-1}_{x_p,x_q}$ and the matrix $W[\{x\}_i]$ is a $(V - k_i) \times (V - k_i)$ matrix

[1] Simon Hands, private communication

identical to the matrix D except that the sites in the set $\{x\}_i$ are dropped from the matrix. All the determinants appearing in Eq.(13) can be shown to be positive (or zero). The simplest way to see this is to consider the matrix W. Since it is exactly the same as the staggered fermion matrix with some sites removed, its eigenvalues come in complex conjugate pairs of the form $m \pm i\lambda$. Unpaired eigenvalues are always m and they too come in pairs when the lattice is bipartite. When $m = 0$ then the determinant can be exactly zero. Thus, $C_i(x_{i_1}, .., x_{i_{k_i}}) \geq 0$. We will use this property to prove the absence of a sign problem in the fermion bag approach.

Instead of integrating out the fermion fields let us integrate out the auxiliary fields first and construct the appropriate four fermion action for the models. Let us first consider the Z_2 model. Each integral over the auxiliary field $\sigma(\tilde{x})$ on the dual site $\tilde{x}$ gives,

$$I_{\tilde{x}} = \int d\sigma(\tilde{x})\ \mathrm{e}^{-S_{AF} - \frac{\sigma(\tilde{x})}{8}\left(\sum_{i,[x,\tilde{x}]} \overline{\chi}_i(x)\chi_i(x)\right)} = \mathcal{N}\mathrm{e}^{-S_I(\tilde{x})}, \tag{14}$$

where $\mathcal{N} = \sqrt{2\pi g^2/N}$ and

$$S_I(\tilde{x}) = -\frac{g^2}{128N}\Big[\sum_{i,[x,\tilde{x}]} \overline{\chi}_i(x)\chi_i(x)\Big]^2, \tag{15}$$

is the effective four-fermion interaction term at each dual site $\tilde{x}$. The symbol $[x,\tilde{x}]$ denotes the set of all lattice sites surrounding the dual site $\tilde{x}$ (see Fig. 1) . Thus, each integral generates many four-fermion couplings of the form $\overline{\chi}_i(x)\chi_i(x)\overline{\chi}_j(y)\chi_j(y)$ where i and j are arbitrary flavor indices and x and y are corners of the cube surrounding the dual site $\tilde{x}$. We can classify the possible couplings into four types based on the bonds $\langle xy\rangle$ connecting the corners x and y. If the two corners are the same we refer to it as a site-bond or a S-bond. If the two corners are the two neighboring sites we get a L-bond (or a link-bond). Similarly, if the two corners are across a face diagonal or a body diagonal, we call the bonds F-bond and B-bond respectively. These four bond types are illustrated Fig. 3.

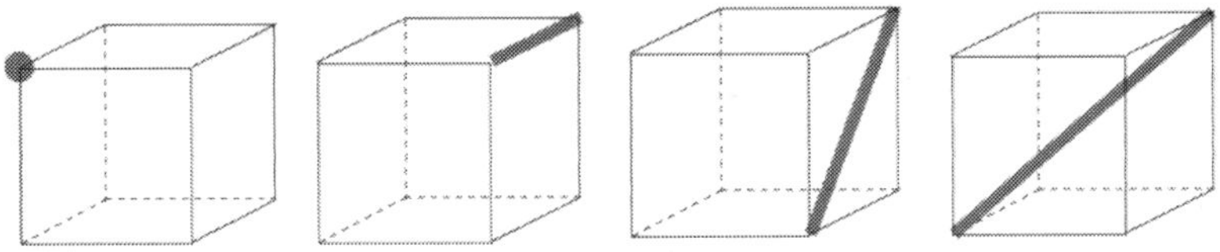

Figure 3. An illustration of the four types of four-fermion couplings (or bonds) generated through the auxiliary field integration. From left to right we have a S, L, F and B bond respectively.

Integration over all the auxiliary field variables yields the four-fermion interaction term of the action $S_{Z_2,\mathrm{int}} = \sum_{\tilde{x}} S_I(\tilde{x})$. Collecting the terms in each of the four types of four fermion couplings separately we see that

$$S_{Z_2,\mathrm{int}} = U_S\mathcal{B}_S + U_L\mathcal{B}_L + U_F\mathcal{B}_F + U_B\mathcal{B}_B \tag{16}$$

where $U_S/4 = U_L/4 = U_F/2 = U_B = g^2/(64N)$ and

$$\mathcal{B}_{\mathrm{bond}} = \sum_{i,j,\langle xy\rangle\in\mathrm{bond}} \overline{\chi}_i(x)\chi_i(x)\overline{\chi}_j(y)\chi_j(y). \tag{17}$$

Based on the above results, the partition function of the Z_2 model can be rewritten as

$$Z_{Z_2} = \int \prod_i [d\overline{\chi}_i d\chi_i]\ \mathrm{e}^{-S_{Z_2}}, \tag{18}$$

where $S_{Z_2} = S_0 + S_{Z_2,\text{int}}$ is the equivalent four-fermion action of the model. Here $S_0 = \sum_{x,y,i} \overline{\chi}_i(x) D_{x,y} \chi_i(y)$ is the free fermion action.

In the fermion bag approach, each four-fermion coupling is represented as a bond and expanded in powers of the coupling. For example the four-fermion coupling of the type $\overline{\chi}_i(x_p)\chi_i(x_p)\overline{\chi}_j(x_q)\chi_j(x_q)$ can be denoted by the bond variable $b_{ij}(x_p, x_q) = 0, 1$, such that if it is 0 then no bond is assumed to exist between the sites x_p and x_q, otherwise the specific four-fermion coupling is inserted in the partition function. Due to the Grassmann nature of the couplings higher powers of the couplings do not exist. More details can be found in [18]. Thus, in the fermion bag formulation, the partition function can be written as a sum over these bond configurations $[b]$, such that

$$\begin{aligned} Z_{Z_2} &= \sum_{[b]} U_S^{n_S} U_L^{n_L} U_F^{n_F} U_B^{n_B} \int \prod_i [d\overline{\chi}_i d\chi_i]\, \mathrm{e}^{-S_0} \prod_i \overline{\chi}_i(x_{i_1})\chi_i(x_{i_2})...\overline{\chi}_i(x_{i_{k_i}})\chi_i(x_{i_{k_i}}) \\ &= \sum_{[b]} U_S^{n_S} U_L^{n_L} U_F^{n_F} U_B^{n_B} \Big\{ \prod_i C_i(x_{i_1}, .., x_{i_{k_i}}) \Big\} \end{aligned} \tag{19}$$

where n_S, n_L, n_F and n_B are the total number of bonds of each type and the correlation function $C_i(x_{i_1}, .., x_{i_{k_i}})$ was defined in Eq.(12). A given bond configuration $[b]$ uniquely determines the k_i sites $x_{i_1} .... x_{i_{k_i}}$ (ordered in a consistent way). Since we argued above that $C_i(x_{i_1}, .., x_{i_{k_i}}) \geq 0$ there is no sign problem in this expansion of the partition function for all non-negative values of U_S,U_L,U_F, U_B, any positive integer N and real mass m.

In the case of the $U(1)$ model, we need to integrate over both the auxiliary fields $\sigma(\tilde{x}), \pi(\tilde{x})$ on every dual site. It is straightforward to verify that

$$\begin{aligned} I_{\tilde{x}} &= \int [d\sigma(\tilde{x}) d\pi(\tilde{x})]\, \mathrm{e}^{-S_{AF} - \frac{\sigma(\tilde{x})}{8}\left(\sum_{i,[\tilde{x},x]} \overline{\chi}_i(x)\chi_i(x)\right)} \\ &\quad \times \mathrm{e}^{-i\frac{\pi(\tilde{x})}{8}\left(\sum_{i,[\tilde{x},x]} \varepsilon(x)\overline{\chi}_i(x)\chi_i(x)\right)} = \mathcal{N} \mathrm{e}^{-S_I(\tilde{x})} \end{aligned} \tag{20}$$

where $\mathcal{N} = (4\pi g^2/N)$ and

$$S_I(\tilde{x}) = \frac{g^2}{64N} \Big\{ \Big[\sum_{i,[x,\tilde{x}]} \overline{\chi}_i(x)\chi_i(x) \Big]^2 - \Big[\sum_{i,[x,\tilde{x}]} \varepsilon(x)\overline{\chi}_i(x)\chi_i(x) \Big]^2 \Big\}, \tag{21}$$

Interestingly, the four-fermion couplings of the type S and F get canceled between the two terms in the above equation. On the other hand couplings of the type L and B survive so that the four-fermion action for the $U(1)$ model turns out to be

$$S_{U(1)} = S_0 + U_L \mathcal{B}_L + U_B \mathcal{B}_B \tag{22}$$

with $U_L/4 = U_B = g^2/(16N)$. Thus, the only difference between the Z_2 and $U(1)$ models is that the couplings $U_S = U_F = 0$ in the $U(1)$ model. Indeed these couplings break the $U(1)$ symmetry to a Z_2 symmetry as can be easily verified. Since we already proved that the sign problem in the Z_2 model was absent for all non-negative values of U_S, U_L, U_F, U_B and N in the fermion bag formulation, the same is true for the $U(1)$ model as well.

5. Results for $U(1)$ model

It has been pointed out in [32] that the Thirring model is different from the Gross-Nevue(GN) model in many ways. The critical exponents in a lattice GN model with a $U_f(1) \times Z_2$ symmetry with staggered fermions have been computed and it was found that $\nu = 1.00(4)$ and $\eta = 0.754(8)$ [31]. In a $U(1)$ lattice GN model simulation [30], $\eta = 0.904(50)$ has been found. However, those simulations are all based on the conventional Monte Carlo methods with auxiliary fields and an extra scale m in the action. It could be tricky to carry out the chiral extrapolation analysis while simulations are close to U_c. The most importantly, as we have pointed out in the previous sections, the original theory which has $U(1)$ flavor symmetry suffers from the sign problem, the conventional methods instead studied the modified theory which has $U(1) \times U(1)$ flavor symmetry with unchanged chiral symmetry. The expanding of the flavor symmetry can potentially affect the critical exponents and turn the model into different universality class. It is desirable to study the original model using a complete new method.

In our previous paper [33], we used fermion bag approach to study the quantum critical exponents of the Thirring model that has the action

$$S_{Thirring} = S_0 + U_L \mathcal{B}_L \tag{23}$$

It has been shown that our results agree with the early studies by the conventional approaches, but our data are more accurate. By formulating the GN model in the fermion bag approach, in fact we can show that on the lattice the difference between the two models is an additional four-fermion coupling along the body diagonal after integrating out the auxiliary fields. It has the action in Eq. (22) with the extra coupling $U_B = U_L/4$ and the original theory can be simulated without sign problem under the fermion bag approach. The simulation of $U(1)$ GN model becomes straightforward, we should be able to compare the Thirring model to GN model at the transition with fairly small amount of modifications of the previous simulation. Before we start our simulations, as one can see, the body diagonal coupling is just quarter of the link coupling, it also preserves the $U(1)$ chiral symmetry. We expect this extra term should not change the universality class of the GN model compared to the Thirring model, both models can have the same critical exponents. In the simulations, we actually let $U_L = U_B = U$, we believe this little modification should not have an affect on the critical exponents.

In order to study the critical exponent, we focused on three observables (Let L be the lattice size): The chiral condensate susceptibility

$$\chi = \frac{1}{2L^3}\sum_{x,y}\langle \overline{\psi}_x \psi_x \overline{\psi}_y \psi_y \rangle, \tag{24}$$

the chiral winding number susceptibility

$$\langle q_\chi^2 \rangle = \langle \frac{1}{3}\sum_\alpha (q_\chi^2)_\alpha \rangle, \tag{25}$$

, and the ratio of fermion two-point correlator

$$R_f = C_F(L/2-1)/C_F(1), \tag{26}$$

The details of those observables are explained in [19]. Since the fermions are exactly massless, in the vicinity of U_c we expect these three observables to satisfy the following simple finite size scaling relations:

$$\begin{aligned}
\chi^{-1} L^{2-\eta} &= \sum_{k=0}^{3} f_k \left[(U-U_c)L^{\frac{1}{\nu}}\right]^k \\
\langle q_\chi^2 \rangle &= \sum_{k=0}^{3} \kappa_k \left[(U-U_c)L^{\frac{1}{\nu}}\right]^k \\
R_f L^{2+\eta_\psi} &= \sum_{k=0}^{3} p_k \left[(U-U_c)L^{\frac{1}{\nu}}\right]^k
\end{aligned} \tag{27}$$

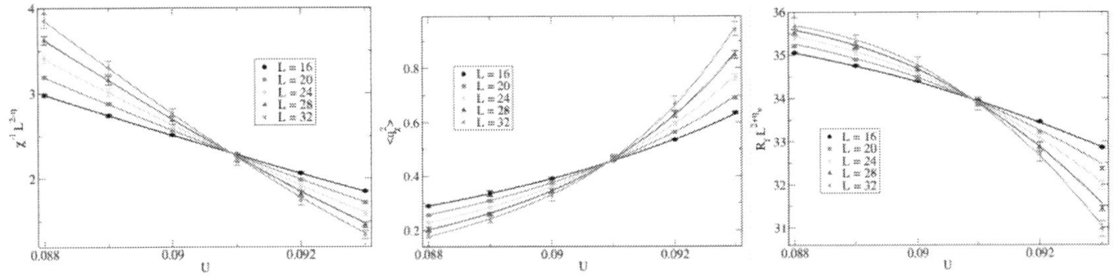

Figure 4. Plots of $\chi^{-1}L^{2-\eta}$, $\langle q_\chi^2\rangle$ and $R_f L^{2+\eta_\psi}$ as a function of U for L from 16 to 40. The solid lines show the combined fit which gives $U_c = 0.0909(1), \nu = 0.88(1), \eta = 0.63(1)$ and $\eta_\psi = 0.37(1)$ with $\chi^2/d.o.f = 0.89$

where we have kept the first four terms in the Taylor series of the corresponding analytic functions. Our goal is to compute the critical exponents η, ν and η_ψ at the quantum critical point. Plots of our data are shown in Fig. 4. Since our data fits very well to the expected scaling form for a whole range of lattice sizes, we feel confident that the corrections to scaling are small.

We compared the critical exponents from $U(1)$ GN model to the Thirring model in Table 1. There are slight deviations between two models, we think it may be the due to the modification on the body diagonal coupling and different statistics between two models. By combining data sets from two models, we can fit the same scaling relations quite well with $\chi^2/d.o.f = 1.19$. The joined fitting results are listed as "Combined" in the table. It is clear that those critical exponents are completely different from mean-field analysis and early MC studies by conventional methods [30]. Our data shows that $U(1)$ GN model and Thirring model belong to the same universality class. It will be very interesting to carry out the same study on the Z_2 GN model and compare to the early results [31]. Since the $U(1)$ chiral symmetry has been broken explicitly by the face diagonal bond, the critical behaviors can be different from the $U(1)$ model.

	ν	η	η_ψ	$\chi^2/d.o.f.$
Thirring	0.85(1)	0.65(1)	0.37(1)	1.3
$U(1)$ GN	0.88(1)	0.63(1)	0.37(1)	0.89
Combined	0.85(1)	0.64(1)	0.37(1)	1.19

Table 1. Results from the fit of the data to Eqs. (27). In addition to fit Thirring and $U(1)$ model separately, we also combined two data sets and fit all the critical exponents using the same scaling relations.

6. Conclusions

The fermion bag approach provides an alternative approach to fermion field theories where solutions to new sign problems emerge naturally. Here we have demonstrated that some sign problems in the auxiliary field formulation of GN models, especially with Z_2 and $U(1)$ chiral symmetries, disappear in the fermion bag approach. By using the fermion bag approach, we have also shown the $U(1)$ GN model and Thirring model belong to the same universality classes. While we have not shown here, we can solve sign problems in some lattice field theories containing both dynamical boson and fermion fields with similar chiral symmetries. In these more complex models, the solutions emerge when bosons are formulated in the world-line approach and the fermions are formulated in the bag approach. Such an approach to quantum field theories was proposed in [34].

Sign problems in other fermion models with more complex symmetries are also solvable in the fermion bag approach. However, in many interesting cases the Boltzmann weight of a fermion bag, although non-negative, turns out to be a fermionant instead of a determinant [35]. Since the computation of the fermionant can be exponentially hard, the fermion bag approach loses its practical appeal in such cases. Still, we believe that there are many other interesting models where the weight of the fermion bag continues to be positive and computable with polynomial effort.

Acknowledgments
I would like to thank my collaborator S. Chandrasekharan for the wonderful collaboration and fruitful discussions. This work was supported in part by the Department of Energy grants DE-FG02-05ER41368 and DE-FG02-00ER41132.

References

[1] Franz M and Tesanovic Z 2001 *Phys. Rev. Lett.* **87**(25) 257003
[2] Ran Y, Hermele M, Lee P A and Wen X G 2007 *Phys. Rev. Lett.* **98**(11) 117205
[3] Meng Z Y, Lang T C, Wessel S, Assaad F F and Muramatsu A 2010 *Nature* **464** 847
[4] Bedaque P F and van Kolck U 2002 *Ann.Rev.Nucl.Part.Sci.* **52** 339–396 (*Preprint* nucl-th/0203055)
[5] Epelbaum E, Krebs H, Lee D and Meissner U G 2010 *Phys.Rev.Lett.* **104** 142501
[6] Rosenstein B, Warr B and Park S 1991 *Phys.Rept.* **205** 59–108
[7] Herbut I F, Juričić V and Roy B 2009 *Phys. Rev. B* **79** 085116
[8] Armour W, Hands S and Strouthos C 2011 *Phys.Rev.* **B84** 075123
[9] Grabowska D, Kaplan D B and Nicholson A N 2012 (*Preprint* 1208.5760)
[10] Wolff U 2008 *Nucl.Phys.* **B789** 258–276 (*Preprint* 0707.2872)
[11] Wolff U 2009 *Nucl.Phys.* **B814** 549–572 (*Preprint* 0812.0677)
[12] Wenger U 2009 *Phys.Rev.* **D80** 071503 (*Preprint* 0812.3565)
[13] Scalettar R T, Scalapino D J, Sugar R L and Toussaint D 1987 *Phys. Rev. B* **36** 8632–8641
[14] Duane S, Kennedy A D, Pendleton B J and Roweth D 1987 *Phys. Lett.* **B195** 216–222
[15] Clark M and Kennedy A 2007 *Phys.Rev.Lett.* **98** 051601 (*Preprint* hep-lat/0608015)
[16] Assad F F and Evertz H G 2008 *Computational Many Particle Physics* vol 739 (Lect. Notes Phys. Springer) chap 10, pp 277–356
[17] Hands S, Kocic A and Kogut J B 1993 *Annals of Physics* **224** 29 – 89 ISSN 0003-4916 URL http://www.sciencedirect.com/science/article/pii/S0003491683710390
[18] Chandrasekharan S 2010 *Phys.Rev.* **D82** 025007 (*Preprint* 0910.5736)
[19] Chandrasekharan S and Li A 2011 *PoS* **LATTICE2011** 058 (*Preprint* 1111.5276)
[20] Chandrasekharan S and Li A 2012 *Phys. Rev. Lett.* **108**(14) 140404 (*Preprint* 1111.7204) URL http://link.aps.org/doi/10.1103/PhysRevLett.108.140404
[21] Chandrasekharan S 2012 *Phys.Rev.* **D86** 021701 (*Preprint* 1205.0084)
[22] Chandrasekharan S and Li A 2012 *Phys.Rev.* **D85** 091502 (*Preprint* 1202.6572)
[23] Chandrasekharan S and Wiese U J 1999 *Phys. Rev. Lett.* **83** 3116–3119
[24] Troyer M and Wiese U J 2005 *Phys.Rev.Lett.* **94** 170201 (*Preprint* cond-mat/0408370)
[25] Sharatchandra H S, Thun H J and Weisz P 1981 *Nucl. Phys.* **B192** 205
[26] Kogut J B and Susskind L 1975 *Phys.Rev.* **D11** 395
[27] Banks T, Susskind L and Kogut J B 1976 *Phys.Rev.* **D13** 1043
[28] Hands S, Kim S and Kogut J B 1995 *Nucl. Phys.* **B442** 364–390 (*Preprint* hep-lat/9501037)
[29] Hands S, Lucini B and Morrison S 2002 *Phys. Rev.* **D65** 036004 (*Preprint* hep-lat/0109001)
[30] Christofi S and Strouthos C 2007 *JHEP* **05** 088 (*Preprint* hep-lat/0612031)
[31] Karkkainen L, Lacaze R, Lacock P and Petersson B 1994 *Nucl. Phys.* **B415** 781–796 (*Preprint* hep-lat/9310020)
[32] Hands S 1995 *Phys. Rev. D* **51** 5816–5826
[33] Chandrasekharan S and Li A 2012 *Phys. Rev. Lett.* **108**(14) 140404 URL http://link.aps.org/doi/10.1103/PhysRevLett.108.140404
[34] Chandrasekharan S 2008 *PoS* **LATTICE2008** 003
[35] Chandrasekharan S and Wiese U J 2011 (*Preprint* 1108.2461)

The sign problem and the Lefschetz thimble

Marco Cristoforetti, Luigi Scorzato[1]
ECT*. Strada delle tabarelle, 286 – I-38123, Trento, Italy.

Francesco Di Renzo
Università di Parma and INFN gruppo collegato di Parma. Viale G.P. Usberti n.7/A – I-43124, Parma, Italy.

Abstract. Recently, we have introduced a novel approach to deal with the sign problem that prevents the Monte Carlo simulations of a class of quantum field theories (QFTs). The idea is to formulate the QFT on a Lefschetz thimble. Here we review the formulation of our approach and describe the *Aurora* Monte Carlo algorithm that we are currently testing on a scalar field theory with a sign problem.

1. Introduction

Recently, we have proposed a novel approach [1] to deal with the sign problem that hinders Monte Carlo simulations of many quantum field theories (QFTs). The approach consists of formulating the QFT on a Lefschetz thimble. In this paper we concentrate on the application to a scalar field theory with a sign problem. In particular, we review the formulation and the justification of the approach, and we also describe the *Aurora* Monte Carlo algorithm that we are currently testing.

2. Formulation

In this paper we consider a scalar field theory with $U(1)$ global symmetry:

$$S = \int d^4x[|\partial\phi|^2 + (m^2 - \mu^2)|\phi|^2 + \mu j_0 + \lambda|\phi|^4], \qquad j_\nu := \phi^* \overleftrightarrow{\partial_\nu} \phi, \tag{1}$$

where $\phi(x)$ is a complex scalar field. This model can be regularized on a lattice as:

$$S[\phi] = \sum_x \left[\left(2d + m^2\right) \phi_x^*\phi_x + \lambda(\phi_x^*\phi_x)^2 - \sum_{\nu=0}^{d-1} \left(\phi_x^* e^{-\mu\delta_{\nu,0}} \phi_{x+\hat{\nu}} + \phi_{x+\hat{\nu}}^* e^{\mu\delta_{\nu,0}} \phi_x \right) \right],$$

and has a sign problem when $\mu \neq 0$, which has been successfully treated in [2] and also in [3].

Our approach consists of defining the observables as:

$$\langle \mathcal{O} \rangle_0 = \frac{1}{Z_0} \int_{\mathcal{J}_0} \prod_{a,x} d\phi_{a,x}\, e^{-S[\phi]} \mathcal{O}[\phi], \qquad Z_0 = \int_{\mathcal{J}_0} \prod_{a,x} d\phi_{a,x}\, e^{-S[\phi]}, \tag{2}$$

[1] Speaker

where the integration domain $\mathcal{J}_0$ is the Lefschetz thimble [4, 5] attached to a global minimum $\phi_{\rm glob}$ of the real part of the action $S_R = \Re S$, when restricted to the original domain $\mathcal{C} = \mathbb{R}^{2V}$. More precisely, $\mathcal{J}_0$ is the manifold of real dimension $n = 2V$, defined as union of all the curves of steepest descent (SD) for S_R, i.e., solutions of:

$$\begin{aligned} \frac{d}{d\tau}\phi^{(R)}_{a,x}(\tau) &= -\frac{\delta S_R[\phi(\tau)]}{\delta\phi^{(R)}_{a,x}}, \quad \forall a, x, \\ \frac{d}{d\tau}\phi^{(I)}_{a,x}(\tau) &= -\frac{\delta S_R[\phi(\tau)]}{\delta\phi^{(I)}_{a,x}}, \quad \forall a, x, \end{aligned} \tag{3}$$

that end in $\phi_{\rm glob}$ for $\tau \to \infty$. Here we have assumed the usual complexification of the complex scalar field (see, e.g., [3]), where both the real (ϕ_1) and imaginary (ϕ_2) part of the original fields become complex $\phi_a = \phi_a^{(R)} + i\phi_a^{(I)}$, $a = 1, 2$.

2.1. Formulation in presence of SSB

Note that the formulation described above is possible only when $\phi_{\rm glob}$ is a minimum of the action with positive definite Hessian. In presence of symmetries that act non-trivially in $\phi_{\rm glob}$ this is not the case.

Gauge symmetries can be treated as described in [6, 5, 1], where, essentially, the thimble is defined modulo gauge transformations. But this is not suitable to study the possibility of spontaneous symmetry breaking (SSB). However, the proper way to study SSB (even in the ordinary formulation of the functional integral) is to introduce a small term of explicit symmetry breaking and study the limit when such term goes to zero. In this way, $\phi_{\rm glob}$ becomes a true minimum and the thimble can be defined.

2.2. Justification of the approach

The functional integral in Eq. (2) does not coincide, in general, with the standard formulation. Morse theory [5] only enables us to say that—under suitable conditions on $S[\phi]$ and $\mathcal{O}$— the standard functional integral coincides with an integral over a particular combination of thimbles $\sum_\sigma n_\sigma \mathcal{J}_\sigma$, for some integer n_σ, and where the sum runs over all the stationary points of the complexified action[2].

However, we have shown in [1] that the model defined by Eq. (2) has the same degrees of freedoms, the same symmetries and symmetry representations, and also the same perturbative expansion as the standard formulation. In this sense, by universality, the formulation of Eq. (2) may be seen as an alternative regularization of the model in Eq. (1).

3. The *Aurora* algorithm

In this section we describe an updated form of the algorithm proposed in [1]. What we want to compute are expectation values like:

$$\frac{1}{Z_0} e^{-iS_I} \int_{\mathcal{J}_0} \prod_x d\phi_x \, e^{-S_R[\phi]} \mathcal{O}[\phi], \tag{4}$$

where the imaginary part of the action S_I is constant in $\mathcal{J}_0$ and can be taken out of the integral. Hence the phase of e^{-S} cancels from all observables. Moreover, the real part S_R is bounded from below by its value at the stationary point[3]. The functional integral in Eq. (4) possesses

[2] An argument in [5] (see Sec. 3.2), can be used to show that only the stationary points near the global minima in $\mathcal{C}$ can bring a non negligible contribution. See [1], Sec. II.B.3.

[3] The integral is convergent, because $S[\phi]$ is a polynomial in ϕ and $\mathcal{J}_0$ is union of the curves of SD.

a bounded real action and hence can be studied, e.g., with a Langevin algorithm[4]. The choice of the Langevin algorithm is particularly convenient because the drift term of the Langevin equation for the real part of the action coincides with Eq. (3), i.e., the SD. Hence, such drift preserves the manifold $\mathcal{J}_0$ by construction. The Langevin noise, however, needs to be projected on the tangent space to $\mathcal{J}_0$, in order to explore the integration domain correctly.

The projection on the tangent space of $\mathcal{J}_0$ at ϕ is challenging, because we seem to have no way to tell which configurations in the neighborhood of $\phi \in \mathcal{J}_0$ will eventually also fall in ϕ_{glob}, by following the SD flow. However, the tangent space at the stationary point ϕ_{glob} can be computed, by computing the Hessian matrix[5]. In order to use this fact to generate a noise vector tangent to $\mathcal{J}_0$ at ϕ, we need a way to transport a vector η along the flow of SD, while keeping it tangent to $\mathcal{J}_0$. One way to achieve this is by requesting that the vector η is parallel transported along the flow of S_R, i.e. requesting that its Lie derivative along ∂S_R to be zero:

$$\mathcal{L}_{\partial S_R}(\eta) = [\partial S_R, \eta] = 0$$

which nicely translates into a first order ordinary differential equation, which is also linear in η:

$$\frac{d}{d\tau}\eta_j(\tau) = \sum_k \eta_k(\tau)\partial_k\partial_j S_R. \tag{5}$$

Note that we use j, k as multi-indices for $(R/I, a, x)$.

Actually, we can do better. Eq. (5) ensures a parallel transport of η along the flow ∂S_R. This is actually more than what we need, because we only need to keep η tangent to $\mathcal{J}_0$. On the other hand, it would be nice to transport η by means of an orthogonal flow, that preserves orthonormality, isotropy, and also enables the use of stable numerical integrators (see Sec. 5.3 of [7]). This can be done quite easily by employing the Iwasawa decomposition of $\partial^2 S_R$ (see, e.g., Sec. VI of [8]), which, at the group level, is a generalization of the Gram-Schmidt orthonormalization procedure, and it can be implemented also at the algebra level. More precisely, if we write:

$$\left(\partial^2 S_R\right)_{i,j} = A_{i,j} + D_{i,j} + N_{i,j},$$

where A is skew-symmetric, D is diagonal and N is upper triangular, then the flow

$$\frac{d}{d\tau}\eta(\tau) = \eta(\tau)A(\tau) \tag{6}$$

defines an orthogonal flow whose solutions η span the same space spanned by the solutions of Eq. (5). This means that the tangent space of $\mathcal{J}_0$ is still preserved. But now, the flow defined by Eq. (6) also preserves orthonormality, isotropy, and such properties can be even ensured numerically by using suitable numerical integrators, such as the implicit midpoint rule [7].

We can now summarize the algorithm as follows. We use t to represent the Langevin (Monte Carlo) time and τ to represent the parameter of the SD. Let us assume that we have already a configuration ϕ_t in $\mathcal{J}_0$, and we also have a path $\phi_t(\tau)$ that fulfills the equations of SD and connects ϕ_t to a configuration $\phi_t^{(\varepsilon)}$ with norm less than ε. The value of ε must be sufficiently small so that the quadratic approximation of the action is valid. One step of Langevin proceeds as follows:

- apply the Langevin force, that consists of taking one step forward along the curve of SD which is already available. The result is the configuration ϕ'_t;

[4] The phase that comes from the measure $d\phi$ will be treated with reweighting, as discussed in Sec. 3.1.

[5] Which is easy to compute also analytically, when $\phi_{\text{glob}} = 0$, and can be computed numerically, once and for all, for general ϕ_{glob}.

- extract a Gaussian noise η;
- project out the components of η orthogonal to $\mathcal{J}_0$ at the origin. More precisely, set $\eta_{\|} = (P-1)\eta$, where $P = \partial^2 S_R/\sqrt{(\partial^2 S_R)}$, and then re-scale the norm of $\eta_{\|}$ according to the χ distribution;
- evolve $\eta_{\|}$ with Eq. (6) from $\phi_t^{(\varepsilon)}$ to ϕ'_t. The orthogonal properties of the flow can also be preserved numerically by employing, e.g., the implicit midpoint rule. This leads to solving a 5-D linear system in $\eta_{\|}(\tau)$;
- define a tentative new configuration as $\phi_{t+dt}^{\text{guess}} = \phi'_t + \sqrt{2t}\,\eta(\tau)$;
- use $\phi_{t+dt}^{\text{guess}}$ as a starting point to find a solution of the SD equation. This can also be integrated with the implicit midpoint rule with the constraint that $(P-1)\phi_{t+dt}^{(\varepsilon)} = 0$. This leads to a non-linear 5-D system that can be solved—if the initial guess is sufficiently good—with a number of Newton-Raphson iterations.

3.1. The residual phase

As noted in the previous section, the integral of Eq. (4) also includes a residual phase that comes from the determinant of the tangent space of $\mathcal{J}_0$ at ϕ, and is not necessarily real and positive. We argue that this should not represent a *sign problem*. In fact, a sign problem means that the average phase $|\langle\prod d\phi\rangle| \ll 1$. However, that phase is constant in the region where the integrand is not very small. In fact, such phase is completely neglected in the saddle point expansion, which, on the other hand, can be identified with some form of perturbative expansion of the QFT. Saying that the average phase could be negligible is equivalent to saying that the "perturbative contribution" (in some sense that we cannot define precisely) could be negligible. We do expect that a non-perturbative contribution may be important, but that the perturbative one could be negligible is contrary to the general expectations.

In this sense, we have good reasons to hope that residual phase can be treated with reweighting, but we have no justification to neglect it. If so, we have to compute it, which is quite expensive. This can be done as follow. If we call T_ϕ the complex $(2V \times 2V)$ matrix that spans the tangent space of $\mathcal{J}_0$ at ϕ, we have:

$$\det(T_{\phi_\tau}) = -\int_\tau^\infty ds\,\mathrm{Tr}[T_{\phi_s}^{-1}\frac{d}{ds}T_{\phi_s}] \simeq \sum_{i=0}^{N_\tau}\mathrm{Tr}[T_{\phi_i}^{-1}\,\Delta T_{\phi_i}] = \sum_{i=0}^{N_\tau}\sum_{k=1}^{N_R}\xi^{(k)^T}T_{\phi_i}^{-1}\,\Delta T_{\phi_i}\,\xi^{(k)}$$

where the $\xi^{(k)}$, $k = 1,\ldots,N_R$ are noisy estimators for the trace. Assuming that the inversion of T_ϕ would require N^{CG} iterations, the computation of the determinant would cost as much as evolving $N_R N^{CG}$ vectors, which is described in the previous section.

Some saving may also come from the fact that the computation of the phase is necessary only for the configurations that are actually used for measurement.

3.2. Preliminary tests

As a first test, we implemented the method by using the naive Euler integration method. Although bound to fail because of the nature of the Euler method, this is an interesting test in order to get a feeling of how difficult it is to keep the system on a thimble and how sensitive the system is to perturbations. Some results are shown in Fig. 1 and 2. These are encouraging, because it seems that even the Euler method is able to keep the system on the thimble long enough to see a thermalization. This lets us hope that a better integrator will have good chances to converge. Observables appear around the expected order of magnitude, but their thermalization is not clear.

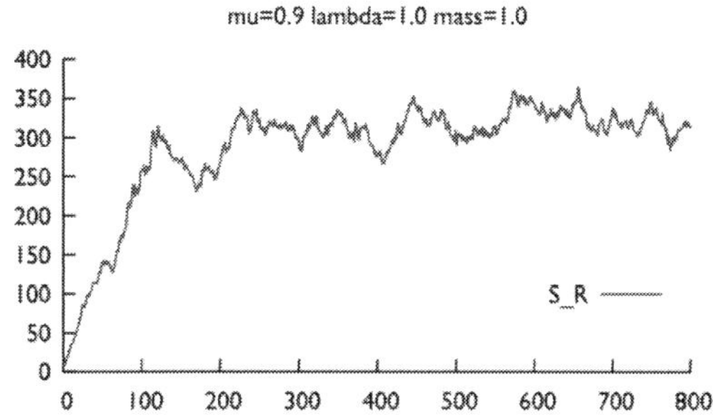

Figure 1. The real part of the action on a $V = 4^4$ lattice with Euler integrator. The system is stable long enough to show signs of a thermalization.

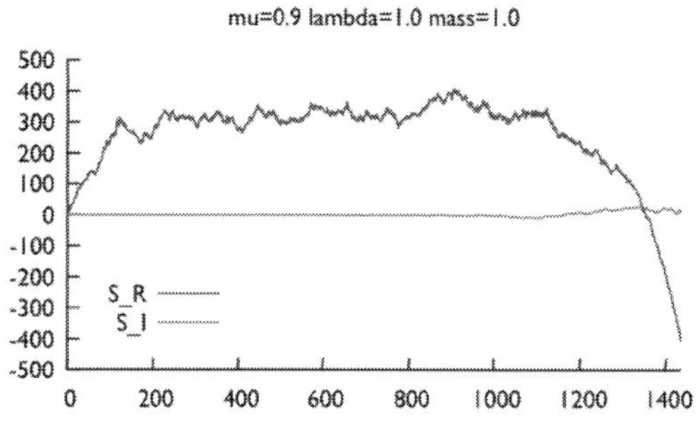

Figure 2. Same as Fig. 1, but for a longer time. The Euler integrator cannot constrain the system on the thimble, and the system drifts away at some point. This is also confirmed by the deviation of the imaginary part of the action from a constant.

4. Conclusions

We have illustrated an new approach to deal with the sign problem that afflicts a class of QFTs. It consists of regularizing the QFT on a Lefschetz thimble. Although it does not coincide with the usual regularization, it is a legitimate one on the basis of universality. In fact, we could prove that QCD on the thimble has the same symmetries, d.o.f. and PT as the usual formulation.

We have also introduced a Monte Carlo algorithm to achieve an importance sampling of the configurations on the thimble. Its numerical implementation will be certainly challenging and expensive, but all the steps of the algorithm are, a priori, feasible and have acceptable scaling. The residual phase should not give a sign problem (unless we believe that the perturbative contribution can be negligible) and hence should be manageable with reweighting, but this must be checked. We are presently testing the method for a scalar QFT on tiny lattices with encouraging results.

Acknowledgments

This research is supported by the AuroraScience project (funded by the Provincia Autonoma di Trento and INFN), and by the Research Executive Agency (REA) of the European Union under Grant Agreement No. PITN-GA-2009-238353 (ITN STRONGnet). L.S. and M.C. are members of LISC. FDR is partially supported by INFN i.s. MI11 and by MIUR contract PRIN2009 (20093BMNPR_004).

References

[1] Cristoforetti M, Di Renzo F and Scorzato L (AuroraScience Collaboration) 2012 *Phys. Rev. D* **86**(7) 074506 (*Preprint* `1205.3996`) URL `http://link.aps.org/doi/10.1103/PhysRevD.86.074506`
[2] Endres M G 2007 *Phys.Rev.* **D75** 065012 (*Preprint* `hep-lat/0610029`)
[3] Aarts G 2009 *Phys.Rev.Lett.* **102** 131601 (*Preprint* `0810.2089`)
[4] Pham F 1983 *Proc. Symp. Pure Math.* **40**(2) 319–333 URL `http://books.google.it/books?id=zGaprN2qx_MC&lpg=PP1&ots=1AFjzEhUBO&pg=PA319&f=true#v=onepage&q&f=true`
[5] Witten E 2010 (*Preprint* `1001.2933`)
[6] Atiyah M and Bott R 1983 *Phil.Trans.Roy.Soc.Lond.* **A308** 523–615
[7] Iserles A 2008 *A First Course in the Numerical Analysis of Differential Equations* 2nd ed Cambridge Texts in Applied Mathematics (Cambridge University Press) ISBN 9780521734905 URL `http://books.google.it/books?id=M0tkw4oUucoC`
[8] Knapp A 2002 *Lie Groups Beyond an Introduction* 2nd ed Progress in Mathematics (Birkhäuser Boston) ISBN 9780817642594 URL `http://books.google.it/books?id=U573NrppkA8C`

Lattice QCD at non-zero isospin chemical potential

Zhifeng Shi[1,2]
[1]Department of Physics, The College of William & Mary, Williamsburg, Virginia, USA
[2]Jefferson Lab, Newport News, Virginia, USA

E-mail: zshi@email.wm.edu

Abstract. Systems of non-zero isospin chemical potential are studied from a canonical approach by computing correlation functions with the quantum numbers of N π^+'s ($C_{N\pi}$). In order to reduce the number of contractions required in calculating $C_{N\pi}$ for a large N in the Wick's theorem, we constructed a few new algorithms. With these new algorithms, systems with isospin charge up to 72 are investigated on three anisotropic gauge ensembles with a pion mass of 390 MeV, and with lattice spatial extents $L \sim 2.0,\ 2.5,\ 3.0$ fm. The largest isospin density of $\rho_I \approx 9\ \mathrm{fm}^{-3}$ is achieved in the smallest volume, and the QCD phase diagram is investigated at a fixed low temperature at varying isospin chemical potentials, $m_\pi \le \mu_I \le 4.5\ m_\pi$. By investigating the behaviour of the extracted energy density of the system at different isospin chemical potentials, we numerically identified the conjectured transition to a Bose-Einstein condensation state at $\mu_I \ge m_\pi$.

1. Introduction

An important goal of nuclear physics is to investigate interactions between hadrons; however studying many body systems are hampered by the factorially growing number of Wick contractions naively required in computing the corresponding correlation function. In order to study multi-meson systems, a few algorithms [1, 2] have been constructed to alleviate the cost of computing larger number of independent contractions. With these new algorithms, many-pion systems have been studied in Ref. [2, 3, 4], and many-kaon systems have been studied in Ref. [5], furthermore systems of mixed species have also been studied in Ref. [6]. In order to study more complicated multi-baryon systems, various methods have also been introduced. With methods constructed in [7], correlation functions of ^{28}Si have been computed within a manageable amount of time.

Knowledge of systems at high density, with non-zero chemical potential, and non-zero temperature are vital to explore the QCD phase diagram. Systems with zero chemical potential have been investigated for a wide range of temperatures, and the transition from confined state at low temperature to the de-confined state at high temperature has been investigated substantially [8]. However, direct simulations of non-zero chemical potential systems are much harder due to the famous sign problem resulting from the non positive fermion determinants. But the sign problem does not exist for isospin chemical potential systems. Although QCD phase diagrams of these two systems are not the same, they do share some common properties, and it is also theoretical interesting to study non-zero isospin chemical potential systems.

In this proceeding, systems of non-zero isospin chemical potential are studied from a canonical approach by directly computing correlation functions of n-π^+ systems. By appling algorithms constructed in [2], we studied systems containing up to 72 π^+'s on three anisotropic $2+1$ flavor dynamical gauge ensembles, $16^3 \times 128$, $20^3 \times 256$ and $24^3 \times 128$, with $\xi = a_s/a_t = 3.5$, $a_s = 0.125$ fm, and with a poin mass of $m_\pi = 390$ MeV and a kaon mass of $m_K = 540$ MeV. The QCD phase diagram at a fixed low temperature, $\mathcal{T} \sim 20$ MeV, for a range of isospin chemical potentials, $m_\pi \leq \mu_I < 4.5\ m_\pi$, are investigated, and evidence for a phase transition from pion gasses to Bose-Einsetin Condensate (BEC) state around $\mu_I = 1.3 m_\pi$ is numerically identified.

The layout of this proceeding is as follows. In Sec.2, we briefly discuss the method applied to study n-π systems. Main results from the simulation are presented in Sec.3, and we conclude in Sec.4.

2. Methodology

Non-zero isospin chemical systems can be investigated from the study of n-π systems by directly computing their correlation functions. Because of the Pauli principle, the largest number of pions can be studied from a single source location is $N_s N_c = 12$. In order to study systems of more than 12 π's, additional source locations are required. The correlation function for a system of total $\overline{n} = \sum_{i=1}^{N} n_i$ π^+'s with n_i π^+'s from source locations $(\mathbf{y}_i, 0)$ is defined as:

$$C_{n_1,\ldots,n_N}(t) \quad = \quad \left\langle \left(\sum_{\mathbf{x}} \pi^+(\mathbf{x},t) \right)^{\overline{n}} \left(\pi^-(\mathbf{y_1},0) \right)^{n_1} \ldots \left(\pi^-(\mathbf{y_N},0) \right)^{n_N} \right\rangle , \tag{1}$$

where the interpolating operator $\pi^+(\mathbf{x},t) = \overline{d}(\mathbf{x},t)\gamma_5 u(\mathbf{x},t)$ and $\pi^-(\mathbf{x},t) = \overline{u}(\mathbf{x},t)\gamma_5 d(\mathbf{x},t)$.

Computing the above correlation function includes $\mathcal{O}(\overline{n}!)$ contractions. For larger $\overline{n}$ and N, number of ways to distribute pions over N sources grows exponentially, which further increase the computational cost of all $C_{n_1,\cdots,n_N}$'s. However, correlation functions of $\overline{n}$ pions, $C_{\overline{n}}$, without knowing details of the distributions of n_i can be constructed, which significantly reduces the number of correlation functions to compute as energies extracted from $C_{\overline{n}}$ is same as those extracted from $C_{n_1,\cdots,n_N}$'s as long as $\overline{n} = \sum_{i=1}^{N} n_i$. The explicit methods constructed in Ref.[2] are used to calculate $C_{\overline{n}}$ in this proceeding. Detailed discussions of the method and evaluations of n-π correlation functions in momentum space can be found in Ref.[2].

3. Lattice Results

Computations are performed on ensembles of $n_f = 2+1$ anisotropic gauge configurations with clover-improved fermions generated by the Hadron Spectrum Collaboration and the Nuclear physics with Lattice QCD collaboration. Pion correlations function, $C_{\overline{n}\pi}(t)$, for $\overline{n} = 1, 2, \cdots, 72$, are computed from three lattice ensembles of $L^3 \times T = 16^3 \times 128, 20^3 \times 256, 24^3 \times 128$. More details of these gauge ensembles can be found in Ref.[9].

The expected form of $C_{n\pi}(t)$ computed on a lattice with temporal extent T is [6]

$$C_{n\pi}(t) \quad = \quad \sum_{m=0}^{\lfloor \frac{n}{2} \rfloor} \binom{n}{m} A_m^n Z_m^n e^{-(E_{n-m}+E_m)T/2} \cosh((E_{n-m} - E_m)(t - T/2)) + \ldots , \tag{2}$$

where $A_m^n = 1$ when $m = n/2$, otherwise $A_m^n = 2$. E_m is the ground state energy of a m-π system, the Z_m^n are overlap factors of thermal states with m π's propagating backward around the temporal boundary [1], and the ellipsis denotes contributions from excited states. For infinite T, all thermal states vanish, and only the ground state survives, however the ground state is always contaminated by thermal states for finite T. Even for $C_{12\pi}$ computed on $T = 128$ ensembles, the ground state does not dominate in any region. Under the current precision, $C_{n\pi}$ computed from ensembles with a temporal extent $T = 256$ shows a clear plateau region even for $n = 72$. Thus, in order to reduce thermal contaminations, the anti-periodic plus periodic

[1] When $m = 0$, Z_0^n denotes the ground state contribution.

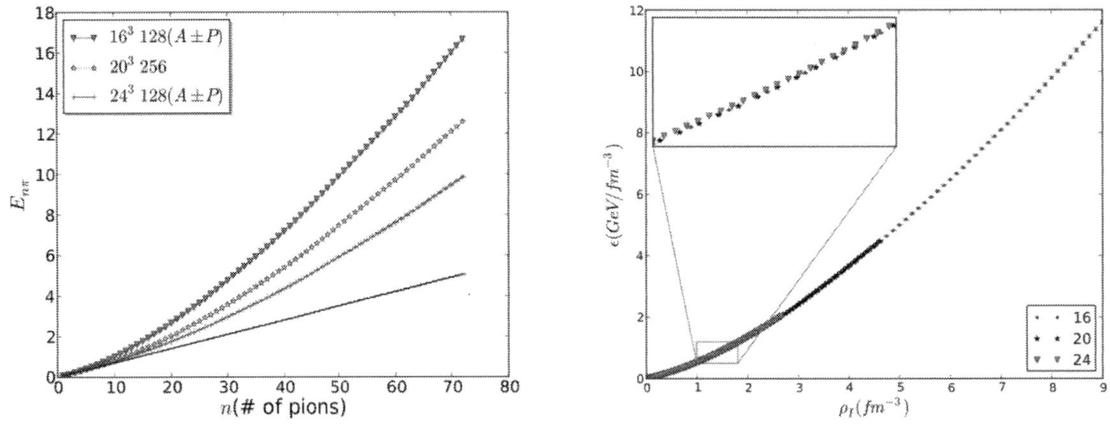

Figure 1. On the left plot, ground state energies of n-$\pi(E_{n\pi})$ systems extracted from ensembles $16^3 \times 128$ (red), $24^3 \times 128$ (blue) and $20^3 \times 256$ (green) are shown. The black line represents the free energy of n non-interacting pions. Energy densities, $\epsilon(\rho_I)$, computed from three ensemble are compared on the right plot.

propagation ($A \pm P$) method [2] has been applied to effectively double the temporal extent of $T = 128$ ensembles.

3.1. Ground state energies

Ground state energies of n-π systems are extracted by fitting a single exponential to $C_{n\pi}(t)$ within time slices where a clear plateau can be identified in the corresponding effective mass plot. Statistical uncertainties are evaluated from the bootstrap method, and systematic uncertainties are from averaging systematic uncertainties computed on each bootstrap sample by moving the fitting window forward and backward two time slices. Extracted ground state energies from three volumes are shown in the left plot of Fig. 1, and the corresponding energy densities, $\epsilon = E/V$, are compared in the right. Because of the repulsive interactions between pions, $E_{n\pi}$ is larger in smaller volumes as pions are closer to each other, however energy densities computed from three volumes are approximately the same.

3.2. Isospin chemical potential

From the ground state energies of n-π systems ($E_{n\pi}$), an effective isospin chemical potential $\mu_I(n) = \frac{dE}{dn}$ can be computed from a finite backward derivative, $\mu_I(n) = E_{n\pi} - E_{(n-1)\pi}$. In order to take into account the correlation between $E_{n\pi}$ and $E_{(n-1)\pi}$ extracted from the same bootstrap sample, $\mu_I(n)$ is also evaluated on each bootstrap sample. The extracted isospin chemical potential as a function of isospin density is plotted in Fig. 2.

At small isospin density, μ_I grows in an accelerating rate, agreeing with the prediction from χ-PT [10], however it starts to deviate from the χ-PT and begins to grow in a decelerating rate at large isospin densities, and at even larger isospin density it begins to flatten out.

3.3. QCD phase diagram

The change of the behavior of the isospin chemical potential at different isospin density signals a possible change of the physical state of the n-π system. In order to investigate this change, we

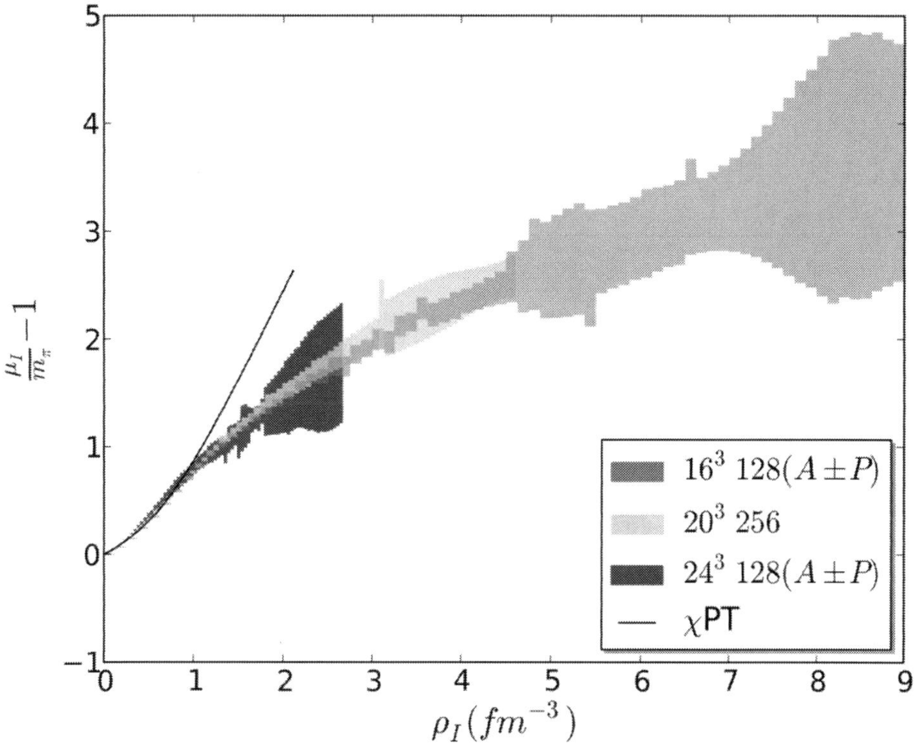

Figure 2. The isospin chemical potential, μ_I, is plotted as a function of the isospin density, ρ_I, from three lattice ensembles, $16^3 \times 128$ (red, $\rho_I = [0, 9]$), $20^3 \times 256$ (green, $\rho_I = [0, 4.7]$) and $24^3 \times 128$ (Blue, $\rho_I = [0, 2.8]$). The solid black line is from χ-PT [10]

have studied the ratio of the energy density, ϵ, to its zero temperature Stefan-Boltzmann limit,

$$\epsilon_{SB} = \frac{N_f N_c}{4\pi^2} \mu_I^4, \tag{3}$$

where $N_f = 3$ and $N_c = 3$ is used in this formula. The ratio is plotted in Fig. 3, where it grows at small μ_I, reaches a peak around 1.3 m_π, starts to decrease after the peak, and begins to flatten out at large isospin chemical potential. With a linear extrapolation of the peak position for different volumes as a function of the inverse of volume, the peak position in the infinite volume limit is $\mu^I_{peak} = 1.30(7)\ m_\pi$.

At low temperature, as soon as μ_I reaches m_π, one pion is produced out of vacuum, and the system is an interacting pion gas. When μ_I reaches the peak at $\mu_I = 1.30(7)\ m_\pi$, pions start to condense and the n-π system becomes Bose-Einsten Condenstate (BEC). Such a transition to BEC state at $\mu_I > m_\pi$ are also observed in two flavor QCD [11], and similar behavior occurs in two-color QCD [12].

4. Conclusion

In this proceeding, we studied n-π systems from a canonical approach by explicitly computing correlation function of the n-π system, $C_n(t)$. Ground state energies of the n-π systems were extracted, and subsequently, the isospin chemical potential was computed as a function of the isospin density.

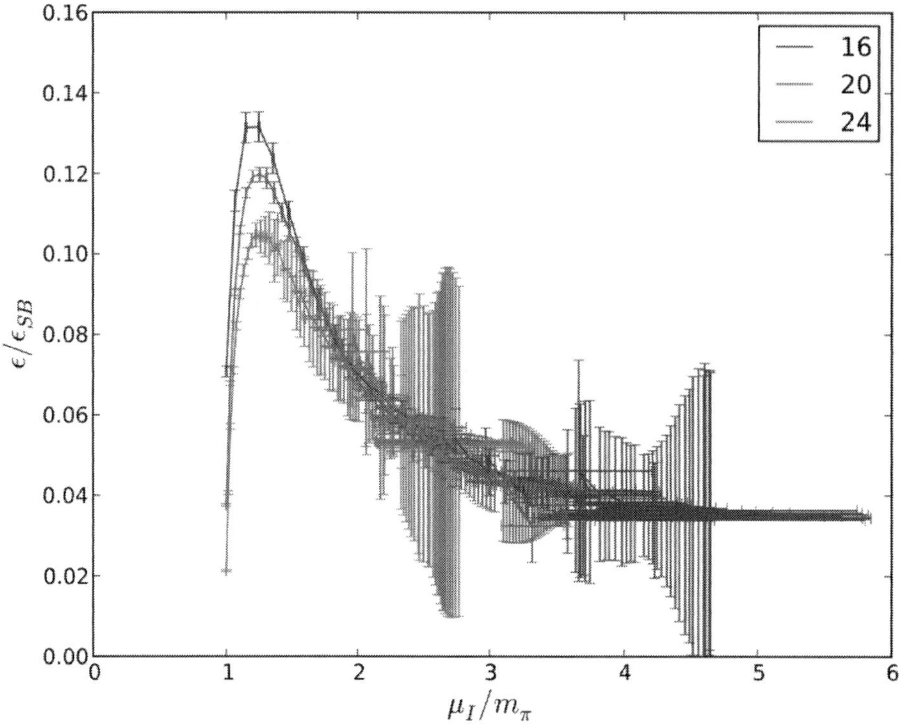

Figure 3. The ϵ/ϵ_{SB} is plotted as a function of μ_I/m_π.

By studying the ratio of the isospin energy density to its zero temperature Stefan-Bolzeman limit, the QCD phase diagram is investigated at a fixed low temperature, $\mathcal{T}$ = 20 MeV for a range of isospin chemical potentials, from $\mu_I = m_\pi$ to $\mu_I = 4.5\ m_\pi$. An evidence of transition from a pion gas to BEC is identified at $\mu_I = 1.30(7)\ m_\pi$.

5. Acknowledgment

The author thanks William Detmold and Kostas Originos for valuable discussions, and is grateful for the support of the DOE NERSC facility, NSF XSEDE resources and in particular TG-PHY080039N, as well as the Sporades cluster at the College of William & Mary.

References

[1] W. Detmold and M. J. Savage, Phys. Rev. D**82**, 014511 (2010).
[2] W. Detmold, K. Orginos, Z. Shi, Phys. Rev. D**86**, 054507 (2012).
[3] S. R. Beane, W. Detmold, etc. Phys. Rev. Lett. **100**, 082004 (2008).
[4] W. Detmold, M. J. Savage, etc. Phys. Rev. D **78**, 014507 (2008).
[5] W. Detmold, K. Orginos, etc. Phys. Rev. D **78**, 054514 (2008).
[6] W. Detmold, B. Smigielski, Phys. Rev. D **84**, 014508 (2011).
[7] W. Detmold, K. Orginos, [arXiv:1207.1452].
[8] O. Philipsen, [arXiv:1207.5999]
[9] B. Sheikholeslami, R. Wohlert, Nucl. Phys. B**259**, 572 (1985).
[10] D. T. Son and M. A. Stephanov, Phys. Rev. Lett. **86**, 592 (2001).
[11] J. B. Kogut and D. K. Sinclair, Phys. Rev. D **70**, 094501 (2004).
[12] S. Hands, S. Kim, J. I. Skullerud, Phys. Rev. D**81**, 091502 (2010).

Exploring QCD phase diagram at vanishing baryon density on the lattice

Heng-Tong Ding
Physics Department, Brookhaven National Laboratory, Upton NY 11973
E-mail: htding@quark.phy.bnl.gov

Abstract.
I report on the current status of QCD phase diagram at vanishing baryon density. I focus on the QCD phase diagram with three degenerate quark flavor using Highly Improved Staggered Quarks on $N_\tau = 6$ lattices. No evidence of a first order phase transition in the pion mass window of $80 \lesssim m_\pi \lesssim 230$ MeV is found. The pion mass at the critical point where the chiral first order phase transition ends is estimated to be $m_\pi^c \lesssim 45$ MeV.

1. Introduction

Due to the asymptotic freedom feature of QCD, it is expected that the QCD matter will undergo a phase transition from hadronic phase to the quark gluon plasma phase at sufficient high energy density. The nature of such a phase transition crucially depends on the values of quark masses (light m_l and strange m_s), the number of quark flavors (N_f) and the baryon density (μ_B) [1,2] [1]. The conjectured QCD phase diagram at vanishing baryon density from effective theories on the basis of the renormalized group together with the universality [3] is summarized in the left plot of Fig. 1. As sketched in Fig. 1, in the pure gauge case, i.e. in the limit of $m_l \to \infty$ and $m_s \to \infty$, QCD has an exact Z(3) symmetry and the phase transition is of first order. In the chiral limit of $N_f = 3$ case, i.e. $m_l = m_s \to 0$, the phase transition is also known to be a first order. In the chiral limit of a $N_f = 2$ theory, i.e. $m_l \to 0$ and $m_s \to \infty$, if $U_A(1)$ symmetry is restored the relevant symmetry is $O(2) \times O(4)$ and the transition is likely first order[2]; if $U_A(1)$ symmetry is broken the relevant symmetry becomes isomorphic to that of the 3-d O(4) spin model and the transition becomes second order belonging to that universality class. In the intermediate quark mass region, there is no real phase transition but a (rapid) crossover takes place from the hadronic phase to the quark gluon plasma phase. All the first order phase transition regions are separated from the crossover by lines of second order phase transitions belonging to the 3-d Ising $Z(2)$ universality class. Thus the first order region for $N_f = 3$ case, the second order O(4) line for $N_f = 2$ case and the second order $Z(2)$ line are supposed to meet at a tri-critical point characterized by a certain value m_s^{tri} of the strange quark mass. For the system with massless $N_f \geq 3$ flavors, the order of its chiral phase transition is always first order and is independent on the fate of $U_A(1)$ symmetry.

[1] We discuss implicitly the case with number of colors $N_c = 3$ in this proceedings.
[2] A second order phase transition is also allowed with a different symmetry breaking pattern [4].

The QCD phase diagram at vanishing baryon density has been studied extensively on the lattice. It has been well established that QCD with $N_f = 0$ has a first order phase transition from a confined phase to a deconfined phase at the transition temperature $T_c \approx 270$ MeV [5]. With physical values of the quark masses $m_u = m_d = m_l^{phys}$ and $m_s = m_s^{phys}$ it is found from lattice simulations with staggered fermions that QCD does not have a true transition but a rapid crossover and the pseudo critical transition temperature $T_{pc} \approx 154$ MeV [6]. The location of the physical point with respect to the tri-critical point is not fully determined. More specifically, it is not yet clear whether $m_s^{phys} > m_s^{tri}$ or $m_s^{phys} = m_s^{tri}$ or $m_s^{phys} < m_s^{tri}$. If $m_s^{phys} > m_s^{tri}$ then in the limit of $m_l \to 0$ one should see a second order transition belonging to the 3-d $O(4)$ universality class, if $m_s^{phys} = m_s^{tri}$ the tri-critical point is a Gaussian fixed point of the 3-dimensional ϕ^6 model and its critical exponents take the mean field values [7] and if $m_s^{phys} < m_s^{tri}$ then in $m_l \to 0$ limit first one should cross through a second order transition belonging to the 3-d $Z(2)$ universality class and then end up in the first order transition region. Recent lattice QCD calculations with improved staggered fermions indicate that QCD transition in the chiral limit of m_l with $m_s = m_s^{phys}$ is likely of second order and belongs to O(4) (O(2)) universality class [8]. To confirm this scenario, further study with better discretization scheme of QCD action and the investigation on the fate of $U_A(1)$ symmetry at finite temperature [9] are crucially needed.

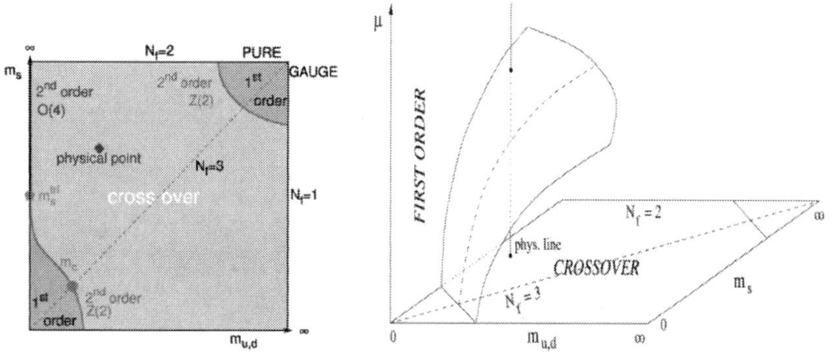

Figure 1. Left: schematic QCD phase transition behavior for different choices of quark masses $(m_{u,d}, m_s)$ at zero chemical potential. Right: The critical surface swept by the chiral critical line at finite chemical potential. A QCD chiral critical point may exist if the surface bends towards the physical point. The left and right plots are taken from Ref. [10] and Ref. [11], respectively.

For QCD with $N_f \geq 3$ a first order phase transition has been observed at small quark mass on coarse lattices using unimproved [12] as well as the p4-improved staggered fermions [13]. However, the value of pion mass at the critical point where the first order phase transition ends, m_π^c, is not yet determined in the continuum limit and its value at finite lattice cutoff varies from ~300 MeV to ~50 MeV, strongly depending on the lattice spacing and discretization schemes of the action [10–12, 14]. The value of m_π^c and the curvature of the surface swept from the chiral first order phase transition region at finite μ_B may be relevant to the (non-)existence of a QCD chiral critical point as sketched in the right plot of Fig. 1 although other possibilities for generating a second order transition at the physical values of quark masses have been discussed in Ref. [15]. Recent lattice QCD studies based on the imaginary chemical potential approach disfavor the existence of such a QCD chiral critical point [16]. This result, however, is obtained from simulations using unimproved staggered fermions on coarse lattices and may be prominently influenced by lattice cutoff effects. Thus it would be interesting to investigate this topic using an improved fermion action on finer lattices. As a starting point to determine the curvature of the chiral critical surface, the value of m_π^c needs to be determined. In this talk we thus focus on the QCD phase diagram with three degenerate Highly Improved Staggered Quarks at vanishing baryon density and discuss the value of m_π^c.

2. Universality class near critical lines

As mentioned above in the chiral limit of three flavor QCD the transition is of first order. This first order phase transition becomes weaker with finite quark mass and ends at a critical quark mass m_c sitting on a second order phase transition line. The universal properties of this critical point belong to a $Z(2)$ symmetry. The proper order parameter could be a mixture of two relevant quantities, e.g. a combination of the chiral condensate with the pure gauge action S_G [10] , $M = (\langle\bar{\psi}\psi\rangle + rS_G)\Big|_{\mathrm{T=T_c}} \propto (m - m_c)^{1/\delta}$ and the susceptibility of the order parameter M, i.e. $\chi_M/T^2\Big|_{\mathrm{T=T_c}} \propto (m-m_c)^{1/\delta-1}$. Here for $Z(2)$ universal class $1/\delta-1$=-0.785. To estimate the value of m_c, we ignore the contribution from S_G and will use the quark chiral condensate as an approximate order parameter for the second order phase transition at the critical point.

lattice dim.	am_q	m_π [MeV]	# β values	max no. of traj.
$16^3\times 6$	0.0075	230	17	8990
$10^3\times 6$	0.00375	160	12	12900
$12^3\times 6$	0.00375	160	12	12900
$16^3\times 6$	0.00375	160	12	12900
$24^3\times 6$	0.00375	160	12	12690
$24^3\times 6$	0.0025	130	5	11900
$24^3\times 6$	0.001875	110	7	8990
$24^3\times 6$	0.00125	90	7	10190
$24^3\times 6$	0.0009375	80	8	11790
$16^3\times 6$	0.0009375	80	6	7670

Table 1. Parameters of the numerical simulations.

As mentioned in Ref. [10] the chiral first order phase transition region from lattice QCD simulations shrinks with reduced cutoff effects. This is also observed by the lattice QCD calculation using the stout action [14]. The Highly Improved Staggered Quark (HISQ) action achieves better taste symmetry than the asqtad and p4 actions at a given lattice spacing [17]. Simulations presented here have been carried out with three degenerate quark flavors using the HISQ action. The quark masses vary from 0.0075 to 0.0009375 corresponding to pion masses in the region of $80 \lesssim m_\pi \lesssim 230$ MeV. In this proceeding we update the results reported in Ref. [10] by performing additional simulations with $am = 0.00375$ on $16^3 \times 6$, $12^3 \times 6$ and $10^3 \times 6$ lattices and simulations with quark mass $am = 0.025$ on $24^3 \times 6$ lattices.

3. Results

We show the volume dependence of quark chiral condensate with $am = 0.0009375$ ($m_\pi \approx 80$ MeV) in the left plot of Fig. 2. The volume dependence at high temperature is small while at low temperature it is relatively large. This is obvious since at high temperature the scale of the system is controlled by the temperature T while at low temperature the scales are the chiral symmetry breaking scales. In the right plot of Fig. 2 we show the volume dependence of chiral susceptibility. If the system with quark mass $am = 0.0009375$ is in the first order phase transition region, the chiral susceptibility should scale with volume. However, such volume scaling behavior is not observed in the system with pion mass $m_\pi \approx 80$ MeV.

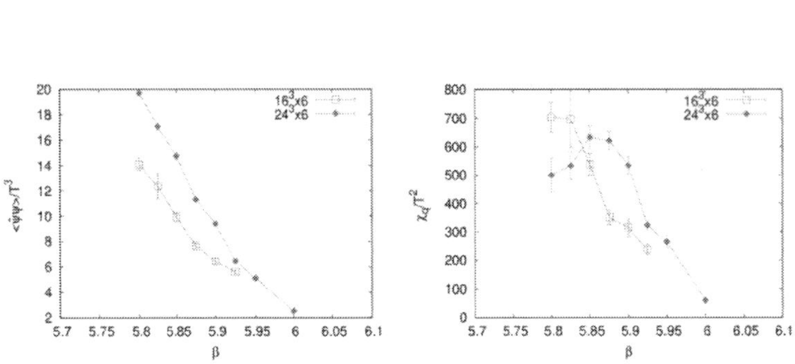

Figure 2. Volume dependences of chiral condensates (left) and chiral susceptibilities (right) with quark mass $am = 0.0009375$.

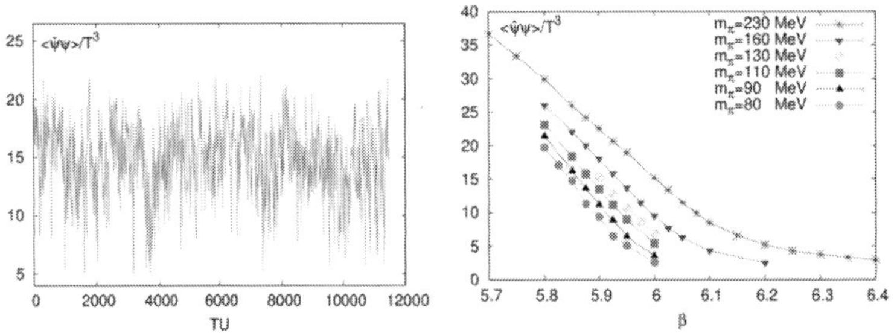

Figure 3. Left: time history of the quark chiral condensate near β_c with $am = 0.0009375$ on $24^3 \times 6$ lattices. Right: Chiral condensates as a function of β.

In the left plot of Fig. 3, we show the time history of quark chiral condensate near the pseudo critical β value for our lowest quark mass on $24^3 \times 6$ lattices. There is no evidence for the coexistence of two phases. We then investigate the temperature dependence of the chiral condensate at different quark masses shown in the right plot of Fig. 3. No evidence for a discontinuity in $\langle\bar{\psi}\psi\rangle$ as function of β at all quark masses is found. Together with the evidence from Fig. 2 we conclude that there is no first order phase transition in the system with pion mass ranging from 230 to 80 MeV.

In the left plot of Fig. 4 we show the disconnected part of the chiral susceptibility as a function of β at all quark masses. The pseudo critical temperature decreases with decreasing quark mass. It is expected as the hadronic degrees of freedom in the system become lighter and thus they become more easily excited in the thermal heat bath. They then can contribute to the energy density of the system and thus trigger the onset of a phase transition already at a lower temperature. The peak height of chiral susceptibility grows with decreasing quark mass as the system is approaching the first order phase transition region. We then performed a scaling fit using an ansatz of $A(m - m_c)^{1-1/\delta}$ to the inverse heights of chiral susceptibility peaks. The results are shown in the right plot of Fig. 4. The intercept of the solid line with the x axis gives an estimate for the critical quark mass. Taking into account that in the infinite volume limit the inverse peak height becomes larger (see Fig. 2 left) an upper bound for the value of the critical pion mass at the critical point can be obtained, i.e. $m_\pi^c \lesssim 45$ MeV.

4. Conclusion

We have performed 3-flavor QCD simulations using the HISQ action on $N_\tau = 6$ lattices with six pion masses in the region of $80 \lesssim m_\pi \lesssim 230$ MeV. Through the study of quark chiral condensates and chiral susceptibilities, we found no evidence of a first order chiral phase transition in this

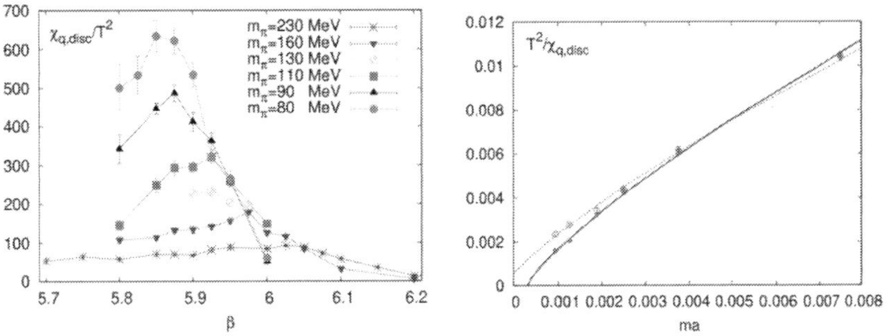

Figure 4. Left: the disconnected part of chiral susceptibility as a function of β. Right: the scaling fit to the inverse height of chiral susceptibility peaks. The filled points denote the inverse peak heights of chiral susceptibilities on largest volume at all quark masses while the open points are rescaled from the filled points by the volume effects. The solid line and dashed line are fitting curves using Z(2) critical exponents.

pion mass window. The upper bound on the pion mass at the critical point is estimated to be around 45 MeV by Z(2) universal scaling analysis considering the finite volume effects. In the quark mass plane this upper bound for the critical point in the 3-flavor QCD is thus at about $(m^s_{\rm phy}/270, m^s_{\rm phy}/270)$, which is far away from the physical point at $(m^s_{\rm phy}/27, m^s_{\rm phy})$. Together with the results from Ref. [14], our results suggest that the first order phase transition region is very small and thus the critical surface swept by the chiral critical line at finite chemical potential has to bend towards the physical point with a very large curvature to affect the nature of the transition at physical values of quark masses at a small chemical potential.

5. Acknowledgements

The numerical simulations were carried out on clusters of the USQCD Collaboration in Jefferson Lab and Fermilab, and on BlueGene computers at the New York Center for Computational Sciences (NYCCS) at Brookhaven National Lab. This manuscript has been authored under contract number DE-AC02-98CH10886 with the U.S. Department of Energy.

References

[1] F. Karsch, Lect. Notes Phys. **583**, 209 (2002).
[2] K. Fukushima, T. Hatsuda, Rept. Prog. Phys. **74**, 014001 (2011).
[3] B. Svetitsky and L. G. Yaffe, Nucl. Phys. B **210** (1982) 423, R. D. Pisarski and F. Wilczek, Phys. Rev. D **29** (1984) 338, S. Gavin, A. Gocksch and R. D. Pisarski, Phys. Rev. D **49** (1994) 3079.
[4] F. Basile, A. Pelissetto and E. Vicari, PoS LAT **2005** (2006) 199 [hep-lat/0509018].
[5] F. Karsch, Nucl. Phys. Proc. Suppl. **83** (2000) 14 [hep-lat/9909006].
[6] Y. Aoki, *et al.*, Nature 443, 675 (2006), A. Bazavov *et al.*, Phys. Rev. D **85** (2012) 054503.
[7] E. K. Riedel and F. J. Wegner, Phys. Rev. Lett. **29** (1972) 349.
[8] S. Ejiri, F. Karsch, E. Laermann *et al.*, Phys. Rev. D **80** (2009) 094505 [arXiv:0909.5122 [hep-lat]].
[9] H. -T. Ding, arXiv:1207.5187 [hep-lat], A. Bazavov *et al.*, Phys. Rev. D **86**, 094503 (2012).
[10] H. -T. Ding *et al.*, PoS LATTICE **2011** (2011) 191 [arXiv:1111.0185 [hep-lat]].
[11] F. Karsch, C. R. Allton, S. Ejiri *et al.*, Nucl. Phys. Proc. Suppl. **129**, 614-616 (2004).
[12] F. Karsch, E. Laermann and C. Schmidt, Phys. Lett. B **520**, 41 (2001), P. de Forcrand and O. Philipsen, Nucl. Phys. B **673**, 170 (2003), P. de Forcrand *et al.*, PoS LAT **2007**, 178 (2007), D. Smith and C. Schmidt, PoS LATTICE **2011**, 216 (2011), X. -Y. Jin and R. D. Mawhinney, PoS LATTICE **2011** (2011) 066.
[13] M. Cheng, *et al.*, Phys. Rev. D **75**, 034506 (2007).
[14] G. Endrodi, Z. Fodor, S. D. Katz, K. K. Szabo, PoS **LAT2007**, 182 (2007).
[15] S. Gupta, J. Phys. G **35** (2008) 104018 [arXiv:0806.2255 [nucl-th]].
[16] P. de Forcrand and O. Philipsen, JHEP **0811**, 012 (2008), PoS LATTICE **2008**, 208 (2008).
[17] E. Follana *et al.* [HPQCD and UKQCD Collaborations], Phys. Rev. **D75**, 054502 (2007).

Extreme QCD 2012 (xQCD) IOP Publishing
Journal of Physics: Conference Series **432** (2013) 012028 doi:10.1088/1742-6596/432/1/012028

Overview of large N QCD with chemical potential at weak and strong coupling

Timothy J. Hollowood[1] **and Joyce C. Myers**[2,3]

[1]College of Science, Department of Physics, Swansea University, Singleton Park, Swansea SA2 8PP, UK
[2]Centre for Theoretical Physics, University of Groningen, Nijenborgh 4, 9747 AG Groningen, The Netherlands
[3]Discovery Center, Niels Bohr Institute, University of Copenhagen, Blegdamsvej 17, 2100 Copenhagen Ø, Denmark

E-mail: t.hollowood@swansea.ac.uk, jcmyers@nbi.dk

Abstract. In this note we summarize the results from a longer article on obtaining the QCD phase diagram as a function of the temperature and chemical potential at large N_c and large N_f in the weak coupling limit $\lambda \to 0$, and the strong coupling limit $\lambda \to \infty$. The weak coupling phase diagram is obtained from the Polyakov line order parameter, and the quark number, calculated using 1-loop perturbation theory for QCD formulated on $S^1 \times S^3$. The strong coupling phase diagram is obtained from the same observables calculated at leading order in the lattice strong coupling and hopping parameter expansions. We show that the matrix models in these two limits agree at temperatures and chemical potentials which are not too high, such that observables in the strongly-coupled theory can be obtained from the observables in the weakly-coupled theory, and vice versa, using a simple transformation of variables.

1. Introduction

QCD at non-zero chemical potential provides a description of systems at large densities, yet it is a description which is currently not directly accessible since it occurs at strong coupling and the non-zero chemical potential leads to a complex action, giving rise to the well-known sign problem. What this means is that the conventional techniques of studying finite temperature QCD: conventional lattice simulations and ordinary perturbation theory, are not applicable. In this note we consider two nonconventional perturbative techniques which allow us to calculate the partition function and related observables such as the Polyakov line and quark number, and from these to map out the phase diagram in an otherwise inaccessible range of temperatures and chemical potentials. This is an executive summary of our longer paper [1].

The two perturbative techniques we employ are 1) weakly-coupled QCD from continuum 1-loop perturbation theory on $S^1 \times S^3$, and 2) strongly coupled lattice QCD with heavy quarks at leading order in a strong coupling, and hopping parameter expansion. In both cases we perform the calculations in the Veneziano large N_c limit, where N_c, $N_f \to \infty$, while $\frac{N_f}{N_c}$ remains fixed. This leads to advantages in both cases (for recent reviews on the progress towards understanding gauge theories and large N_c see [2, 3]). In the weakly coupled theory on $S^1 \times S^3$, the large N_c limit is required to have sharp phase transitions, since the calculation is only valid in very small volumes, such that $R \ll \Lambda_{QCD}^{-1}$, where R is the radius of the S^3. For the lattice strong coupling expansion, the spatial volume is large, but large N_c factorization and translational invariance lead to a simplification of the action in that terms which include correlations between different lattice sites drop out. In both cases the large N_c limit allows for a description of the theory in terms of the distribution of the Polyakov line eigenvalues such that the theory reduces to an analytically solvable matrix model. For temperatures and chemical potentials which are not too high we will show that there is an exact correspondence of the matrix models of the weakly-coupled and strongly-coupled theory, under a simple change of parameters.

2. QCD on $S^1 \times S^3$ with $\lambda \to 0$ vs. lattice QCD with heavy quarks as $\lambda \to \infty$

The action for continuum 1-loop QCD with constant A_0 was derived in [4] for theories with a matter content of scalars, vectors, and/or fermions. For QCD with N_f quarks of mass m, chemical potential μ, and at temperature $T = \frac{1}{\beta}$, the action in terms of the Polyakov line observable $\rho_n \equiv \frac{1}{N_c} \operatorname{Tr} \mathscr{P} e^{n \int_0^\beta \mathrm{d}t A_0(x)} = \frac{1}{N_c} \sum_{i=1}^{N_c} e^{in\theta_i}$ takes the form [4, 5]

$$\begin{aligned} S_{S^1\times S^3} - S_{Vdm} = &- N_c^2 \sum_{n=1}^{\infty} \frac{1}{n} z_{vn} \rho_n \rho_{-n} \\ &+ N_f N_c \sum_{n=1}^{\infty} \frac{(-1)^n}{n} z_{fn} \left(e^{n\beta\mu} \rho_n + e^{-n\beta\mu} \rho_{-n} \right), \end{aligned} \tag{1}$$

where S_{Vdm} is the contribution from the Vandermonde determinant, and z_{vn}, z_{fn} refer to the single particle partition functions for vectors and fermions,

$$z_{vn} = 2 \sum_{l=1}^{\infty} l(l+2) e^{-n\beta(l+1)/R} = \frac{2e^{-2n\beta/R}(3 - e^{-n\beta/R})}{(1 - e^{-n\beta/R})^3}, \tag{2}$$

$$z_{fn} = 2 \sum_{l=1}^{\infty} l(l+1) e^{-n\frac{\beta}{R}\sqrt{(l+\frac{1}{2})^2 + m^2R^2}}. \tag{3}$$

The action for large N_c, large N_f lattice QCD in terms of the Polyakov line $W(x) = \operatorname{Tr} \prod_{t=0}^{N_\tau - 1} U_{t,i}$, at leading order in the strong coupling and hopping parameter expansion is given by [6, 7]

$$\begin{aligned} S_{lat} - S_{Vdm} = &- JD \sum_x \left[\langle W \rangle W^\dagger(x) + \langle W^\dagger \rangle W(x) \quad \langle W \rangle \langle W^\dagger \rangle \right] \\ &- hN_c \sum_x \left[e^{\mu\beta} W(x) + e^{-\mu\beta} W^\dagger(x) \right], \end{aligned} \tag{4}$$

where $J \equiv 2 \left(\frac{\beta_{lat}}{2N_c^2} \right)^{N_\tau}$ for inverse coupling $\beta_{lat} = \frac{2N_c}{g^2}$ and number of temporal slices N_τ, and $h \equiv 2\frac{N_f}{N_c} \kappa^{N_\tau}$ is the hopping parameter with $\kappa \equiv \frac{1}{am+1+D}$ for lattice spacing a and number of spatial dimensions D.

The actions in (1) and (4) appear fairly similar with the exception of the sum over n in the former, and the sum over x in the latter (the term $\langle W \rangle W^\dagger(x) + \langle W^\dagger \rangle W(x) - \langle W \rangle \langle W^\dagger \rangle$ in (4) compared to $\rho_1 \rho_{-1}$ in (1) leads to the same equations of motion). The lack of terms with correlations between different lattice sites in (4) means that observables of the form $\langle F(W, W^\dagger) \rangle$ will undergo large cancellations such that

$$\begin{aligned} \langle F(W, W^\dagger) \rangle &= \frac{1}{N_x Z} \int \prod_x \mathrm{d}W(x) e^{-S[W(x), W^\dagger(x)]} \sum_{x'} F[W(x'), W^\dagger(x')], \\ &= \frac{\int \mathrm{d}W e^{-S(W,W^\dagger)} F(W, W^\dagger)}{\mathrm{d}W e^{-S(W,W^\dagger)}}. \end{aligned} \tag{5}$$

Therefore, when it is possible to truncate the sum over n in (1) to the $n = 1$ term there is an exact correspondence of matrix models resulting from (1) and (4) under the transformations

$$\begin{aligned} \rho_1 &\leftrightarrow \frac{1}{N_c}\langle W\rangle\,, \\ \rho_{-1} &\leftrightarrow \frac{1}{N_c}\langle W^\dagger\rangle\,, \\ z_{v1} &\leftrightarrow JD\,, \\ z_{f1}\frac{N_f}{N_c} &\leftrightarrow h\,. \end{aligned} \tag{6}$$

Truncation to the $n = 1$ contribution in (1) is valid when the temperature is not too high ($z_{v1},\ z_{f1}e^{\mu\beta} \gg z_{v2},\ z_{f2}e^{2\mu\beta}$), and the chemical potential is not too high ($\mu \lesssim \varepsilon_{f1} = \frac{1}{R}\sqrt{(l+\frac{1}{2})^2 + m^2R^2}\big|_{l=1,mR\to 0}$). This region of validity includes the line of transitions extending from the temperature-axis to the chemical potential-axis, but corrections from z_{fn} for $n > 1$ would need to be included to go to higher chemical potentials, as done in [5].

3. Large N_c formalism

To calculate observables in the large N_c limit we adopt the methods in [5], which adapts the Gross-Witten-Wadia [8–10] formalism to handle theories with complex actions. Consider a contour $\mathcal{C}$ with the Polyakov line eigenvalues $z_j = e^{i\theta_j}$ distributed along it with density $\varrho(z)$ defined according to the map

$$\frac{1}{N_c}\sum_{i=1}^{N_c} \xrightarrow[N_c\to\infty]{} \int_{-\psi}^{\psi}\frac{\mathrm{d}s}{2\pi} = \int_{\mathcal{C}}\frac{\mathrm{d}z}{2\pi i}\varrho(z)\,, \tag{7}$$

where $\psi = \pi$ in the confined phase, such that the contour is closed, and $\psi < \pi$ in the deconfined phase, such that the contour is open and the distribution of eigenvalues $z(s)$ has a gap. This is illustrated in Figure 1 as the deconfinement transition is crossed for $\mu \neq 0$.

The blue curve is a distribution $z(s)$, obtained from $i\int \mathrm{d}s = \int \mathrm{d}z\varrho(z)$, where $\varrho(z)$ is obtained in the confined phase at a point close to the deconfinement transition. Notice that the eigenvalue distribution lies significantly away from the unit circle (green dotted curve) in the $-z$ direction. The red curve is the distribution $z(s)$ obtained from $\varrho(z)$ in the deconfined (gapped) phase, at a point just past the deconfinement transition. These distributions correspond to configurations with complexified gauge fields (the θ_j are complex), which is necessary to obtain the correct stationary solution since the action is complex.

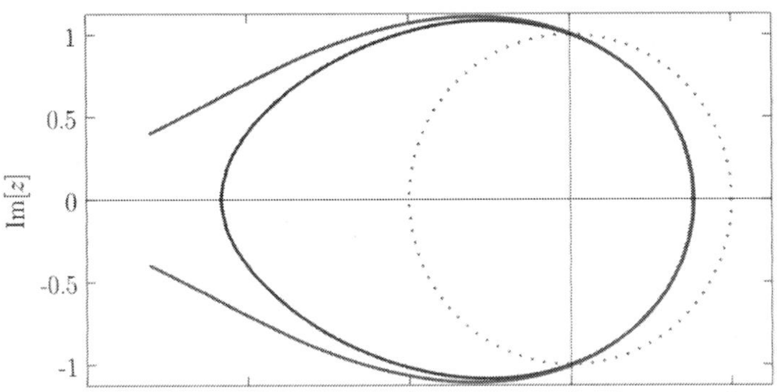

Figure 1. Distribution $z(s)$ of the Polyakov line eigenvalues in the confined phase (blue, $\mu R = 0.74$), and the deconfined phase (red, $\mu R = 0.76$), for $TR = 0.3$, $\frac{N_f}{N_c} = 1$, $mR = 0$.

Another requirement for obtaining the correct saddle point solutions is applying the $SU(N_c)$ constraint. Without this constraint taking $\mu \neq 0$ trivially shifts A_0 by a constant such that the free energy is independent of μ and the quark number is always zero. The $SU(N_c)$ constraint is incorporated by adding an appropriate term to the action with a Lagrange multiplier $\mathcal{N}$,

$$S \to S + i\mathcal{N}N_c \sum_{i=1}^{N_c} \theta_i \,, \tag{8}$$

where $\mathcal{N} = \frac{1}{N_c^2}N_q$ is the effective quark number in the large N_c limit [5]. The density $\varrho(z)$ is obtained by solving the equation of motion from $\frac{\partial S}{\partial \theta_i} = 0$, which, using (8) with (1) or (4) takes the form

$$\mathfrak{P}\int_{\mathcal{C}} \frac{\mathrm{d}z'}{2\pi i}\varrho(z')\frac{z'+z}{z'-z} = \alpha_{-1}z - \alpha_1 z^{-1} - \mathcal{N}\,, \tag{9}$$

where $\mathfrak{P}$ indicates principal value and $\alpha_{\pm1} \equiv \boldsymbol{z}_{v1}\rho_{\pm1} + \frac{N_f}{N_c}\boldsymbol{z}_{f1}e^{\mp\mu\beta}$. The procedure for solving the equation of motion depends on whether the contour $\mathcal{C}$ is open, as in the deconfined phase, or closed, as in the confined phase.

3.1. Confined (ungapped) phase

When $\mathcal{C}$ is considered to be a closed contour, as in the confined phase, the equation of motion (9) can be solved for $\varrho(z)$ using Cauchy's theorem. In the confined phase it is sufficient to consider the Fourier expansion of the density

$$\varrho(z) = \sum_{n=-\infty}^{\infty} \rho_n z^{-n-1}\,, \tag{10}$$

and solve for the ρ_n. The density is constrained to satisfy the identity constraint

$$\frac{1}{N_c}\sum_{i=1}^{N_c} \xrightarrow[N_c\to\infty]{} \int_{\mathcal{C}} \frac{\mathrm{d}z}{2\pi i}\varrho(z) = 1\,, \tag{11}$$

as well as the $SU(N_c)$ constraint

$$\sum_{i=1}^{N_c}\theta_i = 0 \xrightarrow[N_c\to\infty]{} \int_{\mathcal{C}} \frac{\mathrm{d}z}{2\pi i}\varrho(z)\log(z) = 0\,. \tag{12}$$

The free energy and other relevant observables can then be calculated by plugging in the stationary point solutions obtained for the ρ_n.

3.2. Deconfined (gapped) phase

When $\mathcal{C}$ is considered to be an open arc, as in the deconfined phase, then the equation of motion (9) must be solved by defining a resolvent and solving the Plemelj formulae. The resolvent is defined from the singular integral contribution in the equation of motion

$$\phi(z) = \int_{\mathcal{C}} \frac{\mathrm{d}z'}{2\pi i}\varrho(z')\frac{z'+z}{z'-z}\,. \tag{13}$$

Following [8–10] the contour $\mathcal{C}$ along which the eigenvalues are distributed is defined as a square root branch cut. The resolvent can then be evaluated using singular integral techniques for open arc contours to obtain [1]

$$\phi(z) = \left(\alpha_{-1}z^1 - \alpha_1 z^{-1}\right) + \sqrt{z^2 + r^2 - 2rxz}\left(\alpha_1 r^{-1}z^{-1} + \alpha_{-1}\right), \tag{14}$$

where the endpoints of the arc occur at radius r and angle $\pm\psi$, with $x \equiv \cos\psi$. Observables can then be calculated using

$$\int_{\mathcal{C}} \frac{dz}{2\pi i}\varrho(z)F(z) = \oint_{\Gamma} \frac{dz}{4\pi i z}\phi(z)F(z)\,.$$

where Γ is defined as a contour around $\mathcal{C}$, which can then be peeled off to surround the residues outside such that Cauchy's theorem can be used.

4. Results

Using the methods of the previous sections we calculated the Polyakov lines in the confined and deconfined regions and mapped out the phase diagrams [1]. Figure 2 shows the phase diagram for the weakly-coupled theory and Figure 3 for the strongly-coupled theory. The phase boundaries are determined by comparing the free energies of the gapped and ungapped eigenvalue distributions in the regions where both are possible. In the ungapped region the effective quark number $\mathcal{N} = 0$ and where the gapped distribution is favored $\mathcal{N} \neq 0$. In both phase diagrams the order of the transition is at least fifth order at $\mu = 0$, and at least 2nd order for $\mu \neq 0$.

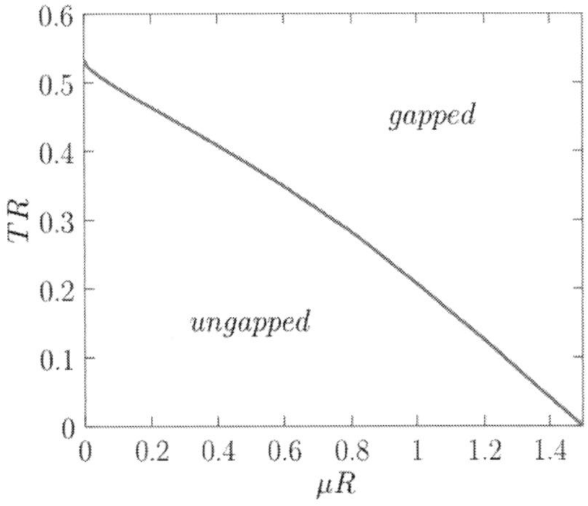

Figure 2. Phase diagram on $S^1 \times S^3$ for $mR = 0$, $\frac{N_f}{N_c} = 1$.

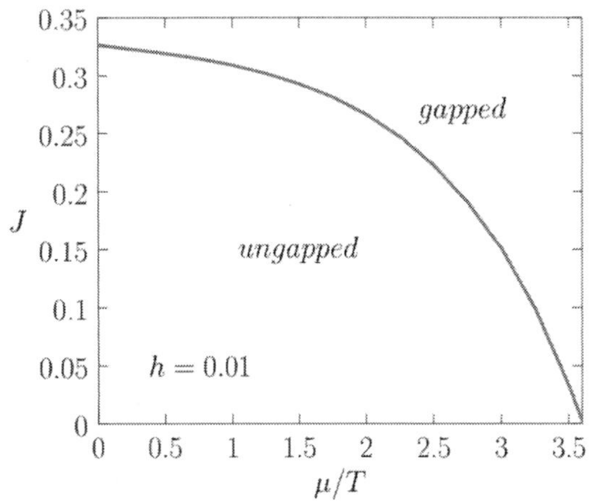

Figure 3. Phase diagram from the strong coupling expansion for $h = 0.01$.

References

[1] Hollowood T J and Myers J C 2012 *JHEP* **1210** 067 (*Preprint* 1207.4605)
[2] Lucini B and Panero M 2012 (*Preprint* 1210.4997)
[3] Ogilvie M C 2012 *J.Phys.* **A45** 483001 (*Preprint* 1211.2843)
[4] Aharony O, Marsano J, Minwalla S, Papadodimas K and Van Raamsdonk M 2004 *Adv.Theor.Math.Phys.* **8** 603–696 (*Preprint* hep-th/0310285)
[5] Hands S, Hollowood T J and Myers J C 2010 *JHEP* **1007** 086 (*Preprint* 1003.5813)
[6] Damgaard P and Patkos A 1986 *Phys.Lett.* **B172** 369
[7] Christensen C H 2012 *Phys.Lett.* **B714** 306–308 (*Preprint* 1204.2466)
[8] Gross D and Witten E 1980 *Phys.Rev.* **D21** 446–453
[9] Wadia S R 1979 *EFI-79/44-CHICAGO*
[10] Wadia S R 1980 *Phys.Lett.* **B93** 403

Quark-gluon thermodynamics with the $\mathbb{Z}_3$ symmetry

Yuji Sakai, Hiroaki Kouno[†], Takahiro Sasaki[‡], Takahiro Makiyama[†], Masanobu Yahiro[‡]

RIKEN Nishina Center, Saitama 351-0198, Japan,
Department of Physics, Saga University, Saga 840-8502, Japan[†],
Department of Physics, Kyushu University, Fukuoka 812-8581, Japan[‡]

E-mail: ysakai@riken.jp

Abstract. We propose a simple model with the $\mathbb{Z}_3$ symmetry in order to answer whether the symmetry is a good concept in QCD with light quark mass. The model is constructed by imposing the flavor-dependent twisted boundary condition (TBC) on the three-flavor Polyakov-loop extended Nambu-Jona-Lasinio model. In the model, the $\mathbb{Z}_3$ symmetry is preserved below some temperature T_c, but spontaneously broken above T_c. Dynamics of the simple model is similar to that of the original PNJL model without the TBC, indicating that the $\mathbb{Z}_3$ symmetry is a good concept. We also investigate the interplay between the $\mathbb{Z}_3$ symmetry and the emergence of the quarkyonic phase.

1. Introduction

In the limit of zero current quark mass, the chiral condensate is an exact order parameter for the chiral restoration. In the limit of infinite current quark mass, on the contrary, the Polyakov loop becomes an exact order parameter for the deconfinement transition, since the $\mathbb{Z}_3$ symmetry is exact there. For the real world in which u and d quarks have small current quark masses, the chiral condensate is considered to be a good order parameter, but it is not clear whether the Polyakov loop is a good order parameter. In this paper, we approach this problem by proposing a simple model with the $\mathbb{Z}_3$ symmetry. This paper is based on our recent papers [1, 2].

We start with the SU(3) gauge theory with three degenerate flavor quarks. The partition function Z in Euclidean spacetime is

$$Z = \int DqD\bar{q}DA\exp(-S_0), \quad S_0 = \int d^4x[\sum_f \bar{q}_f(\gamma_\nu D_\nu + m_f)q_f + \frac{1}{4g^2}{F^a_{\mu\nu}}^2] \tag{1}$$

with the temporal boundary condition $q_f(x, \beta = 1/T) = -q_f(x, 0)$ where T is the temperature. The $\mathbb{Z}_3$ transformation changes the ferimon boundary condition as [3, 4]

$$q_f(x, \beta) = -\exp(-i2\pi k/3)q_f(x, 0) \tag{2}$$

for integer $k = 0, 1, 2$, while the action keeps the original form. The $\mathbb{Z}_3$ symmetry thus breaks down by the fermion boundary condition. Now we assume the twisted boundary conditions (TBC)

$$q_f(x, \beta) \quad = \quad -\exp(i\theta_f)q_f(x, 0) \equiv -\exp[i2\pi(f-1)/3]q_f(x, 0) \tag{3}$$

for flavors $f = 1, 2, 3$. QCD with the TBC has the $\mathbb{Z}_3$ symmetry. Actually the $\mathbb{Z}_3$ transformation changes f into $f - k$, but $f - k$ can be relabeled by f since the action is invariant under the relabeling. The TBC is useful to understand the color confinement.

When the fermion field q_f is transformed as $q_f = \exp(-i\theta_f T\tau)q'_f$ for Euclidean time τ, the action S_0 is changed into

$$S(\theta_f) = \int d^4x[\sum_f \bar{q}'_f(\gamma_\nu D_\nu - \mu_f\gamma_4 + m_f)q'_f + \frac{1}{4g^2}F^2_{\mu\nu}] \tag{4}$$

with the imaginary quark number chemical potential $\mu_f = iT\theta_f$, while the TBC is transformed back to the standard boundary condition. The action S_0 with the TBC is thus equivalent to the action $S(\theta_f)$ with the standard one. The partition function $Z(T, \theta)$ has the Roberge-Weiss (RW) periodicity [3]: $Z(T, \theta) = Z(T, \theta + 2\pi k/3)$ for any integer k. The Polyakov-loop extended Nambu-Jona-Lasinio (PNJL) model is a good model to understand QCD at finite imaginary chemical potential [4].

In this paper, we consider QCD with the TBC in order to answer whether the $\mathbb{Z}_3$ symmetry is a good concept in QCD with light quark mass. Dynamics of the theory is studied concretely by imposing the TBC on the PNJL model, i.e., the TBC model.

2. PNJL and TBC models

The three-flavor PNJL Lagrangian is

$$\begin{aligned}\mathcal{L} &= \sum_f \bar{q}_f(\gamma_\nu D_\nu - \mu_f\gamma_4 + m_f)q_f - G_S\sum_f\sum_{a=0}^{8}[(\bar{q}_f\lambda_a q_f)^2 + (\bar{q}_f i\gamma_5\lambda_a q_f)^2] \\ &+G_D\left[\det_{ij}\bar{q}_i(1+\gamma_5)q_j + \det_{ij}\bar{q}_i(1-\gamma_5)q_j\right] + \mathcal{U}(\Phi[A], \Phi^*[A], T).\end{aligned} \tag{5}$$

For the Polyakov potential $\mathcal{U}$ as a function of the Polyakov-loop Φ and its conjugate Φ^*, we take the potential of Ref. [5]. Now we consider the PNJL model with the TBC (3), that is, the TBC model. The thermodynamic potential of the TBC model is nothing but that of the PNJL model with the flavor-dependent imaginary chemical potential $\mu_f = i\theta_f T$. In the mean-field approximation, the thermodynamic potential Ω of the TBC model is obtained as

$$\Omega = -2\sum_f\int\frac{d^3p}{(2\pi)^3}\left[3E_f + \frac{1}{\beta}\left(\ln\mathcal{F}_f + \ln\mathcal{F}_{\bar{f}}\right)\right] + \sum_f 2G_S\sigma_f^2 - 4G_D\sigma_u\sigma_d\sigma_s + \mathcal{U}(\Phi, \Phi^*, T), \tag{6}$$

where $\sigma_f \equiv \langle\bar{q}_f q_f\rangle$, $E_f \equiv \sqrt{\mathbf{p}^3 + M_f{}^2}$, $M_f = m_f - 4G_S\sigma_f + 2G_D|\epsilon_{fgh}|\sigma_g\sigma_h$ with the antisymmetric symbol ϵ_{fgh}. The function $\mathcal{F}_f$ and $\mathcal{F}_{\bar{f}}$ are obtained as

$$\mathcal{F}_f = 1 + 3\Phi e^{-\beta E_f^-} + 3\Phi^* e^{-\beta 2E_f^-} + e^{-3\beta E_f^-}, \tag{7}$$

$$\mathcal{F}_{\bar{f}} = 1 + 3\Phi^* e^{-\beta E_f^+} + 3\Phi e^{-\beta 2E_f^+} + e^{-3\beta E_f^+}. \tag{8}$$

When $\Phi = 0$, Ω has no flavor dependence, since the factors $e^{\pm 3i\theta_f}$ do not depend on flavor. The flavor symmetry is thus preserved in the confinement phase with $\Phi = 0$.

The vacuum term in (6) is regularized with the three-dimensional cutoff Λ. For the parameter set (G_S, G_D, m_l, m_s, Λ), we take the set of Ref. [6], except that the s-quark mass m_s is taken to be the same as the light quark mass $m_l \equiv m_u = m_d$.

3. Numerical results

Figure 1 shows T dependence of order parameters σ, Φ and $a_0 \equiv \sigma_u - \sigma_d = \sigma_u - \sigma_s$ in (a) the PNJL model with $(\theta_u, \theta_d, \theta_s) = (0, 0, 0)$ and (b) the TBC model with $(\theta_u, \theta_d, \theta_s) = (0, 2\pi/3, 4\pi/3)$; note that a_0 is an order parameter of the flavor symmetry. In the PNJL model, both the chiral restoration and the deconfinement transition are of a crossover type. In the TBC model, the first-order deconfinement transition takes place at $T = T_c \approx 195$ MeV. Below T_c, a_0 and Φ are zero, as expected. The flavor symmetry is thus preserved by the color confinement. Above T_c, a_0 and Φ become finite, indicating that the flavor and $\mathbb{Z}_3$ symmetries break simultaneously. The chiral restoration is very slow in the TBC model because of the breaking of flavor symmetry.

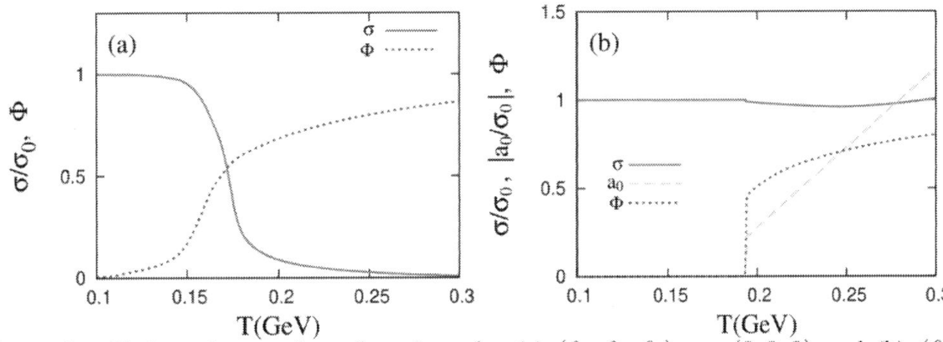

Figure 1. T dependence of σ, Φ and a_0 for (a) $(\theta_u, \theta_d, \theta_s) = (0, 0, 0)$ and (b) $(\theta_u, \theta_d, \theta_s) = (0, 2\pi/3, 4\pi/3)$. σ is normalized by the value σ_0 at $T = 0$.

Next we consider the entanglement-PNJL (EPNJL) model [7, 8]. The four-quark vertex G_S is originated in a gluon exchange between quarks and its higher-order diagrams. If the gluon field A_ν has a vacuum expectation value $\langle A_0 \rangle$, A_ν is coupled to $\langle A_0 \rangle$ and hence to Φ [9]. This effect allows G_S to depend on Φ: $G_S = G_S(\Phi)$ [9]. In this paper, we simply assume the following $G_S(\Phi)$ by respecting the chiral symmetry, the charge-conjugation symmetry [10] and the extended $\mathbb{Z}_3$ symmetry [4]:

$$G_S(\Phi) = G_S[1 - \alpha_1 \Phi\Phi^* - \alpha_2(\Phi^3 + \Phi^{*3})]. \tag{9}$$

In principle, G_D can also depend on Φ. However, Φ-dependence of G_D yields qualitatively the same effect on the phase diagram as that of G_S [8]. We can then neglect Φ-dependence of G_D. The parameters α_1 and α_2 in (9) are so determined as to reproduce two results of LQCD at finite T; one is the result of 2+1 flavor LQCD at $\mu = 0$ [11] that the chiral transition is crossover at the physical point and another is the result of degenerate three-flavor LQCD at $\theta = \pi$ [12] that the order of the RW endpoint is first-order for small and large quark masses but second-order for intermediate quark masses. The parameter set (α_1, α_2) satisfying these conditions is located in the triangle region [8]

$$\{-1.5\alpha_1 + 0.3 < \alpha_2 < -0.86\alpha_1 + 0.32,\ \alpha_2 > 0\}. \tag{10}$$

As a typical example, we take $\alpha_1 = 0.25$ and $\alpha_2 = 0.1$, following Ref. [8].

In Fig. 2, σ, a_0 and Φ are calculated as a function of T with the EPNJL model for (a) $(\theta_u, \theta_d, \theta_s) = (0, 0, 0)$ and (b) $(\theta_u, \theta_d, \theta_s) = (0, 2\pi/3, 4\pi/3)$. In panel (a), the chiral restoration and the deconfinement transition are first-order, because the current quark mass (5.5MeV) is small and the correlation between σ_f and Φ is strong [8]. In panel (b), one can see that $\Phi = a_0 = 0$ in the confinement phase. The EPNJL model with the TBC yields similar T dependence to that without the TBC for both the chiral restoration and the deconfinement transition, since the flavor-symmetry breaking above T_c is weakened by the strong correlation between σ_f and Φ.

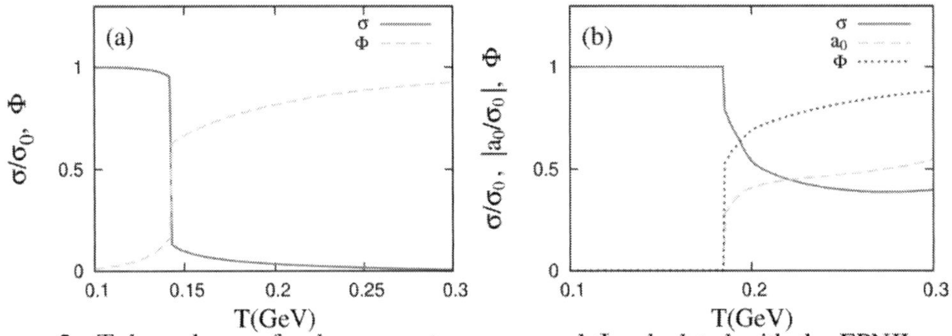

Figure 2. T-dependence of order parameters σ, a_0 and Φ calculated with the EPNJL model for (a) $(\theta_u, \theta_d, \theta_s) = (0, 0, 0)$ and (b) $(\theta_u, \theta_d, \theta_s) = (0, 2\pi/3, 4\pi/3)$. Here σ and a_0 are normalized by σ_0. Note that $a_0 = 0$ in panel (a) and $a_0 \geq 0$ in panel (b), while $\sigma < 0$ in both panels.

Now we consider the PNJL model with the flavor-dependent complex chemical potentials $\mu_f = \mu + iT\theta_f$ in which the θ_f are defined by

$$(\theta_f) = (0, \theta, -\theta). \tag{11}$$

The present model with the μ_f is reduced to the standard PNJL model with the flavor-independent real chemical potential μ when $\theta = 0$ and to the TBC model with real μ when $\theta = 2\pi/3$. Varying θ from 0 to $\theta = 2\pi/3$, one can see how the phase diagram changes between the approximate color-confinement in the standard PNJL model and the exact color-confinement in the TBC model.

Figure 3 shows the phase diagram in the μ-T plane. Panels (a)-(c) correspond to three cases of $\theta = 0$, $8\pi/15$ and $2\pi/3$, respectively. When $\theta = 0$, both the chiral and deconfinement transitions are of a crossover type at smaller μ, but the chiral transition becomes first-order at larger μ. When $\theta = 2\pi/3$, the deconfinement transition is the first-order at any μ, whereas the chiral transition line becomes first-order only at $\mu \approx M_f = 323$ MeV. The region labeled by "Qy" at $\mu \gtrsim M_f$ and small T is the quarkyonic phase [13], since $\Phi = 0$ but the quark number density n is finite there. The region labeled by "Had" is the hadron phase, because the chiral symmetry is broken there and thereby the equation of state is dominated by the pion gas [14]. The region labeled by "QGP" corresponds to the quark gluon plasma (QGP) phase, although the flavor symmetry is broken there by the TBC. As θ decreases from $2\pi/3$ to zero, the first-order chiral transition line declines toward smaller μ and the critical endpoint moves to smaller μ. Once θ varies from $2\pi/3$, the quarkyonic phase defined by $\Phi = 0$ and $n \neq 0$ shrinks on a line with $T = 0$ and $\mu \gtrsim M_f$ and a region at small T and $\mu \gtrsim M_f$ becomes a quarkyonic-like phase with small but finite Φ and $n \neq 0$; the latter region is labeled by "Qy-like".

4. Summary

We have proposed a simple model with the $\mathbb{Z}_N$ symmetry in order to answer whether the $\mathbb{Z}_N$ symmetry is a good concept in QCD with light quark mass. The model called the TBC model is constructed by imposing the flavor-dependent twisted boundary condition (3) on the PNJL model.

In the TBC model, the $\mathbb{Z}_3$ symmetry is preserved below T_c, but spontaneously broken above T_c. Below T_c, the color confinement preserves the flavor symmetry. Above T_c, meanwhile, the flavor symmetry is broken explicitly by the TBC. The flavor-symmetry breaking makes the chiral restoration slower, but the entanglement interaction between σ and Φ makes the restoration faster. Consequently, we can except that QCD with the TBC is similar to original QCD with the standard quark boundary condition.

We have also investigated the interplay between the $\mathbb{Z}_N$ symmetry and the emergence of the quarkyonic phase, considering the complex chemical potentials $\mu_f = \mu + iT\theta_f$ with $(\theta_f) = (0, \theta, -\theta)$

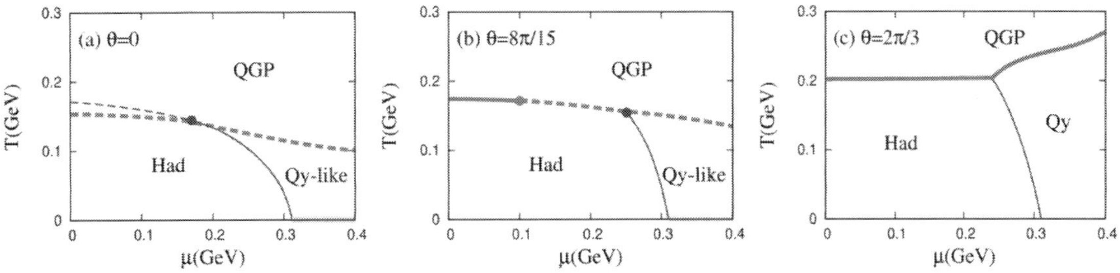

Figure 3. Phase diagram in the μ-T plane. Panels (a)-(c) correspond to three cases of $\theta = 0,\ 8\pi/15$ and $2\pi/3$, respectively. The thick (thin) solid curve denotes the first-order deconfinement (chiral) phase transition line, while the thick (thin) dashed curve the deconfinement (chiral) crossover line. The closed circles stand for the endpoints of the first-order deconfinement and chiral phase transition lines. In panels (a) and (b), the thick-solid line at $T = 0$ and $\mu \gtrsim M_f = 323$ MeV represents the quarkyonic phase.

in the PNJL model. The PNJL model with the μ_f is reduced to the PNJL model with real μ for $\theta = 0$, but to the TBC model with real μ for $\theta = 2\pi/3$. When $\theta = 2\pi/3$, the quarkyonic phase defined by $\Phi = 0$ and $n > 0$ exists at small T and large μ. Once θ varies from $2\pi/3$ to zero, the $\mathbb{Z}_{N_c}$ symmetry is broken and thereby the quarkyonic phase exists only on a line of $T = 0$ and $\mu \gtrsim M_f$. The region at small T and large μ is dominated by the quarkyonic-like phase characterized by small but finite Φ and $n > 0$. The quarkyonic-like phase at $\theta = 0$ is thus a remnant of the quarkyonic phase at $\theta = 2\pi/3$. Since the $\mathbb{Z}_N$ symmetry is explicitly broken at $\theta = 0$, it is natural to expand the concept of the quarkyonic phase and redefine it by a phase with small Φ and finite n. For this reason, the quarkyonic-like phase is often called the quarkyonic phase. The gross structure of the phase diagram thus has no qualitative difference between $\theta = 2\pi/3$ and zero, if the concept of the quarkyonic phase is properly expanded. In this sense, the $\mathbb{Z}_3$ symmetry is a good approximate concept for the case of $\theta = 0$, even if the current quark mass is small.

[1] Kouno H, Sakai Y, Makiyama T, Tokunaga K, Sasaki T and Yahiro M 2012 *J. Phys. G: Nucl. Part. Phys.* **39** 085010
[2] Sakai Y, Kouno H, Sasaki T and Yahiro M 2012 *Phys. Lett.* B **718** 130
[3] Roberge A and Weiss N 1986 *Nucl. Phys.* B **275** 734
[4] Sakai Y, Kashiwa K, Kouno H and Yahiro M 2008 *Phys. Rev.* D **77**, 051901
Sakai Y, Kashiwa K, Kouno H and Yahiro M 2008 *Phys. Rev.* D **78**, 036001
[5] Rößner S, Ratti C and Weise W 2007 *Phys. Rev.* D **75**, 034007
[6] Rehberg P, Klevansky S P and Hüfner J 1996 *Phys. Rev.* C **53** 410
[7] Sakai Y, Sasaki T, Kouno H and Yahiro M 2010 *Phys. Rev.* D **82** 076003
[8] Sasaki T, Sakai Y, Kouno H and Yahiro M 2011 *Phys. Rev.* D **84** 091901
[9] Kondo K -I 2010 *Phys. Rev.* D **82** 065024
[10] Kouno H, Sakai Y, Kashiwa K and Yahiro M 2009 *J. Phys. G: Nucl. Part. Phys.* **36** 115010
Kouno H, Sakai Y, Sasaki T, Kashiwa K and Yahiro M 2011 *Phys. Rev.* D **83** 076009
Kouno H, Kishikawa K, Sasaki T, Sakai Y and Yahiro M 2012 *Phys. Rev.* D **85** 016001
[11] Aoki Y, Endrödi G, Fodor Z, Katz S D and Szabó K K 2006 *Nature* **443** 675.
[12] de Forcrand P and Philipsen O 2010 *Phys. Rev. Lett.* **105** 152001
[13] McLerran L and Pisarski R D 2007 *Nucl. Phys.* A **796** 83
[14] Sakai Y, Sasaki T, Kouno H and Yahiro M 2012 *J. Phys. G: Nucl. Part. Phys.* **39** 035004

Extreme QCD 2012 (xQCD) IOP Publishing
Journal of Physics: Conference Series **432** (2013) 012030 doi:10.1088/1742-6596/432/1/012030

Anomaly-induced charges in nucleons

Yu Maezawa

Physics Department, Brookhaven National Laboratory, Upton, New York 11973, USA

E-mail: ymaezawa@bnl.gov

Abstract. We suggest a novel charge structure inside nucleons in electromagnetic field due to the chiral anomaly. We use Skyrmions, where nucleons appear as solitons of mesons, to calculate the charge distributions in a single nucleon and find that an additional non-integer charge proportional to the magnetic field would be produced. This might look surprising, but the magnitude of the induced charge is evaluated to be tiny enough to have not been observed yet.

1. Introduction

[1]Recent advance in observations and experiments explores new effects of strong electromagnetic fields on fundamental particles. Since matter consists of baryons, electromagnetic properties of protons and neutrons are of most importance. Under the strong magnetic fields such as in neutron stars, supernovae and heavy ion collisions, tiny quantum effect of quantum chromodynamics may lead to an unveiled and significant consequence.

In this article, we investigate baryons under external electromagnetic fields. For describing the baryons, we use the Skyrme model [2] with Wess-Zumino-Witten (WZW) term [3, 4] including electromagnetism. The consequence is amazing: Nucleons in the external electromagnetic fields have anomalous charge distribution due to the chiral anomaly. Nonzero net charge, which is generally non-integer, is induced *even for neutrons*. Correspondingly, we will show that the Gell-Mann-Nishijima formula, $Q = I_3 + N_B/2$ (Q: electric charge, I_3: the third component of isospin, N_B: baryon number), has an additional term due to the quantum anomaly.

Figure 1 shows a schematic description of the phenomenon. Due to the anomalous interaction with a quark loop through the WZW term, nucleons (= Skyrmions) have an additional interaction to the electromagnetic field A_μ. Under the external electromagnetic fields, the anomalous coupling induces the additional electric charge.

2. Skyrmions and anomaly

We adopt the Skyrme model for a concrete illustration in this article. The essential idea of the Skyrme model is to unify baryons and mesons: baryons are described as topological solitons of mesons. This model, known to reproduce experimental data of nucleons within 30% accuracy, is suitable for our purpose. This is because we concentrate on the anomalous contribution to baryons, which is described by the coupling between mesons and photons shown in Fig. 1. Any baryons wearing mesonic clouds will follow our mechanism of anomalous charge generation.

[1] This is study in collaboration with M. Eto (Yamagata Univ.), K. Hashimoto (RIKEN & Osaka Univ.), H. Iida (YITP) and T. Ishii (Seoul National Univ.), and summarized in Ref. [1]

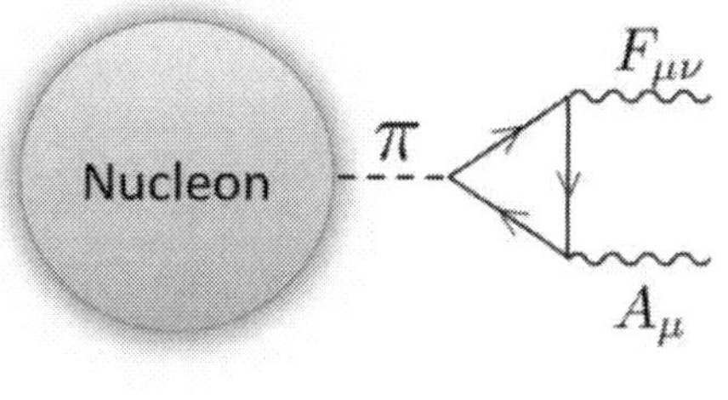

Figure 1. A schematic figure for electric charge generation of a nucleon.

The action of two-flavor Skyrme model [2, 5, 6] coupled with electromagnetic field is

$$S = \int d^4x \left(\mathcal{L}_{\text{kin}} + \mathcal{L}_{\text{mass}} - \frac{1}{4} F_{\mu\nu} F^{\mu\nu} \right),$$
$$\mathcal{L}_{\text{kin}} = -\frac{F_\pi^2}{16} \text{tr}(\hat{R}_\mu \hat{R}^\mu) + \frac{1}{32 e_s^2} \text{tr}([\hat{R}_\mu, \hat{R}_\nu][\hat{R}^\mu, \hat{R}^\nu]),$$
$$\mathcal{L}_{\text{mass}} = \frac{F_\pi^2}{16} m_\pi^2 \text{tr}(U + U^\dagger - 2), \quad \hat{R}_\mu = D_\mu U U^\dagger,$$

where m_π and F_π are the pion mass and the pion decay constant, respectively. e_s is a dimensionless constant and $D_\mu U \equiv \partial_\mu U + ieA_\mu[q, U]$ with $q \equiv \text{diag}(2/3, -1/3)$. The pseudo-scalar field U is an SU(2) matrix which transforms as $U \to G_L U G_R^\dagger$ with $G_L \in \text{SU}(2)_L$ and $G_R \in \text{SU}(2)_R$. In the following, we use dimensionless variables: $r \to r/(e_s F_\pi)$ and $m_\pi \to (e_s F_\pi) m_\pi$. In the Skyrme model, a general hedgehog-type ansatz in the absence of the electromagnetic background is written as

$$U = GU_0G^\dagger = G \exp(if(r)\hat{\boldsymbol{x}} \cdot \boldsymbol{\tau})G^\dagger, \qquad (1)$$
$$G = a_0 + i\boldsymbol{a} \cdot \boldsymbol{\tau} \in \text{SU}(2)_{L+R}, \quad (a_0^2 + \boldsymbol{a}^2 = 1),$$

where $\hat{\boldsymbol{x}} \equiv \boldsymbol{x}/|\boldsymbol{x}|$, $\boldsymbol{\tau}$ are Pauli matrices, and $(a_0, \boldsymbol{a})$ are moduli parameters spanning $\text{SU}(2)_{L+R} \simeq S^3$. We treat electromagnetic effects as a perturbation in terms of e. The equation of motion gives

$$\left(\frac{r^2}{4} + 2\sin^2 f \right) f'' + \frac{r}{2} f' + \sin(2f) \left(f'^2 - \frac{1}{4} - \frac{\sin^2 f}{r^2} \right) - \frac{m_\pi^2 r^2}{4} \sin f = 0. \qquad (2)$$

Solving this under the boundary conditions, $f(0) = \pi$ and $f(r \to \infty) = 0$, one obtains a solution with baryon number $B = 1$. The solution is a topological soliton, called Skyrmion.

We focus on the coupling between mesons and photons in the WZW term. In the two-flavor case, this can be given by [7, 4, 3]

$$S_{\text{WZW}} = -\int d^4x \, A_\mu \left(\frac{eN_c}{6} j_B^\mu + \frac{1}{2} j_{\text{anm}}^\mu \right), \qquad (3)$$
$$j_B^\mu = \frac{1}{24\pi^2} \epsilon^{\mu\nu\rho\sigma} \text{tr}[R_\nu R_\rho R_\sigma], \qquad (4)$$
$$j_{\text{anm}}^\mu = -\frac{ie^2 N_c}{96\pi^2} \epsilon^{\mu\nu\rho\sigma} F_{\nu\rho} \text{tr}[\tau_3(L_\sigma + R_\sigma)], \qquad (5)$$

where j_B^μ is a baryon current giving an integer baryon number, $L_\mu = U^\dagger \partial_\mu U$, $R_\mu = \partial_\mu U U^\dagger$, and $\epsilon^{0123} = -1$ in this article.

In the presence of background electromagnetic fields, not only the first term but also the second term in Eq. (3) is important. The electric charge Q with the contribution from anomaly ($N_c = 3$) is written as

$$Q = I_3 + \frac{N_B}{2} + \frac{Q_{\rm anm}}{2}, \tag{6}$$

where $N_B = \int d^3x j_B^0$ and $Q_{\rm anm} = \int d^3x j_{\rm anm}^0$. Thus, the Gell-Mann-Nishijima formula is corrected under background electromagnetic fields. Substituting Eq. (1) into Eq. (5), we obtain

$$\begin{aligned} j_{\rm anm}^\mu &= \frac{ie^2 N_c}{96\pi^2} \epsilon^{\mu\nu\rho\sigma} F_{\nu\rho} P_\sigma, \\ P_\mu &= 4i \left[(\partial_\mu f) \hat{x}_3^{\rm rot} + \frac{\sin(2f)}{2} \partial_\mu (\hat{x}_3^{\rm rot}) \right], \\ \hat{x}_3^{\rm rot} &= (a_0^2 + a_3^2 - a_1^2 - a_2^2)\hat{x}_3 + 2(a_1 a_3 + a_0 a_2)\hat{x}_1 + 2(a_2 a_3 - a_0 a_1)\hat{x}_2. \end{aligned}$$

This is a classical anomalous current for the general hedgehog solutions.

3. Induced charges from quantized Skyrmion

To obtain physical values of the anomalous charge depending on the baryon states, we need to quantize the Skyrmion. By solving a quantum mechanics of the S^3 moduli parameters on the Skyrmion, quantum states of a nucleon with spin quantized along x^3 are given by $\psi_{p\uparrow} = (a_1 + ia_2)/\pi$, etc.[5]. We evaluate matrix elements of the anomalous current,

$$\langle j_{\rm anm}^\mu \rangle_{I_3,S_3} \equiv \int d\Omega_3 \psi_{I_3,S_3}^* \, j_{\rm anm}^\mu(a_0, \boldsymbol{a}) \, \psi_{I_3,S_3},$$

where $\int d\Omega_3$ denotes the integration over the S^3, and ψ_{I_3,S_3} is the baryon states labeled by the third components of isospin and spin. The matrix elements are calculated as

$$\begin{aligned} \langle j_{\rm anm}^\mu \rangle_{I_3,S_3} &= \frac{ie^2 N_c}{96\pi^2} \epsilon^{\mu\nu\rho\sigma} F_{\nu\rho} \langle P_\sigma \rangle_{I_3,S_3}, \\ \langle P_0 \rangle_{I_3,S_3} &= 0, \\ \langle P_a \rangle_{I_3,S_3}^{a=1,2} &= -\frac{16i}{3} I_3 S_3 \left(f' - \frac{\sin(2f)}{2r} \right) \hat{x}_a \hat{x}_3, \\ \langle P_3 \rangle_{I_3,S_3} &= -\frac{16i}{3} I_3 S_3 \left[\left(f' - \frac{\sin(2f)}{2r} \right) \hat{x}_3^2 + \frac{\sin(2f)}{2r} \right]. \end{aligned}$$

In the following, we concentrate on the case with magnetic-field backgrounds B_i. The anomalous charge density is indeed induced in nucleons:

$$\langle j_{\rm anm}^0 \rangle_{I_3,S_3} = \frac{ie^2 N_c}{48\pi^2} B_i \langle P_i \rangle_{I_3,S_3}. \tag{7}$$

Figure 2 shows the baryon number distribution, and the anomalous charge distribution under magnetic field along the 3rd- and the 1st-axes of quantized Skyrmions at $m_\pi = 0$. The configurations of the charge distribution look like wave functions of an electron in a hydrogen atom.

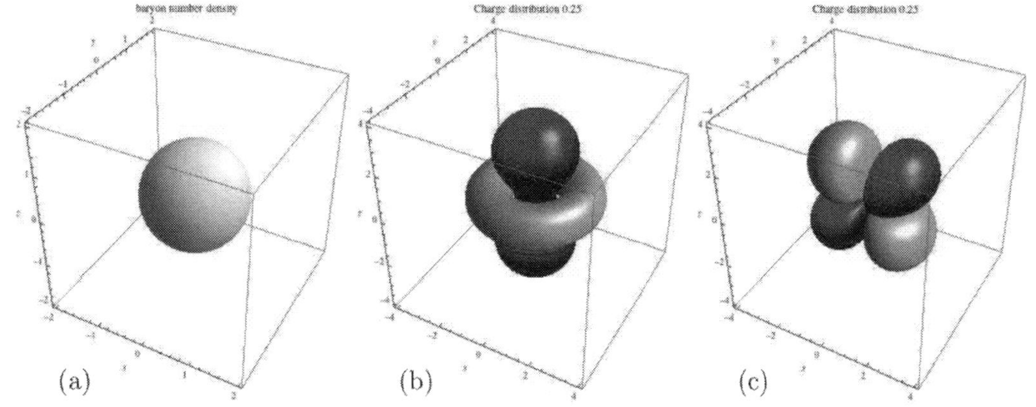

Figure 2. The constant-height surfaces of (a) density distribution of baryon number, (b) electric charge under magnetic field along the 3rd-axis, and (c) electric charge under magnetic field along the 1st-axis.

In contrast, we find that the matrix element of the spatial component of the current density vanishes, $\langle j^i_{\rm anm}\rangle_{I_3,S_3} = 0$. Thus, the electric current is not induced [1].

Let us calculate the total electric charge from the anomalous effect of $\langle P_i\rangle_{I_3,S_3}$ over the whole space gives

$$\int d^3x \langle P_i\rangle_{I_3,S_3} = \begin{cases} 0 & (i=1,2), \\ -\frac{16\pi i}{9}(4I_3S_3)c_0 & (i=3), \end{cases}$$

where $c_0 = \int dr\{r^2 f' + \sin(2f)\}$. Its numerical value is $c_0 = (-5.32, -12.3, -10.2, -7.32)$ for pion masses $m_\pi = (0, m_\pi^{\rm phys}/2, m_\pi^{\rm phys}, 2m_\pi^{\rm phys})$, respectively ($m_\pi^{\rm phys} \equiv 0.263$ is the physical value of the pion mass in the unit of eF_π, determined from the mass splitting between nucleon and Δ [5]). We obtain the anomalous charge for nucleons

$$Q_{\rm anm} = \frac{4e^2N_c}{27\pi} I_3S_3 \frac{c_0 B_3}{(e_s F_\pi)^2}. \tag{8}$$

Equation (8) shows that an electric charge is actually induced by the anomalous effect *even for a neutron.* We further find that dipole moment vanishes while quadrupole moment appears as a leading multipole [1].

4. Observation

Let us argue possibilities to observe the anomalous charge. First, we estimate the amount of induced charge in a nucleon. Using Eq. (8), we obtain $Q_{\rm anm} \sim e \times 10^{-20} I_3S_3[{\rm G}^{-1}] \times B_3[{\rm G}]$. Under the terrestrial magnetic field $B \sim 1[{\rm G}]$, the induced charge $Q_{\rm anm}$ is about $10^{-20}e$, which would be too small to observe. On the surface of a magnetar, which is a neutron star with very strong magnetic field of order $10^{15}[{\rm G}]$, $Q_{\rm anm}$ is about $10^{-5}e$. In heavy ion collisions, magnetic field of order $10^{17}[{\rm G}]$ would be created. However, $Q_{\rm anm}$ is about $10^{-3}e$ even for such an extremely strong magnetic field. Hence, it is natural that the electric charge of neutrons has never been detected until now.

Next, electric dipole moment (EDM) of nucleons is not induced from the anomaly. This is consistent with the experimental results that there is no evidence for the existence of neutron EDM (see, e.g., [8]), which is performed under a magnetic field. In our study, the leading multipole is a quadrupole, $Q_{33} = -2Q_{11} = -2Q_{22} \sim e \times 10^{-19} I_3S_3[{\rm fm}^2{\rm G}^{-1}] \times B_3[{\rm G}]$. Its experimental measurement would be interesting.

5. Summary

We found that an additional non-integer charge would be induced on nucleons in external magnetic fields, based on the Skyrme model with Wess-Zumino-Witten term. We have also found nonzero multipoles such as a quadrupole moment. The detailed analysis of the validity of the computation will be given in [1], as the generation of the charge might sound surprising and counter-intuitive. Here we presented the evaluation of the magnitude of the induced charge, and found it was tiny enough to have not been observed yet. To see the universality of the generation of the anomalous charge, confirmation in other approaches is desirable. For instance, lattice QCD simulation with external electromagnetic fields is a reliable approach. Holographic QCD is also helpful for gaining insights. The novel mechanism due to the anomaly is intriguing, regardless of the smallness of the induced charges. Neutrons play an important role on the frontiers of hadron physics, such as neutron stars and heavy-ion collisions, where strong magnetic fields exist. Such neutrons would have anomalous charges which may be physically significant. Our results will bring new aspects of the dynamics of hadrons.

References

[1] M. Eto, K. Hashimoto, H. Iida, Y. Maezawa and T. Ishii, Phys. Rev. D **85**, 114038 (2012).
[2] T. H. R. Skyrme, Proc. Roy. Soc. Lond. A260, 127, (1961).
[3] J. Wess and B. Zumino Phys. Lett. **B37**, 95 (1971).
[4] E. Witten, Nucl. Phys.**B223**, 422 (1983); Nucl. Phys. **B223**, 433 (1983).
[5] G. S. Adkins, C. R. Nappi and E. Witten, Nucl. Phys. B **228**, 552 (1983).
[6] G. S. Adkins and C. R. Nappi, Nucl. Phys. B **233**, 109 (1984).
[7] D. T. Son and M. A. Stephanov, Phys. Rev. **D77**, 014021 (2008).
[8] C. A. Baker *et al.*, Phys. Rev. Lett. **97**, 131801 (2006).

Journal of Physics: Conference Series **432** (2013) 012031 doi:10.1088/1742-6596/432/1/012031

Meson mass and the sign problem at finite theta

Takahiro Sasaki[1], Hiroaki Kouno[2] and Masanobu Yahiro[1]

[1] Department of Physics, Graduate School of Sciences, Kyushu University, Fukuoka 812-8581, Japan

[2] Department of Physics, Saga University, Saga 840-8502, Japan

E-mail: sasaki@email.phys.kyushu-u.ac.jp

Abstract. We propose a practical way of circumventing the sign problem in lattice QCD simulations with the theta-vacuum term. This method is the reweighting method for QCD Lagrangian after the $U_A(1)$ transformation. In the Lagrangian, the P-odd mass term as a cause of the sign problem is minimized. In order to find out a good reference system in the reweighting method, we estimate the average reweighting factor by using the two-flavor NJL model and eventually find a good reference system.

1. Introduction

The existence of instanton solution requires QCD Lagrangian with the theta vacuum:

$$\mathcal{L} = \sum_f q_f(\gamma_\nu D_\nu + m_f)q_f + \frac{1}{4g^2}F^a_{\mu\nu}F^a_{\mu\nu} - i\theta\frac{1}{64\pi^2}\varepsilon_{\mu\nu\sigma\rho}F^a_{\mu\nu}F^a_{\sigma\rho}, \quad (1)$$

in Euclidean spacetime. Hereafter, we will consider two-flavor QCD and assume isospin symmetry, $m_u = m_d = m_0$. Though the angle θ can take any arbitrary value theoretically, experimental measurements of neutron dipole moment give the upper limit, $|\theta| < 10^{-9}$[1, 2]. Why should θ be so small? This long-standing puzzle is called the strong CP problem.

Since the upper limit is determined only at zero temperature, the behavior is nontrivial for finite temperature. Hence the first-principle lattice simulation is needed, but it has the sign problem for finite θ. After making $U_A(1)$ transformation

$$q = e^{i\gamma_5\frac{\theta}{4}}q', \quad (2)$$

θ dependence appears only through the mass term

$$m_0(\theta) = m_0\cos(\theta/2) + m_0 i\gamma_5\sin(\theta/2), \quad (3)$$

in the transformed Lagrangian

$$\mathcal{L} = \sum_f \bar{q}_f(\gamma_\nu D_\nu + m_0(\theta))q_f + \frac{1}{4g^2}F^a_{\mu\nu}F^a_{\mu\nu}. \quad (4)$$

The P-odd mass term including $i\gamma_5$ makes the fermion determinant complex.

Because of the sign problem, we should perform a reweighting method in lattice simulations. The vacuum expectation value of operator $\mathcal{O}$ is obtained by

$$\langle \mathcal{O} \rangle = \int \mathcal{D}A\mathcal{O}\mathrm{det}\mathcal{M}(\theta)e^{-S_g} \tag{5}$$

$$= \int \mathcal{D}A\mathcal{O}'\mathrm{det}\mathcal{M}_{\mathrm{ref}}(\theta)e^{-S_g} \tag{6}$$

with the gluon part S_g of the QCD action and

$$\mathcal{O}' \equiv R(\theta)\mathcal{O}, \tag{7}$$

$$R(\theta) \equiv \frac{\mathrm{det}\mathcal{M}(\theta)}{\mathrm{det}\mathcal{M}_{\mathrm{ref}}(\theta)}, \tag{8}$$

where $R(\theta)$ is the reweighting factor and $\det \mathcal{M}_{\mathrm{ref}}(\theta)$ is the Fermion determinant of the reference theory that has no sign problem. The simplest candidate of the reference theory is the theory in which the θ-odd term is neglected in the mass term (3). We refer to this reference theory as reference A in this paper. As discussed in Ref. [3], reference A may be a good reference theory for small and intermediate θ, but not for large θ near π. In reference A, the limit of $\theta = \pi$ corresponds to the chiral limit for $\mathrm{det}\mathcal{M}_{\mathrm{ref}}$ that is hard for LQCD simulations to reach.

The expectation value of $R(\theta)$ in the reference theory is obtained by

$$\langle R(\theta) \rangle = \frac{Z}{Z_{\mathrm{ref}}} \tag{9}$$

where Z (Z_{ref}) is the partition function of the original (reference) theory. The average reweighting factor $\langle R(\theta) \rangle$ is a good index for the reference theory to be good; the reference theory is good when $\langle R(\theta) \rangle = 1$.

In this paper, we estimate $\langle R(\theta) \rangle$ with the two-flavor NJL model in order to find out a good reference theory. We find that reference A is good only for small θ, so propose a good reference theory that satisfies $\langle R(\theta) \rangle \approx 1$. This work is based on the Ref. [4].

2. Model setting

The two-flavor NJL Lagrangian with the θ-dependent term is obtained by

$$\mathcal{L} = \bar{q}(\gamma_\nu \partial_\nu + m_0)q - G_1 \sum_{a=0}^{3} \left[(\bar{q}\tau_a q)^2 + (\bar{q}i\gamma_5\tau_a q)^2 \right] - 8G_2 \left[e^{i\theta}\mathrm{det}\bar{q}_{\mathrm{R}}q_{\mathrm{L}} + e^{-i\theta}\mathrm{det}\bar{q}_{\mathrm{L}}q_{\mathrm{R}} \right], \tag{10}$$

in Euclidean spacetime where m_0 is the current quark mass and τ_0 and $\tau_a (a = 1, 2, 3)$ are the 2×2 unit and Pauli matrices in the flavor space, respectively. The parameter G_1 denotes the coupling constant of the scalar and pseudoscalar-type four-quark interactions, while G_2 stands for that of the Kobayashi-Maskawa-'t Hooft determinant interaction [5, 6] where the matrix indices run in the flavor space. Under the $U_A(1)$ transformation (2), the Lagrangian density is then rewritten with q' as

$$\mathcal{L} = \bar{q}'(\gamma_\nu\partial_\nu + m_0(\theta))q' - G_+ \left[(\bar{q}'q')^2 + (\bar{q}'i\gamma_5\vec{\tau}q')^2 \right] - G_- \left[(\bar{q}'\vec{\tau}q')^2 + (\bar{q}'i\gamma_5 q')^2 \right], \tag{11}$$

where $G_\pm = G_1 \pm G_2$.

Applying the saddle-point approximation to the path integral in the partition function, one can get the average reweighting factor $\langle R(\theta) \rangle$,

$$\langle R(\theta) \rangle \approx R_A R_B \tag{12}$$

$$R_A = \sqrt{\frac{\mathrm{det}H_{\mathrm{ref}}}{\mathrm{det}H}}, \quad R_B = e^{-\beta V(\Omega - \Omega_{\mathrm{ref}})}, \tag{13}$$

where $\beta = 1/T$ and Ω ($\Omega_{\rm ref}$) is the thermodynamic potential at the mean-field level in the original (reference) theory [4]. H ($H_{\rm ref}$) is the Hessian matrix in the original (reference) theory defined by [7, 8]

$$H_{ij} = \frac{\partial^2\Omega}{\partial\phi'_i\partial\phi'_j}, \quad \{\phi'_i\} = \{\sigma', \eta', \vec{a}', \vec{\pi}'\}, \tag{14}$$

with the quark-condensates

$$\sigma' = \langle \bar{q}q \rangle \ , \ \eta' = \langle \bar{q}i\gamma_5 q \rangle \ , \ \vec{a}' = \langle \bar{q}\vec{\tau}q \rangle \ , \ \vec{\pi}' = \langle \bar{q}i\gamma_5\vec{\tau}q \rangle \,. \tag{15}$$

The four-dimensional volume βV is obtained by $\beta V = (N_x/N_\tau)^3 T^{-4}$ for the $N_x^3 \times N_\tau$ lattice. Here we consider $N_x/N_\tau = 4$ as a typical example, following Refs. [7, 8].

We consider the following reference theory that has no sign problem:

$$\mathcal{L} = \bar{q}'(\gamma_\nu\partial_\nu + m_{\rm ref}(\theta))q' - G_+\left[(\bar{q}'q')^2 + (\bar{q}'i\gamma_5\vec{\tau}q')^2\right] - G_-\left[(\bar{q}'\vec{\tau}q')^2 + (\bar{q}'i\gamma_5 q')^2\right]. \tag{16}$$

Here $m_{\rm ref}(\theta)$ is θ-even mass defined below. We consider three examples as $m_{\rm ref}(\theta)$.

3. Numerical results

If some reference system satisfies the condition $\langle R(\theta)\rangle \approx 1$, one can say that the reference system is good. As a typical example of the condition, we consider the case of $0.5 \lesssim \langle R(\theta)\rangle \lesssim 2$. This condition seems to be the minimum requirement. The discussion made below is not changed qualitatively, even if one takes a stronger condition.

The first example is reference A defined by

$$\begin{aligned} m_{\rm ref}(\theta) &\equiv m_{\rm A}(\theta) \\ &= m_0\cos(\theta/2). \end{aligned} \tag{17}$$

In this case, the P-odd mass is simply neglected from the original Lagrangian (11).

Figure 1(a) shows θ dependence of $\langle R(\theta)\rangle$ at $T = 100$ MeV. The solid line stands for $\langle R(\theta)\rangle$, while the dashed (dotted) line corresponds to $R_{\rm A}$ ($R_{\rm B}$). This temperature is lower than the chiral transition temperature in the original theory that is 206 MeV at $\theta = 0$ and 194 MeV at $\theta = \pi$. As θ increases from zero, $\langle R(\theta)\rangle$ also increases and exceeds 2 at $\theta \approx 1.2$. Reference A is thus good for $\theta \lesssim 1.2$.

Figure 1(b) shows θ dependence of pion mass $\tilde{M}_\pi$ at $T = 100$ MeV. Since P symmetry is broken at finite θ, P-even modes and P-odd modes are mixed with each other for each meson. Hence, $\tilde{M}_\pi$ is defined by the lowest pole mass of the inverse propagator in the isovector channel[4]. The solid (dashed) line denotes $\tilde{M}_\pi$ in the original (reference A) system. At $\theta = \pi$, $\tilde{M}_\pi$ is finite in the original system, but zero in reference A. As a consequence of this property, $R_{\rm A}$ and $\langle R(\theta)\rangle$ vanish at $\theta = \pi$; see Fig. 1(a). This indicates that reference A breaks down at $\theta = \pi$.

The second example is reference B defined by

$$\begin{aligned} m_{\rm ref}(\theta) &\equiv m_{\rm B}(\theta) \\ &= m_0\cos(\theta/2) + \frac{1}{\alpha}\{m_0\sin(\theta/2)\}^2. \end{aligned} \tag{18}$$

In this case, we have added the m_0^2-order correction due to the P-odd quark mass. Here α is a parameter with mass dimension, so we simply choose $\alpha = M_\pi$. The coefficient of the correction term is $m_0^2/M_\pi = 0.129$ MeV.

The same analysis is made for reference B in Fig. 2. As shown in panel (b), $\tilde{M}_\pi$ in reference

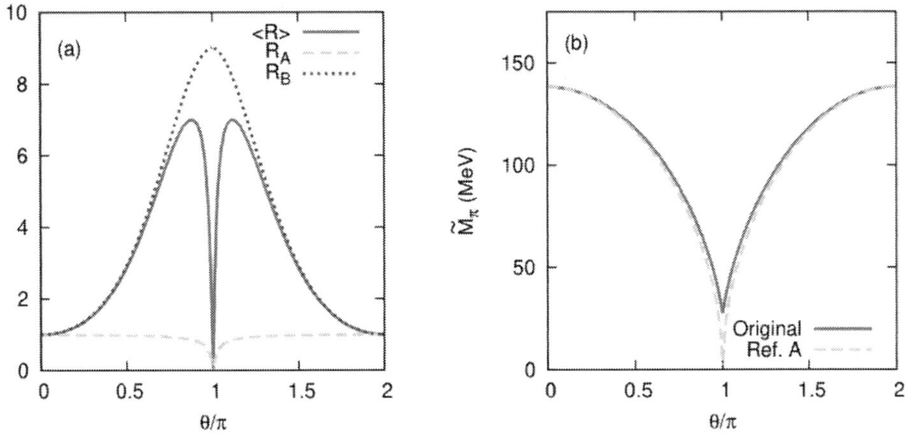

Figure 1. θ dependence of (a) the average reweighting factor and (b) $\tilde{M}_\pi$ at $T = 100$ MeV for the case of reference A.

B well reproduces that in the original theory for any θ. As shown in panel (a), however, the reliable θ region in which $0.5 \lesssim \langle R(\theta)\rangle \lesssim 2$ is located only at $\theta \lesssim 1.3$. Therefore reference B is still not good.

Finally we consider reference C. The pion mass $\tilde{M}_\pi(\theta)$ at finite θ is estimated from the chiral Lagrangian and $1/N_c$ analysis [9]:

$$\tilde{M}_\pi^2(\theta) = \frac{|\sigma_0|}{f_\pi^2}\left[m_0|\cos(\theta/2)| + \frac{m_0 M_\pi^2}{M_{\eta'}^2}\sin^2(\theta/2)\right]. \tag{19}$$

where σ_0 is the chiral condensate at $T = \theta = 0$. Interpreting a θ dependent mass from this result, reference C is defined by

$$\begin{aligned} m_{\rm ref}(\theta) &\equiv m_{\rm C}(\theta) \\ &= m_0\cos(\theta/2) + \frac{m_0 M_\pi^2}{M_{\eta'}^2}\sin^2(\theta/2). \end{aligned} \tag{20}$$

This case also has the m_0^2-order correction, but α is different from reference B. The coefficient of the correction term is $m_0 M_\pi^2/M_{\eta'}^2 = 0.114$ MeV.

As shown in Fig. 3(b), $\tilde{M}_\pi$ in reference C slightly underestimates that of the original theory at small and intermediate θ. As shown in Fig. 3(a), however, $\langle R(\theta)\rangle$ satisfies the condition $0.5 \lesssim \langle R(\theta)\rangle \lesssim 2$ for all θ. Therefore we can think that reference C is a good reference system for any θ.

4. Summary

We have proposed a practical way of circumventing the sign problem in LQCD simulations with finite θ. This method is the reweighting method for the transformed Lagrangian (4). In the Lagrangian, the sign problem is minimized, since the P-odd mass is much smaller than the dynamical quark mass of order $\Lambda_{\rm QCD}$. Another key is to find out which kind of reference system satisfies the condition $\langle R(\theta)\rangle \approx 1$. For this purpose, we have estimated $\langle R(\theta)\rangle$ by using the two-flavor NJL model and eventually found that reference C is a good reference system in the reweighting method.

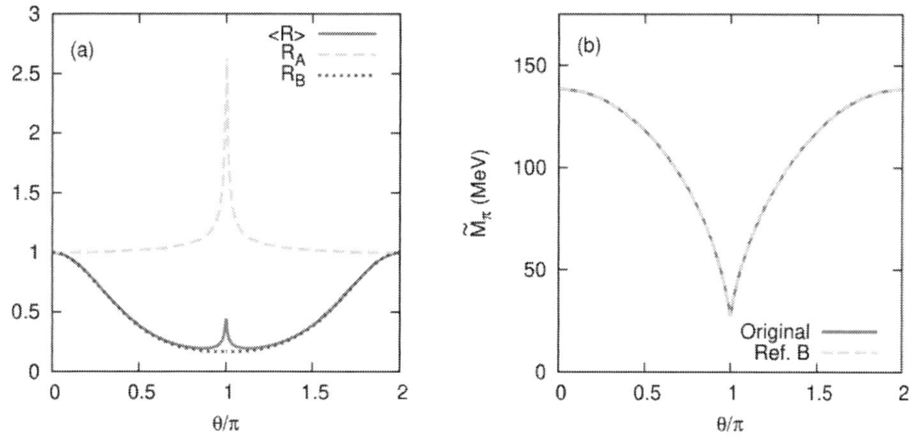

Figure 2. θ dependence of (a) the average reweighting factor and (b) $\tilde{M}_\pi$ at $T = 100$ MeV for the case of reference B.

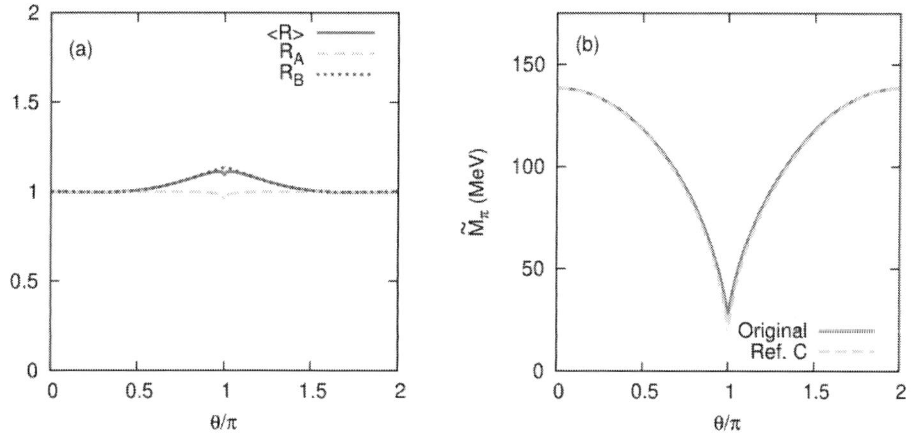

Figure 3. θ dependence of (a) the average reweighting factor and (b) $\tilde{M}_\pi$ at $T = 100$ MeV for the case of reference C.

Acknowledgments
This work was supported by JSPS KAKENHI Grant Number 23-2790.

References
[1] Baker C A, *et al.* 2006 *Phys. Rev. Lett.* **97** 131801.
[2] Kawarabayashi K and Ohta N 1980 *Nucl. Phys.* B **175** 477; 1981 *Prog. Theor. Phys.* **66** 1789; Ohta N 1981 *Prog. Theor. Phys.* **66** 1408; 1982 *Prog. Theor. Phys.* **67** 993.
[3] Sasaki T, Takahashi J, Sakai Y, Kouno H and Yahiro M 2012 *Phys. Rev.* D **85** 056009.
[4] Sasaki T, Kouno H and Yahiro M 2012 *Preprint* arXiv:1208.0375 [hep-ph].
[5] Kobayashi M and Maskawa T 1970 *Prog. Theor. Phys.* **44** 1422, Kobayashi M, Kondo H and Maskawa T 1971 *Prog. Theor. Phys.* **45** 1955.
[6] 't Hooft G 1976 *Phys. Rev. Lett.* **37** 8; 1976 *Phys. Rev.* D **14** 3432; 1978 **18** 2199(E).
[7] Andersen J O, Kyllingstad L T and Splittroff K 2010 *J. High Energy Phys.* **01** 055.
[8] Sakai Y, Sasaki T, Kouno H and Yahiro M 2010 *Phys. Rev.* D **82** 096007.
[9] Metlitski M A and Zhitnitsky A R 2005 *Nucl. Phys.* **B731** 309; 2006 *Phys. Lett.* B **633** 721.

Journal of Physics: Conference Series **432** (2013) 012032 doi:10.1088/1742-6596/432/1/012032

Elucidating the sign problem through noise distributions

Amy N. Nicholson[1], **Dorota Grabowska**[2] **and David B. Kaplan**[2]

[1] Department of Physics, University of Maryland, College Park MD 20742-4111, USA

[2]Institute for Nuclear Theory, University of Washington, Seattle, WA 98195-1550, USA

E-mail: amynn@umd.edu, grabow@uw.edu, dbkaplan@uw.edu

Abstract. Due to the presence of light pions in the theory, lattice QCD at finite densities suffers from issues with noise in both grand canonical and canonical formulations. We study two different formulations of the Nambu-Jona-Lasinio model reduced to 2+1 dimensions at large N, where N is the number of flavors. At finite chemical potential one formulation has a severe sign problem and a fermion correlator which displays a broad probability distribution with small mean. In the other we find no sign problem and a distribution amenable to the cumulant expansion techniques developed in Ref. [1, 2, 3].

1. Introduction

While lattice QCD is currently the only tool for studying QCD from first principles in the non-perturbative regime, the study of QCD at finite densities on the lattice has been greatly hindered by two related problems. In the first, lattice QCD at fixed baryon chemical potential suffers from what is known as a sign problem, which arises due to the complexity of the fermion determinant one wishes to use as a probability measure for Monte Carlo calculations. A possible solution is to employ phase reweighting, in which the phase of the determinant is absorbed into the observable and the magnitude of the determinant is used as a probability measure; however, the expectation value of the phase using the same ensemble must also be calculated in order to reconstruct the original observable. One finds that for μ above some critical value, this expectation value vanishes exponentially and becomes swamped by statistical noise. The second problem is encountered in a canonical formulation, in which a fixed number of quark sources and sinks are separated by a long Euclidean time to project out the ground state. Again, the signal-to-noise ratio is found to vanish exponentially with time.

A notable property of these two problems is that they are both related to the same dynamical property of QCD; namely, they both arise due to the presence of light pions and thus are intimately related to the phenomenon of chiral symmetry breaking. In the case of finite chemical potential, this can be seen by noting that for two degenerate flavors of quark the magnitude of the fermion determinant corresponds to chemical potential for isospin. Thus, at low temperatures for $\mu \geq m_\pi/2$ pion condensation would tend to occur, and the role of the phase must be to cancel this pion condensation [4, 5].

For the other case where finite baryon density is achieved using correlation functions of quark propagators, the role of the pion in the signal-to-noise problem may be noted by considering the variance of an operator used to create nucleons. While at large Euclidean times the correlator for three appropriately contracted quark propagators projects out the ground state of the nucleon, M_N, the variance will be composed of three quark and three anti-quark propagators, projecting out the pion mass at large times [6, 7]. The signal-to-noise ratio will thus be exponentially vanishing with the factor $(M_N - \frac{3}{2}m_\pi)\tau$, where τ is the Euclidean time, for late Euclidean times.

Both of the noise problems presented above can be shown to arise from probability distributions which are broad, symmetric, and centered about zero. For the nucleon correlator, one may see this by looking at higher moments of the distribution function. Even moments consist of equal numbers of quark and antiquark propagators which may pair up to form pions, while odd moments have three unpaired quark propagators which will form a baryon [8]. Thus, because odd moments will always contain a baryon mass in the exponential, they will be suppressed relative to the even moments, giving a symmetric distribution with exponentially small mean. A similar argument may be made for the expectation value of the phase at finite chemical potential, where all odd moments are equivalent to the mean, and therefore exponentially small for $\mu \geq m_\pi/2$, while all even moments give exactly 1 (for a discussion of the distribution of the phase, see [9]).

In [1, 2, 3] it was shown that another type of noise problem exists in which the distribution displays a long tail, resulting in a sample mean which may be far from the true mean with deceptively small error bars. This problem is often referred to as an overlap problem, and has been known to occur for observables when reweighting is performed. This was discussed in a canonical formulation using lattice calculations of unitary fermions, a strongly-interacting nonrelativistic system in which the fermion determinant is real and positive for even numbers of fermion flavors. The probability distributions for fermion correlators were shown to be nearly log-normal, and a possible solution to the overlap problem was postulated in the form of a truncated cumulant expansion for the correlator, $\ln\langle\mathcal{O}(\phi)\rangle = \sum_{n=1}^{N_\kappa} \frac{\kappa_n(\ln\mathcal{O})}{n!}$, where $\mathcal{O}(\phi)$ is some operator of interest on a background field ϕ, $\kappa_n(\ln\mathcal{O})$ is the nth cumulant of the logarithm of the operator, and N_κ is some suitable truncation order, determined by examining the convergence of the series. For a distribution that is sufficiently close to log-normal, this series may converge at a low enough order to be useful in practical applications. Recently it has also been reported that nearly log-normal distributions are also seen in lattice QCD calculations of correlation functions for intermediate times [10].

2. Two formulations for a large-N NJL model in $2+1$-dimensions

To understand the origin of these two types of noise problem in QCD and to further explore the link between signal-to-noise and the presence of light pions, we turn now to a QCD-like model which displays chiral symmetry breaking without the added complication of confinement. For more details behind these calculations, see [11]. The model we will consider is the NJL model in $2+1$-dimensions for large number of fermion flavors, N,

$$\mathcal{L} = N\left(\overline{\psi}_a(\partial\!\!\!/ - m)\psi_a - \frac{C}{2}\left[(\overline{\psi}_a\psi_a)^2 + (\overline{\psi}_a i\gamma_5\psi_a)^2\right]\right) , \tag{1}$$

where $a, b\ldots$ are a flavor index summed over $1,\ldots,N$, indices, $i, j\ldots$ are summed over 3D coordinates $1, 2, 3$, while Greek indices $\mu, \nu\ldots$ are summed over 4D coordinates $1,\ldots,4$. The gamma matrices are the usual 4×4 matrices used in 4D, and so the Lagrangian represents $2N$ flavors of 3D Dirac fermions. In the limit $m \to 0$ this theory has a $U(1)$ chiral symmetry in 4D, which becomes a flavor symmetry in 3D; this "chiral" symmetry is spontaneously broken as in $3+1$-dimensions, giving rise to a Goldstone boson.

To study this model numerically one may introduce auxiliary fields to generate a four-fermion interaction, which can then be integrated over using Monte Carlo methods. We find two physically equivalent methods for doing so. The first, which we will call the σ/π formulation, is the conventional method which introduces scalar fields that are singlet under the $SU(2N)$ flavor symmetry,

$$\mathcal{L} = N\left(\frac{1}{2C}\left(\sigma^2+\pi^2\right) + \bar{\psi}_a\left[\not{\partial} - m + \sigma + i\pi\gamma_5\right]\psi_a\right) . \tag{2}$$

The second formulation, which we will call the A/V formulation, follows if one performs a Fierz rearrangement of the four-fermion interaction in eq. (1) before introducing auxiliary fields. This results in the introduction of $N \times N$ matrix valued vector and axial vector auxiliary fields V and A, and the equivalent theory

$$\mathcal{L} = N\left(\frac{1}{C}\,\mathrm{Tr}\left(V_\mu V_\mu + A_\mu A_\mu\right) + \bar{\psi}_a\left[\not{\partial} - m + i\not{V} + \not{A}\gamma_5\right]_{ab}\psi_b\right) . \tag{3}$$

To determine whether these two formulations will cause a Monte Carlo sign problem at nonzero chemical potential, we should investigate the positivity of the fermion determinants. For the σ/π formulation, it is possible to define a real symmetric charge conjugation matrix C satisfying $C^2 = 1$, $C\gamma_\mu C = \gamma_i^*$ for $i = 1,2,3$ and $C\gamma_5 C = -\gamma_5^*$. Then the fermion operator for a single flavor in the grand canonical formulation satisfies $D^* = CDC$, where $D = (\not{\partial} - m + \sigma + i\pi\gamma_5 + \mu\gamma_1)$, and complex eigenvalues of D must come in conjugate pairs. Thus $(\det D)^N$ is real, and positive for even N. This implies that there is no sign problem at finite density [12].

In the A/V formulation, the fermion matrix at finite chemical potential is given by $D(\mu) = (\not{\partial} - m + i\not{V} + \not{A}\gamma_5 + \mu\gamma_1)$, which is similar in structure to the QCD Dirac matrix with nonzero μ, and its determinant is similarly complex. In fact, as in QCD, the magnitude of the fermion determinant for two degenerate families, $|\det D(\mu)|^2$, corresponds to isospin chemical potential, so that for $\mu \geq m_\pi/2$, the phase is responsible for eliminating pion condensation in the ground state.

Chiral symmetry breaking is simple to see in the σ/π formulation, where the large N expansion is equivalent to the semiclassical expansion. Upon integrating out the fermion fields we have the following effective action:

$$S[\sigma,\pi] = N\int \mathrm{d}^3x\left[\frac{1}{2C}(\sigma(x)^2 + \pi(x)^2) - \mathrm{Tr}\ln(\not{\partial} - m + \sigma(x) + i\pi(x)\gamma_5)\right] . \tag{4}$$

Performing a mean field calculation leads to (using dimensional regularization and the MS subtraction scheme), $\langle\sigma\rangle = \frac{f}{2}\left[1 + \sqrt{1+4m/f} + 2m/f\right]$, $M \equiv \langle\sigma\rangle - m = \frac{f}{2}\left[1+\sqrt{1+4m/f}\right]$, where we have defined $\langle\sigma\rangle = f$ to be the chiral symmetry breaking minimum when $m = 0$, and M is the constituent fermion mass. We may also find the σ and π dispersion relations $D_\sigma(k), D_\pi(k)$ by expanding the effective action to second order about our mean field solution.

In the A/V formulation we cannot use mean field theory; instead, to leading order in $1/N$ we may find the fermion propagator by solving the Schwinger-Dyson equation and finding the same nonzero fermion mass, M, as in the σ/π formalism. Furthermore, one may derive the σ and π meson propagators by solving the matrix equation shown in Fig. 1. One finds $\mathcal{M}_{ij;kl}(k) = -\left(\delta_{ij}\delta_{kl}/D_\sigma(k) + (i\gamma_5)_{ij}(i\gamma_5)_{kl}/D_\pi(k)\right)$. Thus we see that an interaction via t-channel exchange of A_μ and V_μ mesons is exactly equivalent to a single meson in the s or u-channel in the σ/π formulation, corresponding to a valence fermion/antifermion pair or two valence fermions or antifermions, respectively.

3. Probability distributions for fermion correlators

We now proceed to calculate the probability distribution for a fermion propagator in both formulations. If $X[\phi]$ is a functional of a stochastic field ϕ corresponding to an observable,

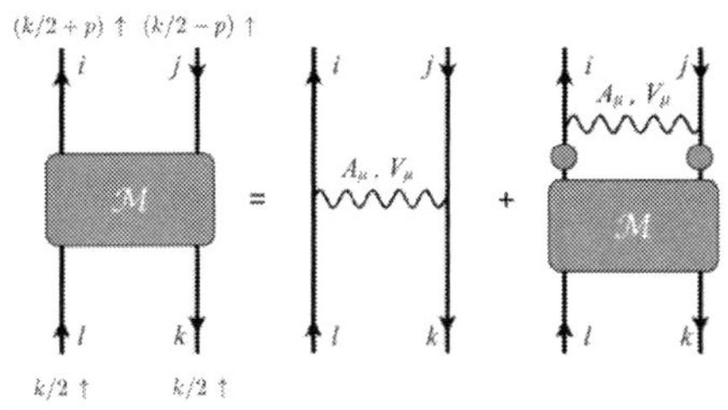

Figure 1. *Graphical representation of the integral equation for the four-point correlator in the A/V formulation for an incoming fermion-antifermion pair of one flavor and an outgoing pair of another. Dirac indices are labeled.*

we define the normalized probability density function for X to be the path integral $\mathcal{P}(x) = \mathcal{N}\int[\mathrm{d}\phi]\, e^{-S[\phi]}\,\delta(X[\phi]-x)$, where we assume $S[\phi]$ is real. This quantity is difficult to analyze field theoretically due to the singular nature of the delta function, so we perform a Fourier transform and work with the characteristic function,

$$\Phi(s,\bar{s}) = e^{-W(s,\bar{s})} = \mathcal{N}\int[\mathrm{d}\phi]\, e^{-S[\phi]+i(sX[\phi]+\bar{s}\bar{X}[\phi])}\ ; \qquad W(s,\bar{s}) = -\sum_{m,n=1}^{\infty}\frac{(is)^m(i\bar{s})^n}{m!\,n!}\kappa_{m,n}\ , \quad (5)$$

where we have generalized the cumulant generating function $W(s,\bar{s}) = -\ln\Phi(s,\bar{s})$ for a complex observable through a double expansion in both s and $\bar{s}$. From this formulation we see that we may compute cumulants using connected Feynman diagrams of the modified action $S[\phi] - i(sX[\phi] + \bar{s}\bar{X}[\phi])$.

For our observables we will use

$$Y_\Gamma = \ln\left[\frac{1}{V}\langle \mathbf{p}=0, t=T/2\,|\,\mathrm{Tr}\,\Gamma\left(\frac{1}{\not{\partial} - m + \sigma + i\pi\gamma_5}\right)|\mathbf{p}=0, t=-T/2\rangle\right] \qquad \sigma/\pi \text{ formulation,} \quad (6a)$$

$$X_\Gamma = \frac{1}{V}\langle \mathbf{p}=0, t=T/2\,|\,\mathrm{Tr}\,\Gamma\left(\frac{1}{\not{\partial} - m + i\not{V} + \not{A}\gamma_5}\right)|\mathbf{p}=0, t=-T/2\rangle \qquad A/V \text{ formulation,} \quad (6b)$$

where Γ is some Dirac matrix of our choosing, and we consider the logarithm of the propagator in the σ/π case to simplify the calculation later on. Measuring the expectation value of this correlator is a procedure for determining the mass m_f of the lightest fermion state allowed for a given Γ using $\lim_{T\to\infty} -\frac{1}{T}\ln\langle X_\Gamma\rangle = \lim_{T\to\infty} -\frac{1}{T}\ln\langle e^{Y_\Gamma}\rangle = m_f$.

The calculation of the variance for the A/V case involves attaching fermion propagators at zero spatial momentum to the legs in the first diagram in Fig. 1. For $\Gamma = 1$ and near the chiral limit the pion is the lightest state which can propagate through such a graph, so that for late Euclidean time we find, $\kappa_{1,1} \simeq \frac{2\pi}{m_\pi f V N}e^{-m_\pi T}$. Higher cumulants can be computed using the equivalent σ/π diagrams. Generically, a graph contributing to $\kappa_{m,n}$ scales as N^{m+n-1}, since we need a minimum of $(m+n-1)$ mesons to make a connected graph. Furthermore, it is straightforward to verify that the minimum mass state that can possibly propagate in a graph for $\kappa_{m,n}$ with $m \geq n$ consists of $(m-n)$ fermions with mass M and n pions. Therefore we expect these cumulant to scale as

$$\kappa_{m,n} \sim \frac{e^{-(m-n)MT}e^{-nm_\pi T}}{N^{m+n-1}} \qquad (m \geq n)\ . \qquad (7)$$

The above scaling implies that the distribution for the real part of the fermion propagator near the chiral limit becomes highly symmetric about zero at late time. That is because odd moments (for which $(m-n) \neq 0$) are seen to fall off much more quickly than even moments. This is consistent with the Lepage-Savage picture for baryon propagator distributions in QCD.

For the σ/π formulation the cumulants are given by the connected graphs derived from the action $S_Y = NS[\sigma,\pi] - isY_\Gamma[\sigma,\pi]$, where $S[\sigma,\pi]$ is given in eq. (4). At leading order in a $1/N$ expansion, κ_n will be given by the sum of all tree level diagrams composed of n external lines and any number of vertices which arise from internal fermion loops. Given our definition of Y_Γ, we find that all external propagators are spatially homogeneous, and, for very large Euclidean time, temporally homogeneous up to edge effects near the source and sink. Because we only consider tree graphs, all internal meson propagators will similarly be spatially and temporally homogeneous; thus, we may once again use mean field theory. Solving for the mean field leads to the following cumulants for the logarithm of the fermion correlator:

$$\kappa_1 \;=\; \ln z - MT\,, \qquad \kappa_{n\geq 2} = -\frac{(2(n-2))!}{(n-2)!}\,(2M-f)T\,\zeta^{n-1}\,, \qquad \zeta \simeq \frac{\pi}{NV(2M-f)^2} \tag{8}$$

where z gives the overlap of our operator with the ground state, and we have approximated the integral over Euclidean time by a step function for $-T/2 \leq \tau \leq T/2$. Note in the limit $N, T, V \to \infty$, $\frac{T}{VN}$ = finite, the cumulants κ_n vanish for $n \geq 3$ and Y_Γ assumes a normal distribution, giving a log-normal distribution for the correlator.

4. Discussion

We found that in the QCD-like "A/V" formulation of the NJL model, the fermion determinant was complex and a Splittorff-Verbaarschot argument [4, 5] could be made to show that the phase of the fermion determinant had to fluctuate wildly for $\mu > m_\pi/2$. When looking at fermion correlators, the distribution evolved to be symmetric about an exponentially small mean relative to its width, implying a severe signal/noise ratio when sampling the correlator using Monte Carlo methods. In contrast the "σ/π" formulation with even N has no sign problem at nonzero μ, and the correlator distribution was log-normal in a certain limit and amenable to the cumulant expansion analysis of Ref. [2, 13, 14].

Our analysis should make it clear that the sign problem encountered in QCD at nonzero chemical potential is not a fermion problem, but instead a consequence of interactions. In particular, if the particles being studied can exist in a tightly bound state of valence fermions, there is going to be a sign problem. We believe that inventing a way to introduce the pion into QCD as a fundamental field could be an important step toward solving the QCD sign problem and beginning to study the properties of ordinary and dense matter from first principles.

Acknowledgments

This work was supported in part by U.S. DOE grants No. DE-FG02-00ER41132 andNo. DE-FG02-93ER-40762.

References

[1] Lee J W, Endres M G, Kaplan D B and Nicholson A N 2011 *PoS* **LATTICE2011** 203 (*Preprint* 1111.3793)
[2] Endres M G, Kaplan D B, Lee J W and Nicholson A N 2011 *Phys.Rev.Lett.* **107** 201601 (*Preprint* 1106.0073)
[3] Endres M G, Kaplan D B, Lee J W and Nicholson A N 2011 *PoS* **LATTICE2011** 017 (*Preprint* 1112.4023)
[4] Splittorff K and Verbaarschot J 2007 *Phys.Rev.Lett.* **98** 031601 (*Preprint* hep-lat/0609076)
[5] Splittorff K and Verbaarschot J 2007 *Phys.Rev.* **D75** 116003 (*Preprint* hep-lat/0702011)
[6] Parisi G 1984 *Phys.Rept.* **103** 203–211
[7] Lepage G P 1989 Invited lectures given at TASI'89 Summer School, Boulder, CO, Jun 4-30, 1989
[8] Savage M J 2010 Private communivation
[9] Lombardo M, Splittorff K and Verbaarschot J 2009 *Phys.Rev.* **D80** 054509 (*Preprint* 0904.2122)
[10] DeGrand T 2012 *Phys.Rev.* **D86** 014512 (*Preprint* 1204.4664)
[11] Grabowska D, Kaplan D B and Nicholson A N 2012 (*Preprint* 1208.5760)
[12] Hands S, Kim S and Kogut J B 1995 *Nucl.Phys.* **B442** 364–390 (*Preprint* hep-lat/9501037)
[13] Endres M G, Kaplan D B, Lee J W and Nicholson A N 2011 *Phys.Rev.* **A84** 043644 (*Preprint* 1106.5725)
[14] Endres M G, Kaplan D B, Lee J W and Nicholson A N 2012 (*Preprint* 1203.3169)

Phase transitions in heavy-quark QCD from an effective theory

M. Fromm, J. Langelage[1], S. Lottini[2,3], M. Neuman, O. Philipsen
ITP, Goethe-Universität, Max-von-Laue-str. 1, 60438 Frankfurt am Main, Germany
E-mail: lottini@th.physik.uni-frankfurt.de

Abstract. With combined hopping parameter and strong coupling expansions, we calculate a dimensionally reduced Polyakov-loop effective theory valid for heavy quarks at nonzero temperature and arbitrary chemical potential. We numerically compute the critical endpoint of the deconfinement transition as a function of quark masses and number of flavours. We also investigate the applicability of the model to the low-T and high density region, specifically in terms of baryon condensation phenomena.

1. Introduction

Despite its brilliant achievements in tackling perturbative problems, analytical investigations of QCD can not do much when it comes to determining the structure of its phase diagram in the (T, μ) plane, with T temperature and μ quark chemical potential. In nonpertubative regimes, the answers numerical lattice QCD calculations can provide are mainly limited to $\mu \simeq 0$: the notoriuos sign problem renders the traditional Monte Carlo sampling approach meaningless and one has to rely on alternative methods, whose efficiency typically degrades about $\mu \sim T$: examples are using an imaginary chemical potential, performing a Taylor expansion around $\mu = 0$ and reweighting of ensembles generated at zero chemical potential.

Yet, the relevant physics for heavy-ion collisions, as well as for compact-star astrophysics, takes place at far-from-zero μ: it would then be desirable to devise strategies able somehow to circumvent the problems. In this category fall the many effective approaches developed throughout the years, such as sigma models, (p)NJL models and so on.

In particular, in characterising the $T \to 0$ region of the (T, μ) phase diagram, the phenomenon of *Silver Blaze* is expected based on very general considerations: there, the partition function – hence all interesting observables – have to be independent of the chemical potential, up to a critical value associated to the lightest baryonic state in the theory. In the path-integral representation of QCD, the Dirac eigenvalues display an explicit dependence on μ, therefore the Silver Blaze phenomenon requires a highly non-trivial cancellation between them. Due to the shortcomings of the numerical methods mentioned above, the phenomenon has not been reproduced successfully in lattice QCD so far; also on the analytical side, an explicit proof exists

[1] Current address: Institut für theoretische Physik, ETH, Wolfgang-Pauli-Str. 27, 8049 Zürich, Switzerland.
[2] Speaker at the conference.
[3] Current address: NIC, DESY, Platanenallee 6, 15738 Zeuthen, Germany.

only for the choice of an isospin chemical potential $\mu_I = \mu_u = -\mu_d$, with a critical onset located at $\mu_I = m_\pi/2$ [1].

It is perhaps possible to avoid the sign problem altogether, thus exploring numerically the cold dense region $\mu \ll T$ of QCD, by employing the so-called Langevin dynamics [2]: recently, numerical results for a scalar complex field have been obtained [3] and later reproduced in the context of a flux representation for the same theory [4]; both techniques, however, are at the moment hardly applicable to QCD, and it is not at all guaranteed that they will ever be.

In this work we present an effective, dimensionally-reduced Polyakov-loop model obtained by applying systematically strong-coupling and hopping-parameter expansions to the original QCD lattice formulation, valid as long as quarks are not too light and the gauge coupling β is not too large; after briefly describing the theory (Section 2), we apply it to the cold dense region (Section 3) and show that the signature of the Silver Blaze phenomenon can be read off the numerical results. A more detailed presentation of the theory can be found in [5, 6]; for the Silver Blaze property in this context, see also [7].

2. The effective theory

The model lives on a three-dimensional lattice, with complex scalars as per-site degrees of freedom, representing the traced Polyakov loops $L_x \equiv \mathrm{Tr}(W_x), W_x \in SU(3)$. The associated effective action consists of a pure-gauge term and a fermionic contribution; the former comes from application of strong-coupling methods to the original four-dimensional Yang-Mills gauge action, and the latter is obtained with a hopping-parameter expansion, starting from the standard Wilson expression for the fermionic term in lattice QCD. In general, then, we have

$$Z_{\mathrm{eff}}(\ldots) = \int \underbrace{\Big(\prod_x \mathrm{d}L_x e^{V(L_x)}\Big)}_{[\mathcal{D}L]} e^{-S_{\mathrm{eff}}[L]} \ , \quad \int \mathrm{d}L e^{V(L)} = \int_{-\pi}^{+\pi} \mathrm{d}\theta \int_{-\pi}^{+\pi} \mathrm{d}\phi \ e^{2V(L(\theta,\phi))} \ , \qquad (1)$$

where the parametrisation of the scalar L is made explicit (see [5]). The partition function Z depends on effective couplings which are in turn some functions of the original couplings appearing in the four-dimensional lattice QCD action.

Let us now sketch how the pure-gauge part of the effective action is obtained [5] starting from the four-dimensional Euclidean Yang-Mills partition function at finite temperature (with time extent N_τ lattice spacings)

$$Z_{SU(3),4d} = \int [\mathcal{D}U_\ell] \exp\Big\{\frac{\beta}{2N}\sum_p (\mathrm{Tr}U_p + \mathrm{Tr}U_p^\dagger)\Big\} \ , \quad \beta = \frac{2N}{g^2} = 18u + \ldots \ . \qquad (2)$$

By integrating out the spatial degrees of freedom and subsequently applying a strong-coupling expansion, one gets a tower of interactions in the effective theory, the leading one of which is simply a "spin-spin" term connecting nearest neighbours in the fundamental representation (the superscript "s" denotes the compliance of the terms to the centre symmetry requirement):

$$-S^s_{\mathrm{eff}} = \lambda_1 S^s_1 + \lambda_2 S^s_2 \cdots \ ; \ \lambda_1 S^s_1 = \lambda_1(\beta, N_\tau) \sum_{\langle ij \rangle} (L_i L_j^* + L_i^* L_j) \ ; \ \lambda_{i>1} = o(\lambda_1) \text{ for } \beta \to 0 \ . \quad (3)$$

The precise form of the map $\lambda_1(\beta, N_\tau)$ comes from an order-by-order enumeration of strong-coupling graphs having the two Polyakov lines as boundary: we calculated the series up to order u^{10} [5].[4] Moreover, one can resum higher powers of (certain classes of) graphs and improve the

[4] When fermions are introduced, a shift in the gauge maps between couplings is induced [6].

effective theory to a "nonlinear" form

$$Z_{\rm eff} = \int [\mathcal{D}L] \prod_{\langle ij\rangle} (1 + 2\lambda_1 {\rm Re}\, L_i L_j^*) \ ; \tag{4}$$

the important point is that this model reproduces the deconfinement transition of the original theory, Eq. (2), and once the critical point (that still falls within the range of applicability of the strong-coupling methods) is known, it can be translated back to a table $\beta_c(N_\tau)$ with a sufficient precision and in a wide enough range of N_τ to allow for a sensible continuum extrapolation of the physical deconfinement point T_c [5].

The next step is the inclusion of fermions in the theory: we perform an expansion in the hopping parameter κ, therefore the masses (we consider N_f degenerate quarks of mass $M = (1-8\kappa)/(2a\kappa)$) must be large enough to guarantee convergence. The Wilson action for quarks is then written as

$$-S_q = -N_f \sum_{\ell=1}^{\infty} \frac{\kappa^\ell}{\ell} {\rm Tr}(H[U]^\ell) \tag{5}$$

with H is the hopping matrix. Its structure is such that the non-zero contributions after integration of the gauge fields are given by various kinds of closed loops of length ℓ; each of their links carries a factor κ (and, if the chemical potential is turned on, an additional factor $e^{\pm a\mu}$ in the temporal direction): the expression above is then also an expansion in powers of κ. The general form of the fermion contribution to the effective action is analogous to an external field in a spin system (here a denotes the explicit breaking of centre symmetry):[5]

$$-S^a_{\rm eff} = -2N_f \sum_{i=1}^{\infty} \Big(h_i(u,\kappa,\mu,N_\tau) S_i^a + \overline{h}_i(u,\kappa,\mu,N_\tau) S_i^{a,\dagger} \Big) \ ; \ \overline{h}_i(-\mu) = h_i(+\mu) \ , \tag{6}$$

with the i-order reflecting increasingly subdominant terms. A closer inspection of the leading term S_1^a reveals that, once again, a partial resummation of higher powers of the same graphs is possible: one finds that

$$-S^a_{\rm eff} = \log \Big[\prod_x \det \Big(1 + h_1 W_x\Big)^{2N_f} \det \Big(1 + \overline{h}_1 W_x^\dagger\Big)^{2N_f} \Big] + \cdots \ ; \ h_1 = (2\kappa e^{a\mu})^{N_\tau}(1+\cdots) \ , \tag{7}$$

The expression for h_1 received also contributions from various types of higher-order graphs; including these, and the first subleading fermionic term S_2^a, one can write the model as [6, 7]:

$$\begin{aligned} Z_{\rm eff} &= \int [\mathcal{D}L] \prod_{<ij>} \Big(1 + 2\lambda {\rm Re}\, L_i L_j^*\Big) \prod_x \det \Big[(1 + h_1 W_x)(1 + \overline{h}_1 W_x^\dagger)\Big]^{2N_f} \\ & \prod_{<ij>} \Big[1 - h_2 {\rm Tr}\frac{W_i}{1 + CW_i} {\rm Tr}\frac{W_j}{1+CW_j}\Big] \Big[1 - \overline{h}_2 {\rm Tr}\frac{W_i^\dagger}{1+\overline{C}W_i^\dagger} {\rm Tr}\frac{W_j^\dagger}{1+\overline{C}W_j^\dagger}\Big] \ , \end{aligned} \tag{8}$$

(we rename $\lambda_1 \to \lambda$ and ignore higher-order corrections). Despite the appearance of W_x and $W_x^\dagger$ in the above, $Z_{\rm eff}$ can still be expressed, in practice, in terms only of the θ, ϕ parameterising L_x in Eq. (1). The effective couplings are given by:

$$h_1 = C\Big(1 + 6\kappa^2 N_\tau \frac{u - u^{N_\tau}}{1-u} + \cdots\Big) \ ; \tag{9}$$

$$h_2 = C^2 \frac{\kappa^2}{3}\Big(1 + 2\frac{u - u^{N_\tau}}{1-u}\Big) \ ; \tag{10}$$

$$C = (2\kappa e^{a\mu})^{N_\tau} \ . \tag{11}$$

[5] The next term seems an ordinary nearest-neighbour interaction, but is actually non-centre-symmetric, Eq. (8).

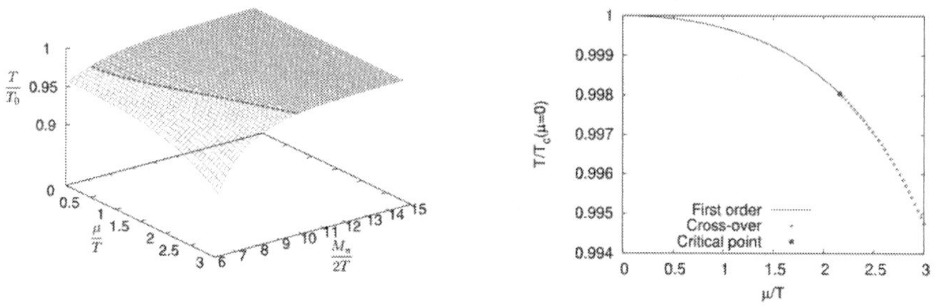

Figure 1. *Left:* $N_f = 2$ phase diagram in the space of temperature, chemical potential and meson mass (here called M_π). T_0 is the pure-gauge transition point; the blue line is the critical line. *Right:* T-μ phase diagram for the choice $N_f = 2$, $M_{ud}/T \simeq 8.68$; the vertical axis is normalised to the $\mu = 0$ transition temperature for this mass choice.

The meson and baryon masses are also evaluated:

$$am_M = -2\ln(2\kappa) - 6\kappa^2 - 24\kappa^2\frac{u}{1-u} + \dots \; ; \; am_B = -3\ln(2\kappa) - 18\kappa^2\frac{u}{1-u} + \dots \,. \quad (12)$$

The model, as formulated in Eq. (8), is well suited to different simulational strategies: we could confirm that the standard Metropolis approach, a flux-representation-based worm algorithm, and a complex-Langevin implementation all agree with each other: however, one has to keep in mind that the first has to rely on a reweighting procedure for $\mu > 0$ (which does not hinder its scope significantly for not too large spatial volumes) and the second ceases to be of use if the h_2 term is included, while the third does not have a sign problem at all and is thus the best choice to deal with large-volume, finite-μ ensembles. All of these algorithms, when compared to traditional lattice QCD approaches, are far less expensive in terms of required system resources and CPU time.

An important application of this effective approach concerns the mapping of the critical surface in the $(M_{ud}, M_s, \mu/T)$ space. Due to the heavy-quark validity of the model, the second-order surface that could be located is the one associated to the upper-right corner of the Columbia plot: there, a nice agreement between the findings at zero, real positive and imaginary μ was verified and the correct universality class of the transition was confirmed. Moreover, the result at zero chemical potential for $N_f = 1, 2, 3$ closely reproduces those found in ordinary QCD simulations e.g. in [8, 9]. The shape of the critical surface at all chemical potentials can be summarised in the parametrisation [6]:

$$(N_f h_1 e^{-\mu/T})_{\rm crit} = \frac{0.00075(1)}{\cosh \mu/T} \; ; \; \kappa_f \simeq e^{-aM_f}/2 \,, \; M_f \to \infty \; ; \quad (13)$$

the phase diagram for heavy-quark QCD can then be determined in $(\frac{\mu}{T}, \frac{M_\pi}{2T}, \frac{T}{T_0})$, and, by taking slices of it, a "heavy fermion" version of the familiar μ-T phase diagram can be drawn (Fig. 1).

3. Low-temperature regime

We now turn to investigate the cold and dense regime of QCD with this effective theory. In the (analytically solvable) static strong-coupling limit, one finds that the Silver Blaze property holds as $T \to 0$, with the quark number density approaching a step function centred at $3\mu = m_B = -3\log(2\kappa)$ (the last equality is exact in this limit):

$$a^3 n \overset{T\to 0}{\longrightarrow} 2N_c\,\Theta(\mu_B - m_B) \; ; \quad (14)$$

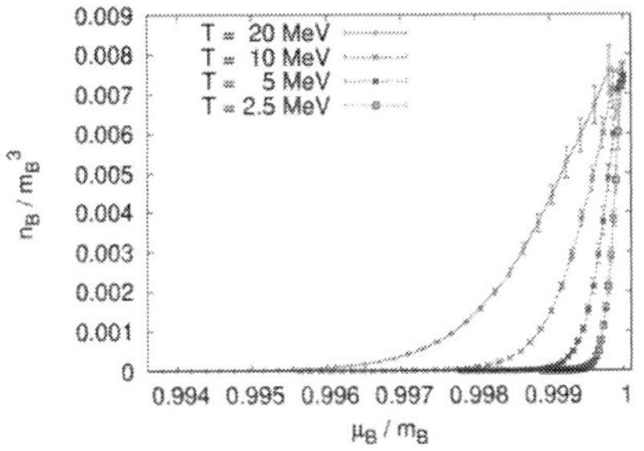

Figure 2. Continuum-extrapolated baryon density, in units of m_B^3, as a function of the baryon chemical potential (also measured in units of m_B) for various temperatures.

one sees then also that a saturation phenomenon is expected (this is connected to the resummations performed in deriving the form of Z_{eff}). The next step is to perform numerical computations with the full expression Eq. (8). Taking advantage of the fact that N_τ is now merely a parameter in the effective couplings' maps, we can tune it to very large values in the hundreds: this effectively realises $\lambda \to 0$ thus simplifying the model; by keeping $\kappa \sim 10^{-3}$, moreover, we can still trust the pure-Yang-Mills scale-setting prescription [10]. In this way, we can measure numerically the baryon density as a function of the baryon chemical potential with $m_B = 30$ GeV and four different temperatures from 20 MeV to 2.5 MeV [7].

The baryon density n_B can be expressed as a meaningful physical quantity after a careful continuum limit is taken: we could generate data at nine different values of the lattice spacings a and perform a fit to the continuum limit; n_b is measured as

$$a^3 n_b = \frac{a^3}{3}\left(\frac{T}{V}\frac{\partial \log Z}{\partial \mu}\right) = -\frac{1}{3N_s^3}\left\langle \frac{\partial S_{\text{eff}}}{\partial(\mu/T)}\right\rangle . \tag{15}$$

The results are shown in Fig. 2: the step-like shape of the density is less and less smoothed-out as $T \to 0$ (and anyway the μ-scale is extremely tiny); besides, as the baryonic chemical potential hits the baryon mass the expected saturation to a common, roughly T-independent density is seen, and this value, when expressed in appropriate units of m_B, is less than a factor two off the real-world nuclear matter, despite this model being limited to extremely massive quarks. This seems to suggest that the core features of nuclear condensation might be largely independent of whether the fundamental constituents of the baryon are heavy or light.

References

[1] Cohen T D 2003 *Phys. Rev. Lett.* **91** 222001 (*Preprint* hep-ph/0307089)
[2] Parisi G, Wu Y -S 1981 *Sci. Sin.* **24** 483
Damgaard P H and Huffel H 1987 *Phys. Rept.* **152** 227
Aarts G and Stamatescu I -O 2008 *JHEP* **0809** 018 (*Preprint* arXiv:0807.1597 [hep-lat])
[3] Aarts G 2009 *Phys. Rev. Lett.* **102**, 131601 (*Preprint* arXiv:0810.2089 [hep-lat])
[4] Gattringer C and Kloiber T 2013 *Nucl. Phys.* B **869** 56 (*Preprint* arXiv:1206.2954 [hep-lat])
[5] Langelage J, Lottini S and Philipsen O 2011 *JHEP* **1102** 057 *and* Erratum 2011 *JHEP* **1107** 014 (*Preprint* arXiv:1010.9051 [hep-lat])
[6] Fromm M, Langelage J, Lottini S and Philipsen O 2012 *JHEP* **1201** 042 (*Preprint* arXiv:1111.4953 [hep-lat])
[7] Fromm M, Langelage J, Lottini S, Neuman M and Philipsen O *Preprint* arXiv:1207.3005 [hep-lat]
[8] Alexandrou C, Boriçi A, Feo A, de Forcrand P, Galli A, Jegerlehner F and Takaishi T 1999 *Phys. Rev.* D **60** 034504 (*Preprint* hep-lat/9811028)
[9] Saito H, Ejiri S, Aoki S, Hatsuda T, Kanaya K, Maezawa Y, Ohno H and Umeda T 2011 *Phys. Rev.* D **84** 054502 *and* Erratum 2012 *Phys. Rev.* D **85** 079902 (*Preprint* arXiv:1106.0974 [hep-lat])
[10] Necco S and Sommer R 2002 *Nucl. Phys.* B **622** 328 (*Preprint* hep-lat/0108008)

Chiral polarization scale of QCD vacuum and spontaneous chiral symmetry breaking

Andrei Alexandru[1] **and Ivan Horváth**[2,3]

[1]George Washington University, Washington, DC, USA
[2]University of Kentucky, Lexington, KY, USA

E-mail: aalexan@gwu.edu, horvath@pa.uky.edu

Abstract. It has recently been found that dynamics of pure glue QCD supports the low energy band of Dirac modes with local chiral properties qualitatively different from that of a bulk: while bulk modes suppress chirality relative to statistical independence between left and right, the band modes enhance it. The width of such chirally polarized zone – *chiral polarization scale* Λ_{ch} – has been shown to be finite in the continuum limit at fixed physical volume. Here we present evidence that Λ_{ch} remains non–zero also in the infinite volume, and is therefore a dynamical scale in the theory. Our experiments in N_f=2+1 QCD support the proposition that the same holds in the massless limit, connecting Λ_{ch} to spontaneous chiral symmetry breaking. In addition, our results suggest that thermal agitation in quenched QCD destroys both chiral polarization and condensation of Dirac modes at the same temperature $T_{ch} > T_c$.

1. Introduction

The work presented in this talk is part of a long–term effort to look at QCD vacuum structure in a model–independent manner. This involves devising meaningful, preferably gauge invariant, characteristics of gauge configurations dominating QCD path integral. These configurations are accessible via first–principles lattice QCD simulations, and are not altered from their equilibrium form. The underlying assumption is that when objective information of this kind sufficiently accumulates, it can be eventually integrated into a coherent picture of vacuum–related QCD phenomena. This is the *bottom–up approach* to QCD vacuum structure [1].

Our focus here is spontaneous chiral symmetry breaking (SChSB), generally viewed as one of the keys to making sense of low–energy hadronic physics. When thinking about SChSB, Dirac eigenmodes immediately pop out. Indeed, spectral representation turns out to be a fruitful way of looking at $\bar{\psi}\psi$, revealing that symmetry breakdown is equivalent to Dirac mode condensation. For example, in N_f=2 theory with quark mass m,

$$\lim_{m \to 0} \langle \bar{\psi}\psi \rangle_m \neq 0 \quad \Longleftrightarrow \quad \lim_{\lambda \to 0} \lim_{m \to 0} \rho(\lambda, m) \neq 0 \tag{1}$$

where $\rho(\lambda, m)$ is the spectral density of modes and we assumed infinite volume [2]. While the above clarifies meaning of chiral symmetry in spectral language, it offers no dynamical detail on the breaking phenomenon. The underlying goal of this project is to identify such detail in local behavior of Dirac modes.

[3] The Speaker.

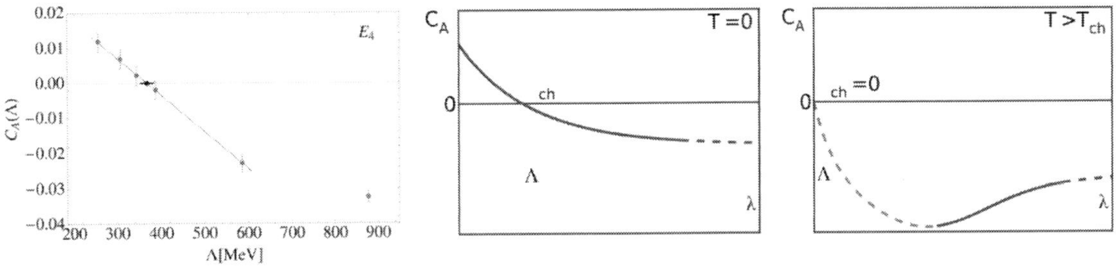

Figure 1. Left: chiral polarization scale as found in ensemble E_4 of Ref. [4]. Middle: qualitative behavior of $C_A(\lambda)$ in zero–temperature QCD. Right: same after chiral symmetry restoration by thermal agitation.

Our initial interest in this direction are *chiral properties* of the modes, owing to the intuitive expectation that breakdown of chiral symmetry should be imprinted in such features. While global chirality of Dirac non–zeromodes vanishes, the local chiral behavior reflects properties of the underlying gauge background [3]. The simplest "local inquiry" of the above type is whether values $\psi(x) = \psi_L(x) + \psi_R(x)$ tend to appear with equal participation of left/right subspaces or with asymmetric one. In other words, whether they tend to be chirally polarized or anti–polarized. Such questions were asked some time ago with different goals in mind [3], but it turns out that for our purposes it is crucial that the corresponding measures be *dynamical*. In the current context, "dynamical measure" is one that it is uniquely defined and involves comparison to statistical independence of left and right. Such measures have been constructed in Ref. [4].

In this work we will only deal with the most basic polarization measure, namely the correlation coefficient of chiral polarization C_A. This quantifies the overall tendency toward local chiral asymmetry, and is constructed as follows. Consider the probability distribution $\mathcal{P}(\psi_L, \psi_R)$ of left–right components for a mode or some group of modes. Associated with $\mathcal{P}$ is the distribution $\mathcal{P}^u(\psi_L, \psi_R) \equiv P(\psi_L)P(\psi_R)$ describing statistically independent components. Here $P(\psi_L) = \int d\psi_R \mathcal{P}(\psi_L, \psi_R)$ and similarly for $P(\psi_R)$, with functional forms being the same due to the symmetry of $\mathcal{P}$. Now, imagine the experiment in which samples are being drawn simultaneously from $\mathcal{P}$ and $\mathcal{P}_u$, and compared by their polarization. The outcomes of such comparisons have no arbitrariness, and their statistics defines the probability Γ_A that sample chosen from $\mathcal{P}$ is more polarized than one chosen from $\mathcal{P}^u$. The correlation coefficient is then $C_A \equiv 2\Gamma_A - 1 \in [-1, 1]$. Consequently, the dynamics enhancing polarization relative to statistical independence is chirally correlated ($C_A > 0$) while the one suppressing it is anti–correlated ($C_A < 0$): the former supports local chirality while the latter local anti–chirality.

The new dynamical information associated with the above approach is encoded in spectral behavior of $C_A \equiv C_A(\lambda)$, with spectral average in finite volume defined by [5]

$$C_A(\lambda, M, V) \equiv \frac{\sum\limits_k \langle\, \delta(\lambda - \lambda_k)\, C_{A,k}\, \rangle_{M,V}}{\sum\limits_k \langle\, \delta(\lambda - \lambda_k)\, \rangle_{M,V}} \tag{2}$$

Here $M \equiv (m_1, m_2, \ldots, m_{N_f})$ is the set of quark masses and $C_{A,k}$ the correlation associated with k-th mode. In Ref. [4] it was found that $C_A(\lambda)$ in quenched QCD has a positive core around zero, and switches to negative values at chiral polarization scale Λ_{ch} (Fig.1, left). It was also shown that Λ_{ch} is non–zero in the continuum limit at fixed physical volume. Here we present evidence that (1) Λ_{ch} remains positive in the infinite volume limit and is thus a dynamical scale in the theory, (2) Λ_{ch} is non–zero in the chiral limit of $N_f = 2+1$ QCD and SChSB thus proceeds via chirally polarized modes, and (3) Λ_{ch} vanishes simultaneously with density of near–zeromodes

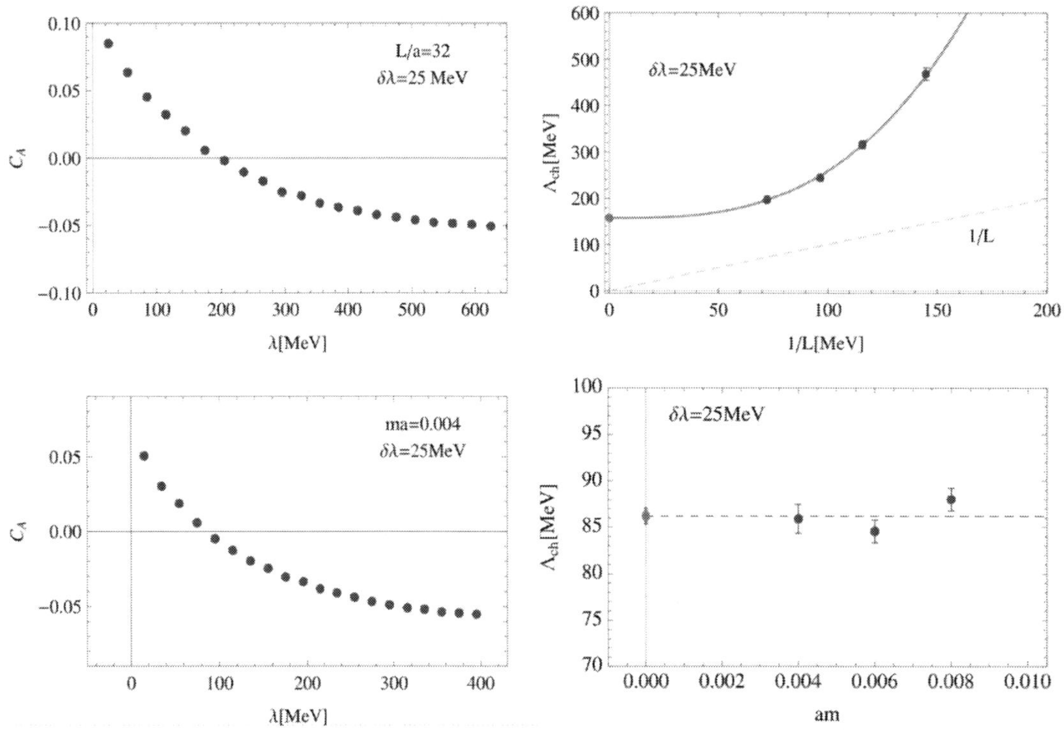

Figure 2. Top: $C_A(\lambda)$ for $L = 32$ lattice of quenched QCD (left) and infinite volume extrapolation (right). Bottom: $C_A(\lambda)$ for m_l=0.004 lattice of N_f=2+1 QCD (left) and chiral limit extrapolation (right).

when temperature is turned on, and is thus a scale closely tied to SChSB. Qualitative behavior of $C_A(\lambda)$ in zero-temperature QCD is shown in the middle plot of Fig.1, while that in the chirally symmetric phase at finite temperature is on the right.

2. Infinite volume and chiral limit.

We start with the question of infinite volume limit. As pointed out in Ref. [4], this is of crucial importance in order to establish that Λ_{ch} represents dynamically generated scale of QCD. We perform the volume study in the context of pure glue lattice QCD with Wilson action at fixed gauge coupling β=6.054. Using the parametrization of Ref. [6] and value r_0=0.5 fm this translates into the lattice spacing a=0.085 fm. Lattices of sizes 16^4, 20^4, 24^4 and 32^4 were studied with 100 independent configurations generated in each case.

To probe local chiral properties of modes, we use overlap Dirac operator with parameters r=1 and ρ=26/19, and these settings were used throughout this study. Approximately 200 conjugate pairs of lowest near–zeromodes were computed on each configuration, and the average $C_A(\lambda)$ was determined in the available region of the spectrum. In Fig.2 (top left) we show the result for the largest volume, exhibiting the anticipated behavior, and with clearly defined chiral polarization scale. Dependence of Λ_{ch} on the infrared cutoff is shown in the top right plot of Fig.2, together with the cutoff $1/L$ itself. The data clearly curves away from the cutoff, indicating a positive infinite volume limit. The fit of the form $\Lambda_{ch}(1/L) = \Lambda_{ch}(0) + b\,(1/L)^3$ was used to extrapolate to infinite volume, yielding $\Lambda_{ch} \approx 160$ MeV at this cutoff. Judging by the behavior of continuum extrapolation in Ref. [4], the continuum limit is estimated to be $\Lambda_{ch} \approx 150$ MeV in Wilson regularization of quenched QCD.

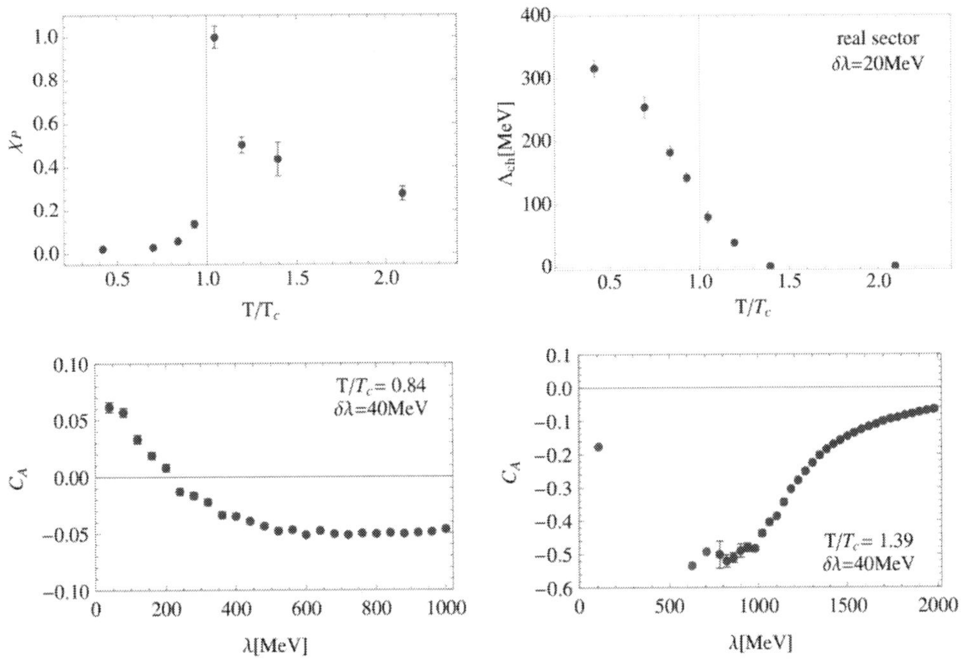

Figure 3. Top: temperature dependence of Polyakov line susceptibility (left) and chiral polarization scale (right) as discussed in the text. Bottom: behavior of $C_A(\lambda)$ at $T/T_c = 0.84$ (left), and at $T/T_c = 1.39$ (right).

To study the effects of dynamical fermions, we use the N_f=2+1 **RBC/UKQCD** domain wall fermion ensembles of Ref. [7]. These $32^3\times64$ lattices are at fixed heavy bare mass $m_h a = 0.03$, approximately corresponding to mass of physical strange quark, and at fixed lattice cutoff $a = 0.085$ fm. The light quark masses $m_l a = 0.004, 0.006, 0.008$ reach the lightest pion mass of 295 MeV. Low–lying overlap Dirac eigenmodes were computed on 50 configurations from each ensemble, with functions $C_A(\lambda)$ evaluated on the spectral region available. The result for lightest quark mass is displayed on the bottom left plot of Fig. 2, showing behavior qualitatively similar to quenched case. Chiral polarization scale has been reduced due to the effects of light dynamical quarks, but the sensitivity to light mass has already become negligible in this region, as seen on the bottom right plot of Fig. 2. Positive chiral limit, indicated via extrapolation by a constant, is thus expected. The preferred value is $\Lambda_{ch} \approx 86$ MeV at this cutoff, with rough estimate of $\Lambda_{ch} \approx 80$ MeV in the continuum. Note that, due to flat mass dependence, the above estimates apply both at physical point and in the chiral limit.

3. Chiral transition

We now wish to examine whether low–energy chiral polarization could be tied to chiral symmetry breaking in yet more fundamental way. In particular, we study whether the condensation of Dirac modes ($\rho(\lambda \to 0) > 0$) and chiral polarization ($\Lambda_{ch} > 0$) tend to occur simultaneously [5]. To that end, we consider quenched QCD at finite temperature T where it is expected that, for $T > T_{ch}$, Dirac modes cease to condense, and "valence" chiral symmetry becomes restored. Note that the chiral transition temperature was denoted as T_{ch}, in order to distinguish it from the deconfinement temperature T_c defined by the breakdown of Z_3 symmetry.

We use Wilson lattice regularization again, at the identical gauge coupling β=6.054. Overlap Dirac eigenmodes were computed on $20^3 \times N_t$ lattices with $N_t = 20, 12, 10, 9, 8, 7, 6, 4$, spanning

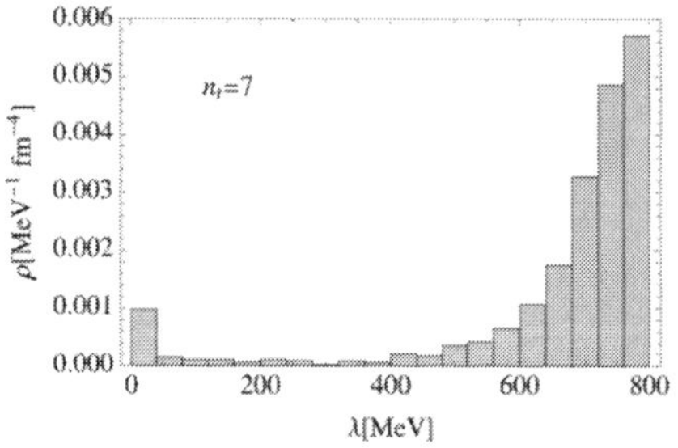

Figure 4. Detail of mode density at low end of the spectrum for N_t=7. Isolated band of near–zero modes is clearly visible. Zero modes are not included.

both confined and deconfined regions. In the top left panel of Fig. 3 we plot the temperature dependence of Polyakov line susceptibility, showing expected behavior with peak at $T/T_c = 1.05$ (N_t=8). The value of T_c, including all extrapolations, is taken from Ref. [8]. Note that for spectral calculations, only configurations from the "real Z_3" Polyakov line sector were used [9], so that the continuation to theory with dynamical quarks is smooth.

In Fig. 3 (bottom) we show the computed $C_A(\lambda)$ at T/T_c=0.84 (N_t=10), and at T/T_c=1.39 ($N_t = 6$). In the former (confined) case there is a positive core of chiral polarization, while only anti–polarization exists in the (deconfined) latter.[4] Closer inspection reveals a discrepancy of chiral polarization behavior with respect to confinement, but not with respect to mode condensation. Indeed, as can be seen in Fig. 3 (top right), chiral polarization is present ($\Lambda_{ch} > 0$) at T/T_c=1.2 ($N_t = 7$), i.e. quite safely in the deconfined phase. At the same time though, as the profile of mode density in Fig. 4 suggests, $\rho(\lambda) > 0$ in the narrow band close to zero, consistently with condensation and breakdown of valence chiral symmetry. The residual density of this kind was first observed in [10]. Our result thus supports both the polarization–condensation equivalence conjecture of Ref. [5], and the possibility that $T_{ch} > T_c$ in quenched QCD.

4. Discussion

"Bottom–up" approach to QCD vacuum structure starts with identifying the features associated with important vacuum effects, such as confinement or SChSB, and characterizing them in a model–independent manner. Here we focused on SChSB and found that it is intimately tied to *dynamical local chirality* of low–lying Dirac modes. In particular, strong interaction supports the band of chirally polarized low–energy modes that condense, and are thus "carriers" of the breaking phenomenon. The width of the band Λ_{ch} provides for a new dynamical scale associated with SChSB. Simultaneous occurrence of mode condensation and chiral polarization at finite temperature supports the possibility that chiral symmetry breakdown (true and "valence") is equivalent to presence of chiral polarization, and thus non–zero Λ_{ch} [5]. This would extend the equivalence of Eq. (1) to include the independent condition $\lim_{m\to 0} \Lambda_{ch}(m) > 0$.

References

[1] Horváth I 2006 *Preprint* hep-lat/0605008
[2] Banks T and Casher A 1980 *Nucl.Phys.* **B169** 125 (*Preprint* hep-lat/9501037)
[3] Horváth I, Isgur N, McCune J, Thacker H B 2002 *Phys.Rev.* **D65** 014502 (*Preprint* hep-lat/0102003)
[4] Alexandru A, Draper T, Horváth I, Streuer T 2011 *Annals Phys.* **326** 1941 (*Preprint* arXiv:1009.4451)
[5] Alexandru A and Horváth I 2012 *Preprint* arXiv:1210.7849
[6] **ALPHA** Collaboration, Guagnelli M, Sommer R and Wittig H 1998 *Nucl.Phys.* **B535** 389 (*Preprint* hep-lat/9806005)
[7] **RBC** and **UKQCD** Collaborations, Mawhinney R 2009 *PoS* **LAT2009** 81 (*Preprint* arXiv:0910.3194)
[8] Karsch F 1998 *Nucl. Phys. Proc. Suppl.* **60A** 169 (*Preprint* hep-lat/9706006)
[9] Chandrasekharan S and Christ N 1996 *Nucl. Phys. Proc. Suppl.* **47** 527 (*Preprint* hep-lat/9509095)
[10] Edwards R G, Heller U M, Kiskis J E, Narayanan R 2000 *Phys.Rev.* **D61** 074504 (*Preprint* hep-lat/9910041)

[4] The red points in the plot for $T/T_c = 1.39$ denote bins with not enough eigenmodes for error to be estimated.

Institute of Physics
Dirac House, Temple Back
Bristol BS1 6BE UK

CURRAN ASSOCIATES INC
proceedings
.com